高等学校创意创新创业教育系列丛书

金百东　刘德山　编著

Android简明程序设计

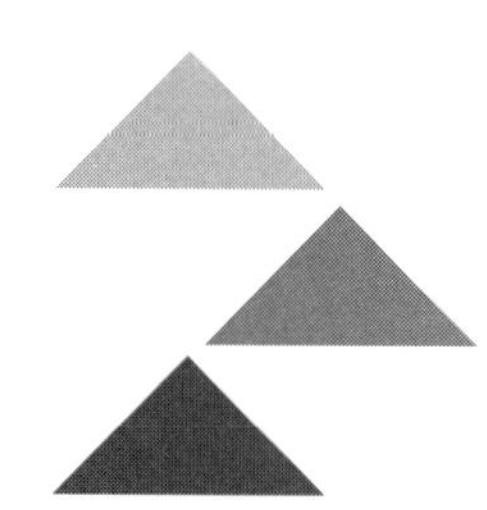

清华大学出版社

北京

内容简介

本书全面而又系统地介绍了 Android 移动开发技术。全书共 12 章，内容包括 Android 简介、Android 布局、Android 控件、对话框与高级控件、Activity、网络通信、广播接收组件、Service、数据存储与共享、图形与动画、设备操作、第三方开发包等。本书注重应用，每章都包含大量示例和详细的结果分析，旨在使读者夯实基础，提高综合运用 Android 各项技术的编程能力，学会软件编程的思考方法。

本书可作为普通高校计算机专业本科生的教材，也可作为专业技术人员、软件工程师、系统架构师等的参考用书。

图书在版编目(CIP)数据

Android 简明程序设计/金百东，刘德山编著. —北京：清华大学出版社，2022.3
(高等学校创意创新创业教育系列丛书)
ISBN 978-7-302-60000-8

Ⅰ. ①A… Ⅱ. ①金… ②刘… Ⅲ. ①移动终端—应用程序—程序设计—高等学校—教材 Ⅳ. ①TN929.53

中国版本图书馆 CIP 数据核字(2022)第 006475 号

责任编辑：谢 琛 常建丽
封面设计：常雪影
责任校对：胡伟民
责任印制：朱雨萌

出版发行：清华大学出版社
网　　址：http://www.tup.com.cn，http://www.wqbook.com
地　　址：北京清华大学学研大厦 A 座　　**邮　　编**：100084
社 总 机：010-83470000　　**邮　　购**：010-83470235
投稿与读者服务：010-62776969，c-service@tup.tsinghua.edu.cn
质量反馈：010-62772015，zhiliang@tup.tsinghua.edu.cn
课件下载：http://www.tup.com.cn，010-83470236
印 装 者：三河市龙大印装有限公司
经　　销：全国新华书店
开　　本：210mm×235mm　　**印　　张**：27.75　　**字　　数**：693 千字
版　　次：2022 年 3 月第 1 版　　**印　　次**：2022 年 3 月第 1 次印刷
定　　价：79.80 元

产品编号：093944-01

前言

创作背景

随着智能手机的飞速发展，越来越多的 App 应用走进人们的生活，App 开发也越发重要，出现了许多优秀的 Android 程序设计书籍。笔者也想为此尽一些微薄之力，以下三点是笔者的创作动机：①利用尽量少的代码，讲清知识内涵；②知识点环环相扣，要达到一定的深度；③体会 Android 编程思想，并将其运用到其他不同语言的编程中。

本书内容

本书共分 12 章，具体内容如下。

第 1 章　Android 简介，介绍 Android 历史，Android Studio 开发环境的安装、配置及应用方法。开发第一个 Android 程序，介绍其相关目录的作用，并简介 Toast 类的功能及两个重要的知识点。

第 2 章　Android 布局，介绍线性布局(LinearLayout)、相对布局(RelativeLayout)、表格布局(TableLayout)、网格布局(GridLayout)、框架布局(FrameLayout)的应用方法。

第 3 章　Android 控件，介绍按钮控件 Button、ImageButton，状态开关控件 ToggleButton、Switch，单选按钮 RadioButton 和多选按钮 CheckBox，图片控件 ImageView，文本控件 TextView、EditText，列表控件 ListView，下拉控件 Spinner 等的创建方法，类中的主要函数及事件处理机制。

第 4 章　对话框与高级控件，介绍系统 AlertDialog 对话框的基本应用方法，讲解日期控件、翻页控件、增强型列表控件 RecyclerView 的具体实现和用法。

第 5 章　Activity，介绍 Activity 的生命周期，创建 Activity 间的通信技术，论述 Activity 隐式启动技术，讲解在 Activity 接口应用 Fragment 技术的方法。

第 6 章　网络通信，介绍 URL、HttpURLConnection 类的基本应用方法，编制最简单的网络通信程序“Hello world”，指出多线程在网络编程中的重要性，并对编码、解码进行详细的讨论。

第 7 章　广播接收组件，介绍广播接收的基本原理，组件的静态注册与动态注册，普通广播与有序广播的不同，并对系统固有广播做了一定的讨论。

第 8 章　Service，介绍 Service 生命周期，启动 Service，绑定 Service 的特点及应用，对跨进程调用 Service 进行了深入的讨论。

第 9 章　数据存储与共享，介绍内部存储、外部存储、资源文件存储；讲解 SharedPreferences 存储、SQLite 数据库存储，并对 ContentProvider 组件进行了深入的论述。

第 10 章　图形与动画，介绍 Android 2D 绘制基本图形、文字、位图的方法，对 Path 路径绘图进行了详细的描述；讲解帧动画、补间动画、属性动画的技术与应用。

第 11 章　设备操作，介绍麦克风、摄像头常规操作，描述传感器的应用方法，对手机定位技术也进行了深入的讨论。

第 12 章　第三方开发包，介绍高德地图在手机定位、搜索、公交查询、天气预报中的应用。

附加说明

(1) 由于篇幅关系，示例均省略了 import 导入包部分，读者自行利用 Android Studio 可视化平台加入即可。

(2) 某些程序运用了 try-catch 异常处理框架，同样，由于篇幅关系，略去了 catch 块中的代码，读者在程序调试时可自行补充。

总之，本书内容循序渐进，采取实例驱动讲授方式，所有实例复制下来编译后就可以运行。许多题目是笔者多年 Android 编程经验的总结，实用性较强。示例前因后果都做了必要的说明，对一些稍难的题目，对其设计思想也做了相应的论述，帮助读者加深理解。

本书第 3、4、6、8、10、11、12 章由金百东完成，其余章由刘德山完成。因本书程序较多，故全书变量均用正体。

由于作者水平有限，书中难免有疏漏之处，恳请广大读者批评指正。

编　者

2021 年 6 月

目　录

第 1 章　Android 简介 …… 1

1.1　Android 历史 …… 1
1.2　开发环境 …… 2
1.3　创建第一个工程 …… 3
1.4　工程主要文件和目录 …… 6
1.4.1　主要目录介绍 …… 6
1.4.2　主要文件介绍 …… 7
1.5　编译与运行 …… 10
1.6　Toast 类 …… 12
1.7　两个知识点 …… 13
1.7.1　接口回调技术 …… 13
1.7.2　适配器技术 …… 15
习题 1 …… 17

第 2 章　Android 布局 …… 18

2.1　Android 布局与 Java 布局的区别 …… 18
2.2　线性布局 …… 18
2.3　相对布局 …… 27
2.3.1　根据父容器定位 …… 27
2.3.2　根据兄弟组件定位 …… 29
2.4　表格布局 …… 32
2.5　网格布局 …… 35
2.6　框架布局 …… 37
2.7　滚动窗口 …… 38
2.8　综合示例 …… 40
2.9　动态控制布局 …… 47

2.10 单位转换 …… 53
习题 2 …… 56

●第 3 章 Android 控件 …… 58

3.1 类层次关系 …… 58
3.2 按钮控件 …… 59
3.2.1 基本按钮 Button …… 59
3.2.2 图像按钮 ImageButton …… 63
3.3 状态开关 …… 65
3.3.1 ToggleButton 开关 …… 65
3.3.2 Switch 开关 …… 67
3.4 单选按钮和多选按钮 …… 70
3.4.1 RadioButton 单选按钮 …… 70
3.4.2 深入探究 …… 73
3.4.3 CheckBox 多选按钮 …… 78
3.5 图片控件 ImageView …… 81
3.5.1 基本函数 …… 81
3.5.2 数学基础 …… 82
3.5.3 典型事例 …… 83
3.6 文本控件 …… 89
3.6.1 TextView …… 89
3.6.2 深入探究 …… 92
3.6.3 EditText …… 99
3.7 列表控件 …… 106
3.7.1 基本函数与事件响应 …… 106
3.7.2 数据适配器 …… 109
3.8 下拉控件 …… 120
3.9 进度条控件 …… 124
3.10 形状文件 …… 126
3.11 状态文件 …… 129
习题 3 …… 133

● 第 4 章 对话框与高级控件 …… 135
4.1 对话框 …… 135
4.1.1 AlertDialog 简介 …… 135
4.1.2 分类介绍 …… 136
4.2 日期控件 …… 141
4.3 翻页控件 …… 148
4.4 计时器控件 …… 151
4.5 增强型列表 RecyclerView 控件 …… 153
4.5.1 简介 …… 153
4.5.2 几个问题 …… 156
4.5.3 布局管理器 …… 159
4.6 菜单控件 …… 163
4.6.1 选项菜单 …… 163
4.6.2 上下文菜单 …… 166
4.6.3 弹出菜单 …… 167
习题 4 …… 169
● 第 5 章 Activity …… 171
5.1 生命周期 …… 171
5.2 建立 Activity …… 173
5.2.1 入口 Activity 类 …… 173
5.2.2 普通 Activity 类 …… 174
5.3 Activity 通信 …… 176
5.4 隐式启动 Activity …… 179
5.4.1 intent-filter …… 179
5.4.2 自定义属性应用 …… 181
5.4.3 系统属性应用 …… 183
5.5 Fragment …… 185
5.5.1 引入 Fragment 的原因 …… 185
5.5.2 静态加载 …… 186
5.5.3 动态加载 …… 188
5.5.4 数据通信 …… 191

5.5.5 生命周期 …… 193
习题 5 …… 194

●第 6 章 网络通信 …… 196

6.1 子线程刷新 UI 问题 …… 196
6.2 Handler 类 …… 197
6.3 URL 类 …… 199
6.4 应用服务器 …… 201
6.5 HttpURLConnection …… 210
6.5.1 简介 …… 210
6.5.2 应用举例 …… 211
6.6 XML 解析 …… 216
6.7 JSON 解析 …… 220
6.8 URL 编码 …… 223
6.9 WebView …… 225
6.9.1 简介 …… 225
6.9.2 应用举例 …… 226
习题 6 …… 229

●第 7 章 广播接收组件 …… 231

7.1 基本原理 …… 231
7.2 基本类 …… 232
7.3 应用示例 …… 232
7.3.1 普通广播＋静态注册 …… 232
7.3.2 普通广播＋动态注册 …… 235
7.3.3 有序广播＋静态注册 …… 236
7.3.4 有序广播＋动态注册 …… 238
7.3.5 其他广播 …… 239
7.4 系统广播 …… 241
习题 7 …… 246

●第 8 章 Service …… 247

8.1 简介 …… 247

8.2 启动 Service …… 248
8.2.1 生命周期 …… 248
8.2.2 几个知识点 …… 251
8.2.3 应用示例 …… 254
8.3 绑定 Service …… 260
8.3.1 生命周期 …… 260
8.3.2 Messenger 技术 …… 266
8.3.3 AIDL 技术 …… 270
习题 8 …… 278

●第 9 章 数据存储与共享 …… 280

9.1 内部存储 …… 280
9.1.1 存储目录 …… 280
9.1.2 存储文件 …… 281
9.2 外部存储 …… 284
9.2.1 存储目录 …… 284
9.2.2 存储文件 …… 285
9.2.3 共享文件夹 …… 287
9.3 资源文件存储 …… 292
9.4 SharedPreferences 存储 …… 294
9.4.1 概述 …… 294
9.4.2 基本用法 …… 294
9.5 数据库存储 …… 296
9.5.1 命令行建库 …… 296
9.5.2 程序建库与操作 …… 298
9.6 ContentProvider 组件 …… 307
9.6.1 简介 …… 307
9.6.2 最简单的示例 …… 307
9.6.3 相关类介绍 …… 311
9.6.4 实现 SharedPreferences 共享 …… 313
9.6.5 实现数据库共享 …… 317
9.6.6 系统数据库共享 …… 322
习题 9 …… 326

●第 10 章 图形与动画 …… 327

10.1 2D 绘图 …… 327
10.1.1 最简单的绘图 …… 327
10.1.2 相关类简介 …… 328
10.1.3 图像变换 …… 330
10.1.4 Path 应用 …… 333
10.1.5 贝塞尔曲线 …… 335
10.1.6 位图操作 …… 338
10.1.7 绘制文字 …… 342
10.2 动画 …… 344
10.2.1 帧动画 …… 344
10.2.2 补间动画 …… 346
10.2.3 属性动画 …… 351
10.2.4 实用动画技术 …… 356
习题 10 …… 360

●第 11 章 设备操作 …… 361

11.1 麦克风 …… 361
11.1.1 SeekBar 类 …… 361
11.1.2 AudioManager 类 …… 361
11.1.3 MediaRecorder 录音类 …… 362
11.1.4 MediaPlayer 类 …… 366
11.2 摄像头 …… 371
11.2.1 相关类简介 …… 371
11.2.2 照相预览功能 …… 372
11.2.3 拍照功能 …… 376
11.2.4 录影功能 …… 378
11.2.5 放映功能 …… 381
11.3 传感器 …… 384
11.3.1 简介 …… 384
11.3.2 编程步骤 …… 385
11.3.3 加速度传感器 …… 385

11.3.4 磁场传感器 …… 389
11.3.5 计步传感器 …… 392
11.4 手机定位 …… 395
11.4.1 定位原理 …… 395
11.4.2 相关类介绍 …… 395
习题 11 …… 399

第 12 章 第三方开发包 …… 401

12.1 签名信息 …… 401
12.1.1 重要性 …… 401
12.1.2 签名查看 …… 402
12.2 构建自定义高德地图工程环境 …… 404
12.3 最简单的高德地图程序 …… 406
12.4 定位功能 …… 409
12.4.1 相关类及接口 …… 409
12.4.2 定位实现 …… 411
12.4.3 基本搜索 …… 418
12.4.4 公交查询 …… 423
12.4.5 天气查询 …… 426
习题 12 …… 429

参考文献 …… 430

第 1 章

Android 简介

本章首先介绍 Android 历史，Android Studio 开发环境的安装、配置及应用方法，接着开发第一个 Android 程序，并介绍其相关目录的作用，最后简介 Toast 类的功能及两个重要的知识点。

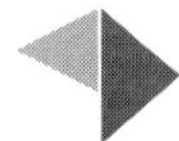

1.1 Android 历史

Android 是一款基于 Linux 的、自由及开放源代码的操作系统，主要用于移动设备，如智能手机和平板电脑，由 Google 公司和开放手机联盟领导及开发。

2003 年 10 月，Andy Rubin 等人创建 Android 公司，并组建 Android 团队，负责 Android 操作系统的开发。2005 年 8 月 17 日，Google 低调收购了成立仅 22 个月的高科技企业 Android 及其团队。Andy Rubin 成为 Google 公司工程部副总裁，继续负责 Android 项目。2007 年 11 月，Google 与 84 家硬件制造商、软件开发商及电信运营商组建开放手机联盟，共同研发改良 Android 系统。随后，Google 以 Apache 开源许可证的授权方式，发布了 Android 的源代码。

第一部 Android 智能手机发布于 2008 年 10 月。随着时间的推进，Google 公司不断地完善并更新 Android 的版本，各个硬件厂家也不断地推出新的基于 Android 的手机。目前 Android 逐渐扩展到平板电脑及其电视、汽车、手表、数码相机、游戏机、智能家电等多个领域。

Android 操作系统的版本号比较有趣，它以甜点作为版本代号，其实最初 Android 的版本号是以著名机器人名称对其进行命名的(例如它的一个内测版本代号是阿童木)，后来由于涉及版权问题，2009 年 4 月，也就是从 1.5 版本发布开始，采用甜点的名字命名。

除 Android 操作系统的版本号外，Android API 也有一个版本号，称为 API 等级 (Android level)。随着 Android 操作系统的升级，API 等级也随之发生改变，对于开发人员来说，经常使用 API 等级描述 App 所对应的 Android 系统。

目前，智能手机已经走进千家万户，各种 App 层出不穷，Android 应用程序开发进入一个黄金时代。

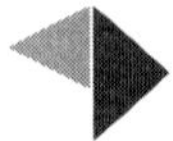

1.2 开发环境

Android 应用程序开发环境需要安装：JDK，Java 基本开发包；Android SDK，Android 应用程序需要的开发包；Android Studio，可视化开发环境。下面论述安装过程。

1. JDK 安装

本书选择的是 JDK 1.8，配置好 JAVA_HOME 环境变量，JAVA_HOME 代表 Java 安装目录。在 PATH 环境变量末尾增加%JAVA_HOME%\bin。

2. Android Studio+ Android SDK 捆绑安装

本部分是安装的重点，安装方法、途径多种多样。本部分安装软件是从安卓中文网（https://developer.android.google.cn/）下载的，文件名是 android-studio-bundle-162.3934792-windows.exe，大小将近 2GB，发布日期是 2017-05-11。它是一个捆绑软件，包含 Android Studio 2.3.2 可视化开发平台及相应的 Android SDK。安装步骤如下所示。

① 建立两个目录：android_studio，用于存放 Android Studio 可视化开发平台系统；android_sdk，用于存放 Android SDK 开发包。当然，上述两个目录可任意取名，不是固定的。

② 运行 android-studio-bundle-162.3934792-windows.exe，按步骤操作即可（在某步骤中可设置①中的两个目录）。安装结束后，选择不启动 Android Studio 可视化开发平台。

③ 设置 Android SDK 环境变量 ANDROID_HOME，即 Android SDK 的安装根目录；在 PATH 环境变量中增加%ANDROID_HOME%\tools。

④ 修改 android_studio\bin\idea.properties，在末尾加一行代码：disable.android.first.run=true。

⑤ 从官网（https://services.gradle.org/distributions/Gradle3.3.all.zip）下载 Gradle 3.3 并解压至 android_studio\gradle\下。

至此，Android Studio 2.3.2 可视化开发平台及 Android SDK 安装完毕。

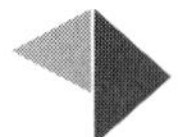

1.3　创建第一个工程

① 运行 Android Studio 安装目录 bin\studio64.exe，界面如图 1-1 所示。

图 1-1　创建新工程初始页面

② 单击 Start a new Android Studio project，出现图 1-2。

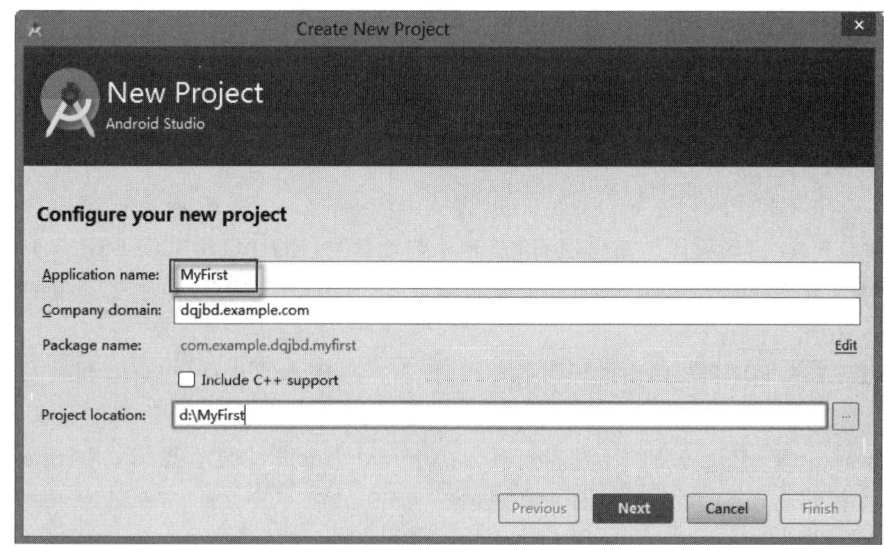

图 1-2　输入工程名页面

③ 输入相关内容后单击 Next 按钮出现选择 Activity 活动界面页面，如图 1-3 所示。

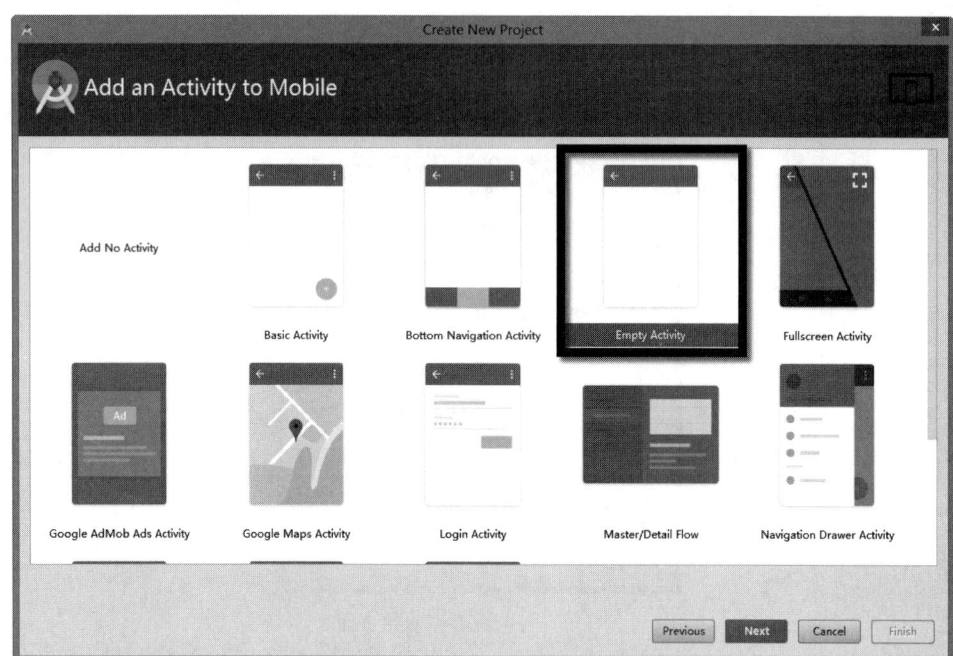

图 1-3 选择 Activity 活动界面页面

④ 之后一直单击 Next 按钮，选择默认设置，最后单击 Finish 按钮，工程进入自动编译状态，但会出现错误信息。这是因为没有设置 Gradle。执行菜单项 File-settings 出现图 1-4 所示界面，将 Gradle 设置成版本 3.3（或其他较高版本，但必须匹配，主要是设置 Gradle 路径）。

⑤ 单击 OK 按钮后，MyFirst 工程进入自动编译状态（或者执行菜单项 Build-Rebuild project 启动编译），当出现图 1-5 所示界面信息（BUILD SUCCESSFUL）时，表示 MyFirst 工程创建成功。若该选项卡页面没显示，则单击下方工具栏右下角的 Gradle Console 选项。

注意，若在上述步骤中出现异常或等待时间过长，则需使计算机与外网相连，通过支持 Android 下载的代理服务器进行下载，本文选择的是东软信息学院的代理服务器 mirrors.neusoft.edu.cn，端口号为 80。通过 Android Studio 的菜单项 File-Settings 设定，如图 1-6 所示。

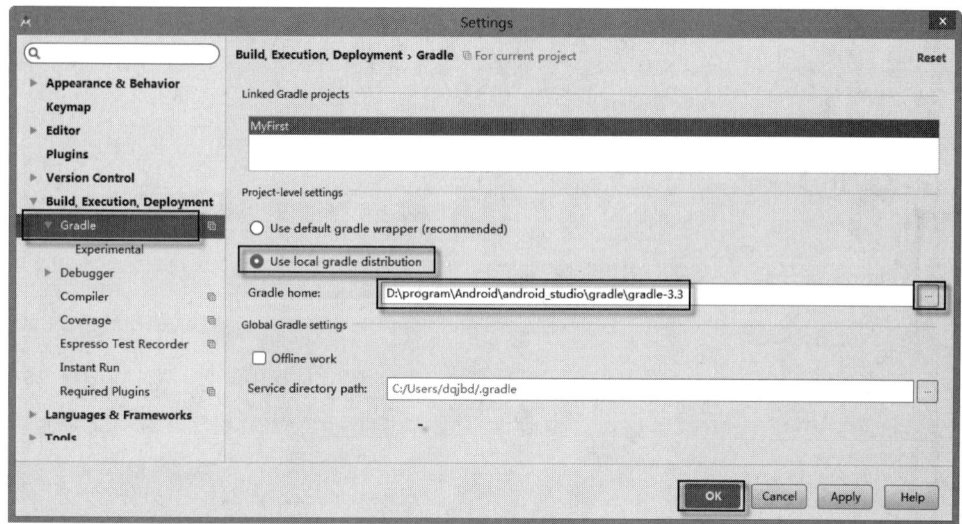

图 1-4　设置 Gradle 界面

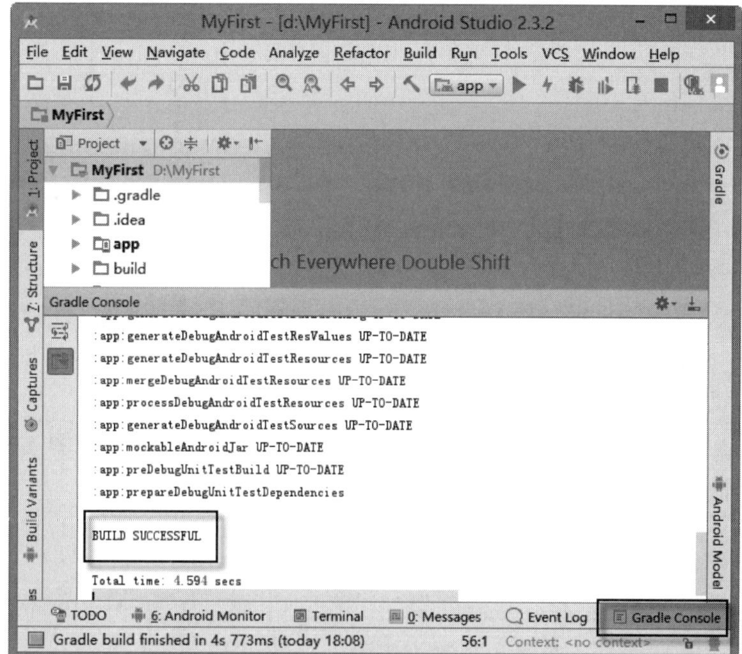

图 1-5　工程编译成功界面

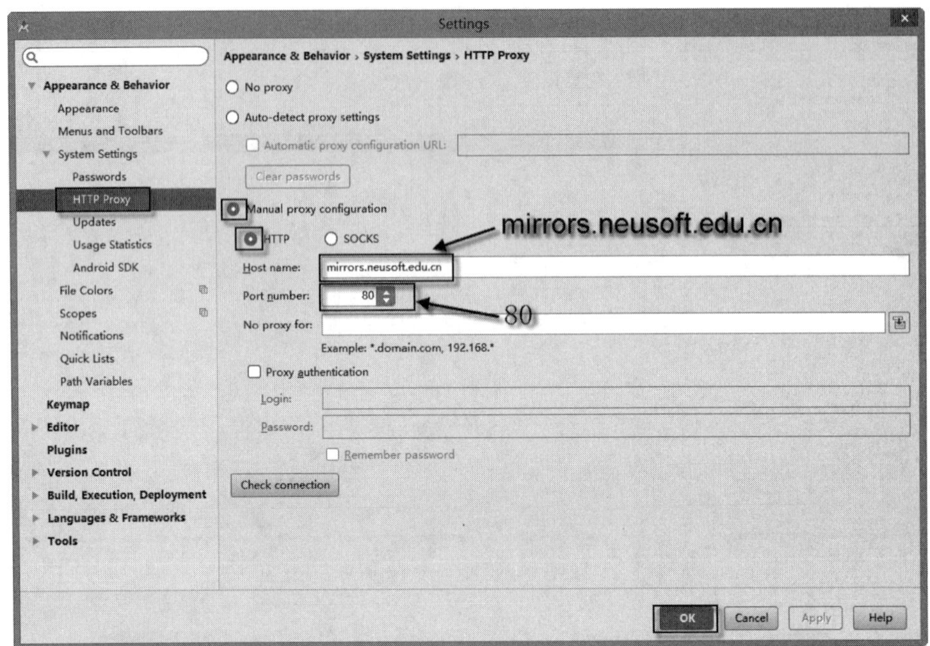

图 1-6 代理设置界面

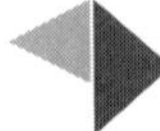

1.4 工程主要文件和目录

以上述 MyFirst 工程为例，选中 Android Studio 左侧的选项卡 Project，再选择上方的工具条显示类型下拉框，选择 Android 模式，出现如图 1-7 所示的界面。

1.4.1 主要目录介绍

当用 Android Studio 产生新工程的时候，产生的主要常用目录如图 1-7 所示。下面分别加以介绍。

manifests 子目录下面只有一个 xml 文件，即 AndroidManifest.xml，它是该工程的配置文件，包含工程的所有功能。

java 子目录，下面有 3 个 com.example.dqjbd.myfirst 包。其中，第 1 个包存放的是工程的 Java 源代码，后两个包存放的是测试用的 Java 源代码。

res 子目录存放的是工程的各种资源文件，其下又有 4 个常用的子目录。

- layout：存放界面的配置文件目录。

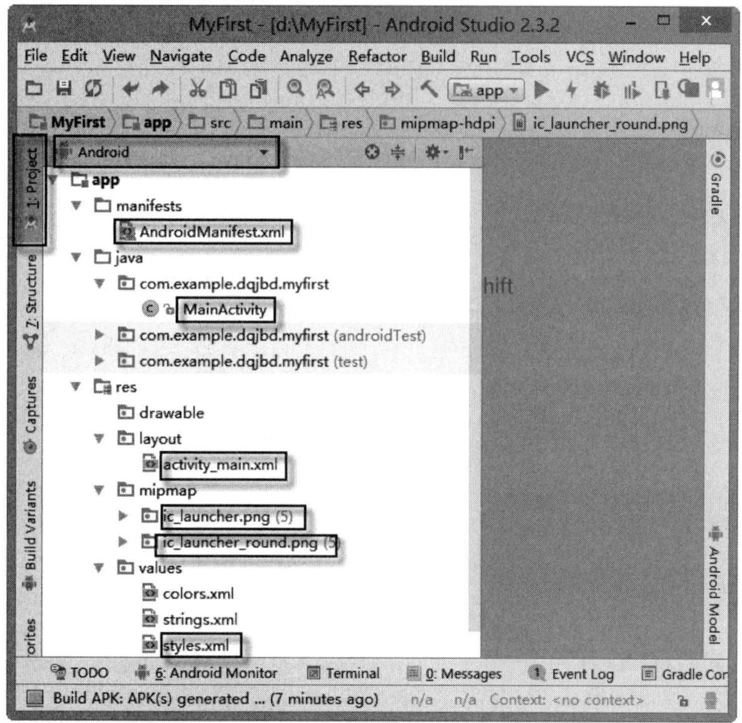

图 1-7　工程主要文件和目录

- drawable：存放图片文件的目录。
- values：存放一些常量的文件。
- mipmap：存放工程图标的目录。

1.4.2　主要文件介绍

1. AndroidManifest.xml

该文件是工程最重要的配置文件，从该文件可以看出工程的所有功能。MyFirst 工程的此文件内容如下，解析见注释。

```
<?xml version="1.0" encoding="utf-8"?>
<manifest xmlns:android="http://schemas.android.com/apk/res/android"
    package="com.example.dqjbd.myfirst" >   //package 用于定义工程默认包名
    <application                            //工程所有配置定义在 application 标签中
        android:allowBackup="true"          //用于指定是否允许备份
```

```
        android:icon="@mipmap/ic_launcher"  //工程图标是 mipmap/ic_launcher.png
        android:label="@string/app_name"
                             //名称见 values/strings.xml 中键 app_name 对应的值
        android:roundIcon="@mipmap/ic_launcher_round"
                                      //工程图标是 mipmap/ic_launcher_round.png
        android:supportsRtl="true"
                  //设置为 true 表示支持阿拉伯语/波斯语这种从右向左的文字排列顺序
        android:theme="@style/AppTheme" >  //主题风格见 values/styles.xml
        <activity android:name=".MainActivity" >
                                           //-定义工程入口类 MainActivity.java
            <intent-filter>
                <action android:name="android.intent.action.MAIN" />
                <category android:name="android.intent.category.LAUNCHER" />
            </intent-filter>
        </activity>
    </application>
</manifest>
```

注释中指向的目标文件在图 1-7 中都能找到。例如，android:label="@string/app_name"，打开 values/strings.xml，内容如下所示。可以看出，键 app_name 对应的值为 MyFirst。

```
<resources><string name="app_name">MyFirst</string></resources>
```

2. MainActivity.java，其内容如下所示。

```
package com.example.dqjbd.myfirst;
import android.support.v7.app.AppCompatActivity;
import android.os.Bundle;
public class MainActivity extends AppCompatActivity {
    protected void onCreate(Bundle savedInstanceState) {
        super.onCreate(savedInstanceState);
        setContentView(R.layout.activity_main);
    }
}
```

可以看出，MyFirst 工程运行时第一个入口函数是 onCreate()，最主要的一行代码是 setContentView(R.layout.activity_main)，与 activity_main.xml 相关，该文件内容如下。

```
<?xml version="1.0" encoding="utf-8"?>
```

```
<android.support.constraint.ConstraintLayout
    xmlns:android="http://schemas.android.com/apk/res/android"
    xmlns:tools="http://schemas.android.com/tools"
    xmlns:app="http://schemas.android.com/apk/res-auto"
    android:layout_width="match_parent"
    android:layout_height="match_parent"
    tools:context="com.example.dqjbd.myfirst.MainActivity">
    <TextView
        android:layout_width="wrap_content"
        android:layout_height="wrap_content"
        android:text="Hello World!"
        app:layout_constraintBottom_toBottomOf="parent"
        app:layout_constraintLeft_toLeftOf="parent"
        app:layout_constraintRight_toRightOf="parent"
        app:layout_constraintTop_toTopOf="parent" />
</android.support.constraint.ConstraintLayout>
```

由 TextView 节点中的 android:text＝“Hello World”知，本示例要在界面中显示“Hello World”字符串。

Android Studio 允许由 layout 目录下的 XML 文件定义所需界面，并且可在线看到。例如，打开 activity_main.xml，单击可视化平台右侧的 Preview 选项卡，即可看见 activity_main.xml 对应的界面视图，如图 1-8 所示。

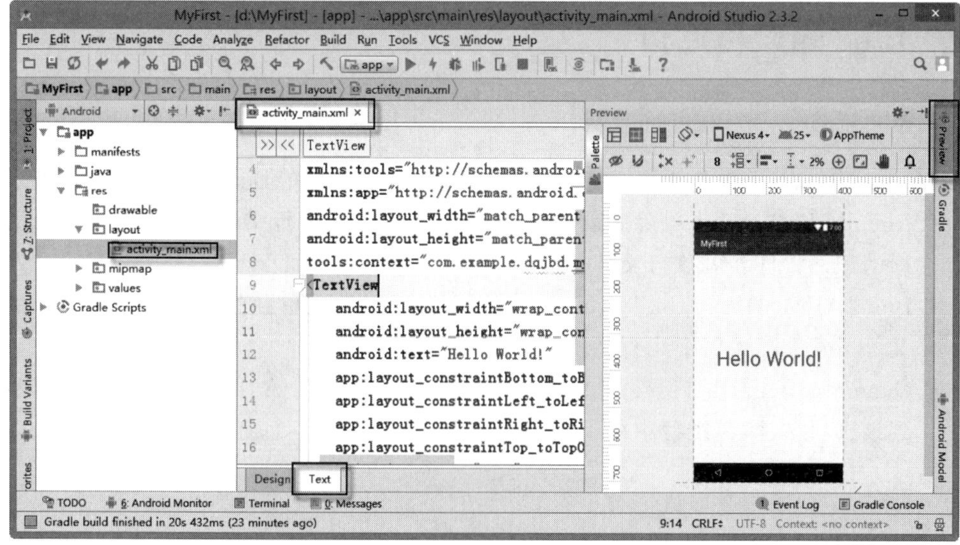

图 1-8　界面配置文件与预览

3. R 文件

MainActivity 类的 onCreate()中有语句：setContentView(R.layout.activity_main)，如何理解 R.layout.activity_main 中的 R 呢？其实 R 是一个类，在工程 MyFirst 中搜索 R.java，会发现它的位置很深，在 MyFirst\app\build\generated\source\r\debug\com\example\dqjbd\myfirst 目录下，而且文件很大，达到几百 K。为了便于理解，仅摘录与图 1-7 相关部分内容的代码如下。

```
package com.example.dqjbd.myfirst;
public final class R {
    public static final class layout {
        public static final int activity_main=0x7f04001b;
    }
    public static final class string {
        public static final int app_name=0x7f060021;
    }
}
```

可以看出，activity_main，app_name 均是整型资源号。也就是说，工程 res 目录下的所有资源均在 R 文件中动态生成一个整型资源号，至于如何根据资源号查找到对应的文件由 Android 系统完成，则无须关心。

1.5 编译与运行

1. 编译命令

Android Studio 可视化平台上有 3 个常用的编译命令均在 Build 菜单下，如下所示。

- Rebuild Project：先删除之前编译的编译文件和可执行文件，然后重新编译生成新的编译文件和可执行文件。
- Build APK：在 app/build/outputs/apk 目录下，生成单独的 APK 文件。该 APK 文件可安装在手机中运行。
- Generated Signed APK：生成有签名的 APK，一般在 App 作为商品发布时才用到该选项，在学习过程中可不用。

2. 运行

方式 1：在手机上运行。将 app/build/outputs/apk/app-debug.apk 复制到手机上，

像其他常规 App 安装方法安装，运行，结果如图 1-9 所示。

图 1-9　手机运行 App 效果图

方式 2：利用 Android Studio 建立手机模拟器，然后运行。当然，建立手机模拟器仅需一次。按菜单步骤"Tools-Android-AVD Manager"运行，需要下载模拟器资源及 HAXM 软件，因此全过程需连接外网。若下载过程中出现异常，则按图 1-6 所示设置通过代理下载。安装成功后，会出现类似图 1-10 的界面。

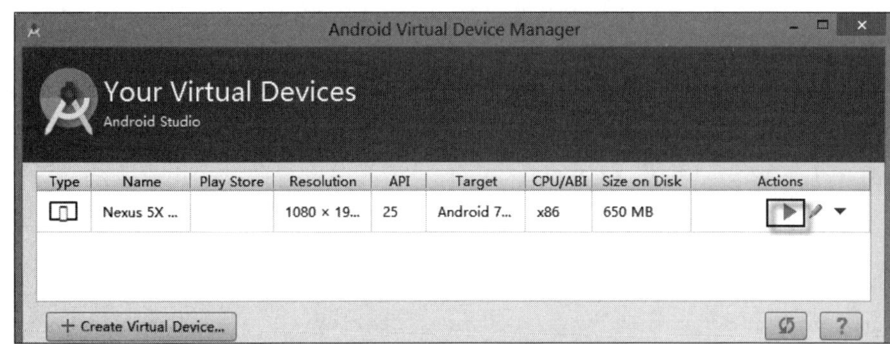

图 1-10　创建手机模拟器过程图

单击图 1-10 中的三角按钮，就会在手机模拟器上运行 MyFirst 应用程序。之后再运行应用时，从菜单下操作"Run-Run 'app'"即可。

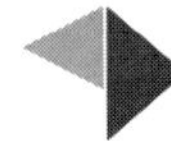

1.6 Toast 类

Toast 是 Android 经常用到的工具类，是一种简易的消息提示框，其显示一定时间后自动消失，作用与 C 语言的 printf()，Java 基础中的 System.out.println()函数类似。

该类常用函数如下所示。

- Toast makeText(Context ctx, CharSequence text, int duration)；静态方法，返回一个 Toast 对象实例。ctx 是上下文对象；text 是待显示的字符串；duration 是消息框显示时间长短标志，一般取 Toast.LENGTH_LONG，表示长时间显示，Toast.LENGTH_SHORT 表示短时间显示。
- void show()；实例方法，显示消息对话框。

【例 1-1】 利用 Toast 显示"Hello World"字符串。

只修改 MyFirst 工程中的文件即可，如下所示。

① 主界面配置文件 activity_main.xml。

```
<?xml version="1.0" encoding="utf-8"?>
<android.support.constraint.ConstraintLayout
    xmlns:android="http://schemas.android.com/apk/res/android"
    android:layout_width="match_parent"
    android:layout_height="match_parent">
</android.support.constraint.ConstraintLayout>
```

MyFirst 工程默认产生的 activity_main.xml 稍显复杂，去掉 TextView 节点所有内容，让界面是空白即可，再去掉多余的属性 xmlIns: tools、xmlIns: app、tools: context，化简成最简式，如上所示。

② 主应用程序 MainActivity.java。

```
package com.example.dqjbd.myfirst;
import android.support.v7.app.AppCompatActivity;
import android.os.Bundle;
import android.widget.Toast;
public class MainActivity extends AppCompatActivity {
    protected void onCreate(Bundle savedInstanceState) {
        super.onCreate(savedInstanceState);
        setContentView(R.layout.activity_main);
        Toast.makeText(this, "I learn Android!",Toast.LENGTH_LONG).show();
```

```
    }
}
```

与原 MyFirst 工程相比，onCreate()函数多了最后一行 Toast 语句。其运行结果如图 1-11 所示。

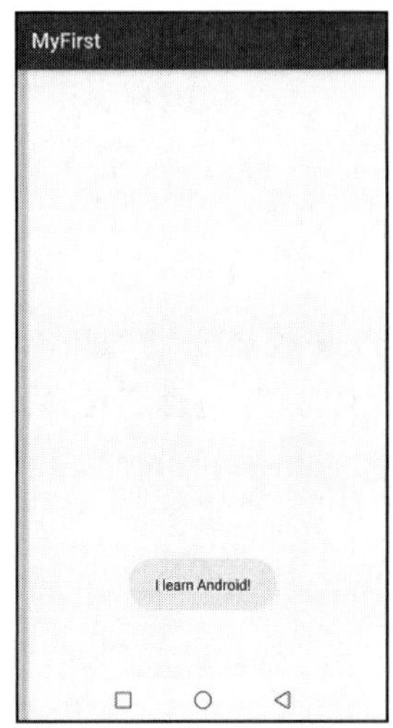

图 1-11　Toast 消息框

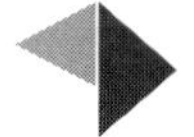

1.7　两个知识点

在学习 Android 过程中发现有两个知识点经常出现：接口回调和适配器技术。本节将简要描述它们的理论基础。

1.7.1　接口回调技术

例如，假设有两个对象 A、B，接口回调是指 A 调用 B，B 再调用 A。很明显，B 是一个中心环节，A 是左右两头。此抽象描述可分为如下文件。

① 定义抽象接口 ICallback。

```
interface ICallback {
    void callfunc();                                    //定义回调接口函数
}
```

② 定义 B 类。

```
class B{
    ICallback callback;
    public B(ICallback callback){
        this.callback = callback;
    }
    public void testB(){
        callback.callfunc();
    }
}
```

B 类中一定有回调接口 ICallback 成员变量 callback，并通过构造方法完成初始化。很明显，callback 是多态对象。testB()函数包含了接口回调功能，目的地是 callback 对象的 callfunc()函数。

③ 定义 A 类。

若先写成如下形式。

```
class A{
    public void testA(){
        B obj = new B(???);
        obj.testB();
    }
}
```

其含义是 testA()调用 B 对象的 testB()函数，testB()函数再回调 A 中的函数。但是，发现“???”处不知如何写了，如何能使 testB()回调 A 中的函数呢？理解两点即可：一是回调函数形式不是随便写的，必须与 ICallback 定义的接口函数一致；二是 testB()要回调 A 中的函数，那么 ICallback 对象一定是 A。基于以上两点，可很快写出如下 A 类的定义。

```
class A implements ICallback{
    public void testA(){
        B obj = new B(this);  //this 是 ICallback 对象,说明 A 类必须实现 ICallback
                                接口,重写 callback()函数
        obj.testB();
    }
    public void callfunc(){
        System.out.println("回调成功!");
```

```
    }
}
```

④ 一个简单的测试类。

```
public class Test {
    public static void main(String []args){
        new A().testA();
    }
}
```

讨论 1：若 A 调用 B,B 不回调 A,而是回调 C,则 A、C 类代码如下所示。

```
class A {                              //回调函数不在 A 中,无须实现 ICallback 接口
    public void testA(){
        B obj = new B(new C());        //C 一定是 ICallback 对象,回调函数在 C 中
        obj.testB();
    }
}
class C implements ICallback{
    public void callfunc(){   }
}
```

讨论 2：在 Android 开发中经常使用匿名类实现函数回调,A 类代码如下所示。

```
class A {
    public void testA(){
        B obj = new B(new ICallback(){ //匿名类实现接口回调
            public void callfunc(){
            }
        });
        obj.testB();
    }
}
```

1.7.2　适配器技术

数据适配器是适配器技术的重要内容,其含义是数据显示和数据管理分离,通过适配器接口进行配合,完成显示等功能。下面以二维数据列表控制台显示为例,其需要的相关文件如下所示。

① 自定义适配器接口 IMyAdapter。

```
interface IMyAdapter{
    int getRows();                              //数据有几行
    int getCols();                              //数据有几列
    String getValue(int row,int col);           //某行、某列的元素值
}
```

该接口定义了画二维表格所需的行、列值及获得具体元素值的方法。该接口的定义与画表格的方法息息相关,请继续向下看。

② 控制台二维表显示类 MyListView。

```
class MyListView{
    void setAdapter(IMyAdapter ad){
        for(int i=0; i<ad.getRows(); i++){
            for(int j=0; j<ad.getCols(); j++){
                String value = ad.getValue(i,j);
                System.out.print(value+"\t");
            }
            System.out.println();
        }
    }
}
```

setAdapter()函数即绘制二维表函数,其仅有一个形参,是 IMyAdapter 对象 ad,内部绘制二维表过程仅与 getRows()、getCols()、getValue()函数相关,因此 IMyAdapter 接口必须定义这 3 个函数。可以看出,setAdapter()与绘制的数据是松耦合的,通过 IAdapter 接口对象,而不是具体的类对象与数据相关联。

③ 自定义适配器类 MyArrayAdapter。

```
class MyArrayAdapter implements IMyAdapter{
    String data[][];
    public MyArrayAdapter(String data[][]){
        this.data = data;
    }
    public int getRows(){  return data.length;  }
    public int getCols(){   return data[0].length;}
    public String getValue(int row, int col){  return data[row][col];  }
}
```

可以看出,适配器类是一个数据管理类,与显示功能是分离的,但它必须提供数据显示所需要相关函数的返回值。

④ 一个简单的测试类。

```
public class Test {
    public static void main(String[] args) throws Exception{
        String d[][]={{"1000","zhang","20"},{"1001","sun","21"},
                       {"1002","wan","20"},{"1001","wu","22"}};
        MyListView view = new MyListView();     //创建显示类对象 view
        IMyAdapter ad = new MyArrayAdapter(d); //创建适配器对象
        view.setAdapter(ad);                   //view 与 ad 绑定
    }
}
```

习题 1

1. 如何理解 Java SDK 与 Android SDK 的作用?
2. Android 工程的 4 个主要目录的用途分别是什么?
3. 分别在手机和模拟器中运行文中的 MyFirst 工程。
4. 利用 Toast 显示“Hello World”字符串,并在手机或模拟器上运行。

第 2 章

Android 布局

Android 布局是一个活动中的用户界面的架构,它定义了布局结构且存储所有显示给用户的元素。有两种方式可以声明布局：在 XML 格式布局文件中声明 UI;在运行时实例化布局元素。Android 常用布局有以下 5 种：线性布局(LinearLayout)、相对布局(RelativeLayout)、表格布局(TableLayout)、网格布局(GridLayout)、框架布局(FrameLayout)。

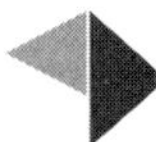

2.1　Android 布局与 Java 布局的区别

在 Java 图形用户界面中,容器和布局是分立的。例如,JFrame 是容器,FlowLayout、BorderLayout、GridLayout 分别是流式、方位、格线布局管理器。可以把 JFrame 设置成流式布局或方位布局或格线布局。当添加子控件时,可按相应布局特点在某位置添加子组件。也就是说,在基本的 Java 图形用户界面中,布局并不是容器。

在 Android 图形用户界面中,容器和布局是统一的。例如,LinearLayout 是线性布局,更形象地说,LinearLayout 是线性布局容器,即它是容器,添加子组件时其具有线性布局的特点。因此,在 Android 中,所说的"XXX"布局就是"XXX"布局容器。

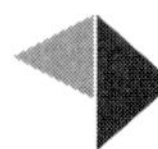

2.2　线性布局

LinearLayout 是 Android 最常用的线性布局,其特点是：子组件依次水平(或垂直)排列。下面分别从方向、填充模型、内外边距、权重、对齐等方面进一步学习 LinearLayout 的使用,其中一部分也适用于其他布局。

1. 方向

通过 android:orientation 设置线性布局的方向。有 horizontal、vertical 两个值可选,

前者表示组件水平排列，后者表示组件垂直排列。android:orientation 属性的默认值是 horizontal。

【例 2-1】　android:orientation 属性示例。

```
<?xml version="1.0" encoding="utf-8"?>
<LinearLayout xmlns:android="http://schemas.android.com/apk/res/android"
    android:orientation="horizontal"
    android:layout_width="match_parent"
    android:layout_height="match_parent">
    <Button
        android:layout_width="wrap_content"
        android:layout_height="wrap_content"
        android:text="OK1"/>
    <Button
        android:layout_width="wrap_content"
        android:layout_height="wrap_content"
        android:text="OK2"/>
    <Button
        android:layout_width="wrap_content"
        android:layout_height="wrap_content"
        android:text="OK3"/>
</LinearLayout>
```

其显示界面如图 2-1 所示。

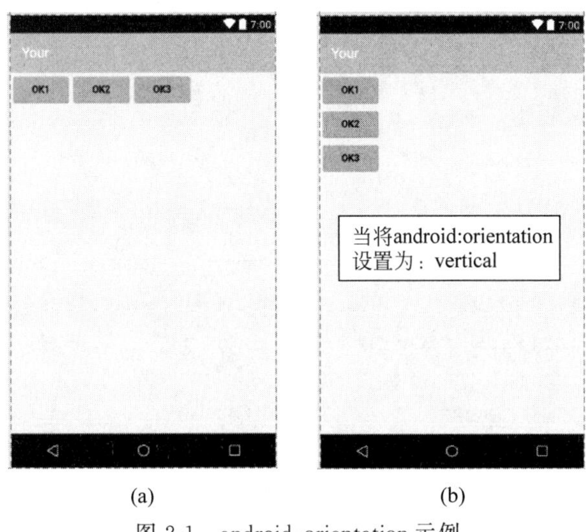

图 2-1　android:orientation 示例

XML 布局文件采用线性布局，依次添加了 3 个按钮组件。当 android:orientation 属性不写（采用默认值）或设置为 horizontal 时，表明 3 个按钮是水平依次添加，如图 2-1(a)；当 android:orientation 设置为 vertical 时，表明 3 个按钮是垂直依次添加，如图 2-1(b)。

2. 填充模型

填充模型主要设置 android:layout_width 和 android:layout_height 两个属性的值，前者设置组件的宽度，后者设置组件的高度。它们均可设置成下列常用参数之一。

- match_parent：当前组件的高度（或宽度）与父组件的高度（或宽度）一致。
- wrap_content：自适应大小，当前组件的高度（或宽度）刚好包含里面的内容，也就是由控件内容决定组件的高度（或宽度）。
- npx：px 即 pixels 像素的简写。当前组件的高度（或宽度）等于 n 个屏幕像素的高度（或宽度）。
- ndp：dp 也叫作 dip，是 device independent pixels（设备独立像素）的简写。当前组件的高度（或宽度）等于 n 个设备独立像素的高度（或宽度）。

【例 2-2】 Android 填充模型属性示例。

```
<?xml version="1.0" encoding="utf-8"?>
<LinearLayout xmlns:android="http://schemas.android.com/apk/res/android"
    android:orientation="vertical"
    android:layout_width="match_parent"
    android:layout_height="match_parent">
    <Button
        android:layout_width="120px"
        android:layout_height="120px"
        android:text="OK1" />
    <Button
        android:layout_width="60dp"
        android:layout_height="60dp"
        android:text="OK2" />
    <Button
        android:layout_width="match_parent"
        android:layout_height="wrap_content"
        android:text="OK3" />
    <Button
        android:layout_width="wrap_content"
        android:layout_height="match_parent"
```

```
        android:text="OK4" />
</LinearLayout>
```

其显示界面如图 2-2 所示。

父容器是 LinearLayout 线性容器。其 android:layout_width="match_parent",表明容器宽度等于手机宽度;其 android:layout_height="match_parent",表明容器高度等于手机高度;其 android:orientation="vertical",表明容器垂直依次添加子组件。子组件 OK1、OK2 是通过像素 px、独立像素 dp 设置宽度和高度的。它们看起来大小一致,但是所设置的值是不同的。那么,px 与 dp 有什么不同,转化关系如何,后文还要继续讨论;子组件 OK3 中 android:layout_width="match_parent",所以它的宽度等于父容器手机的宽度;子组件 OK4 中 android:layout_height="match_parent",所以它的高度等于父容器手机的高度减去 OK1、OK2、OK3 子组件的高度之和。

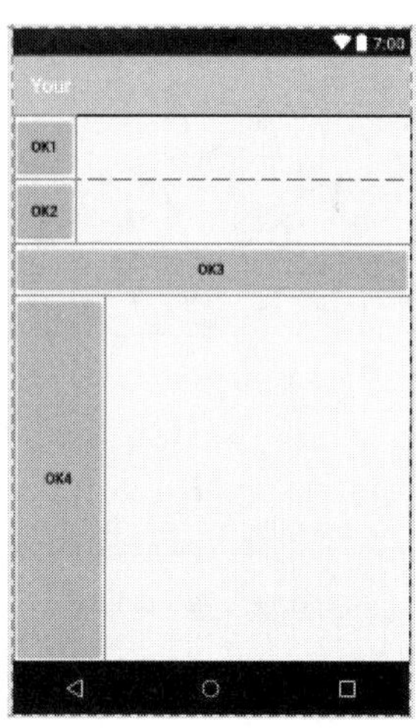

图 2-2 Android 填充模型属性示例

3. 内外边距

内外边距主要是设置 android:padding 和 android:layout_margin 属性的值,即设置组件间的相对距离。不同之处在于,android:padding 指父子组件间的相对距离,android:layout_margin 指同级组件间的相对距离。一般来说,"距离"有上、下、左、右 4 个值。对父子组件来说,对应着 android:paddingTop(paddingBotton、paddingLeft、paddingRight) 4 个值,可以分别设置。当 4 个值相同时,直接设置 android:padding 属性即可;对同级组件来说,对应着 android:layout_marginTop(layout_marginBottom、layout_marginLeft、layout_marginRight)4 个值,可以分别设置。当 4 个值相同时,直接设置 android:margin 属性即可。

【例 2-3】 Android 内外边距属性示例。

```
<?xml version="1.0" encoding="utf-8"?>
<LinearLayout xmlns:android="http://schemas.android.com/apk/res/android"
    android:orientation="vertical"
    android:layout_width="match_parent"
    android:layout_height="match_parent"
    android:paddingLeft="50dp"
    android:paddingTop="100dp">
```

```
    <Button
        android:layout_width="wrap_content"
        android:layout_height="wrap_content"
        android:text="OK1" />
    <Button
        android:layout_width="wrap_content"
        android:layout_height="wrap_content"
        android:text="OK2"
        android:layout_marginBottom="70dp"/>
    <Button
        android:layout_width="wrap_content"
        android:layout_height="wrap_content"
        android:text="OK3" />
</LinearLayout>
```

其界面显示如图 2-3 所示。

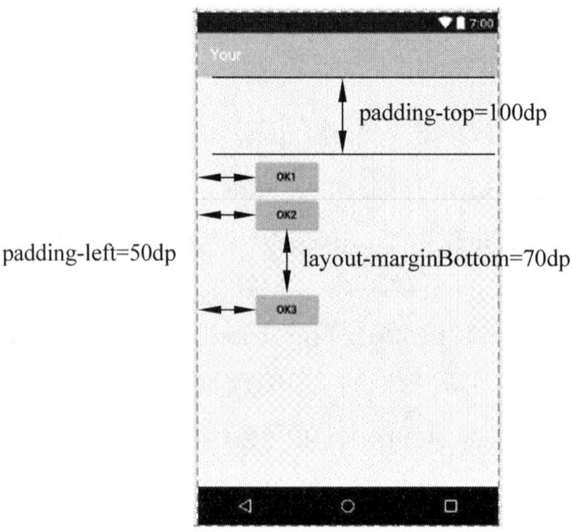

图 2-3 Android 内外边距示例

本示例中，对 LinearLayout 父容器来说：由于 android:paddingLeft＝50dp，所以所有子组件距离父容器的左边距是 50dp，如图 2-3 中的子组件按钮“OK1、OK2、OK3”所示；由于 android:paddingTop＝100dp，表明最初添加的子组件距离父容器上边距为 100dp，如图 2-3 中子组件按钮 OK1 所示；对同级组件来说，如图 2-3 中的子组件按钮“OK2、O3”所示，由于 OK2 中设置了 android:layout_marginBottom＝70dp，所以 OK2、OK3 的间距是 70dp。当然，设置同级组件“OK2、OK3”间距为 70dp 还有其他方法，若在

OK2 中不设置，则只在 OK3 中设置 android:layout_marginTop="70dp"即可。

4. 权重

当需要子组件高度(或宽度)按比例显示时，就需要设置每个组件的 android:layout_weight 属性，一般是一个整数。例如，若 3 个子组件的高度比为 1∶2∶3，则设每个组件的 android:layout_weight 属性为 1、2、3 即可。当然，也可将每个组件的 android:layout_weight 属性设置为 2、4、6，因为 2∶4∶6 等于 1∶2∶3。

【例 2-4】 Android 权重属性示例。

```
<?xml version="1.0" encoding="utf-8"?>
<LinearLayout xmlns:android="http://schemas.android.com/apk/res/android"
    android:orientation="vertical"
    android:layout_width="match_parent"
    android:layout_height="match_parent">
    <Button
        android:layout_width="wrap_content"
        android:layout_height="200dp"
        android:text="OK1" />
    <Button
        android:layout_width="wrap_content"
        android:layout_height="wrap_content"
        android:text="OK2"
        android:layout_weight="1"/>
    <Button
        android:layout_width="wrap_content"
        android:layout_height="wrap_content"
        android:text="OK3"
        android:layout_weight="2"/>
    <Button
        android:layout_width="wrap_content"
        android:layout_height="wrap_content"
        android:text="OK4"
        android:layout_weight="3"/>
</LinearLayout>
```

其界面显示如图 2-4 所示。

图 2-4(a)对应上文代码，OK1 按钮的固定高度为 200dp，“OK2、OK3、OK4”按钮的高度比为 1∶2∶3；当将代码中的 OK1、OK2 代码对调，先产生 OK2，再依次产生“OK1、

OK3、OK4"按钮,所得界面如图 2-4(b),"OK2、OK3、OK4"按钮的高度比仍为 1∶2∶3。也就是说,只要某些组件的高度(或宽度)比固定,在这些组件中再加入其他固定高度(或宽度)的组件,原先定比组件的高度(或宽度)的比值是不会改变的。

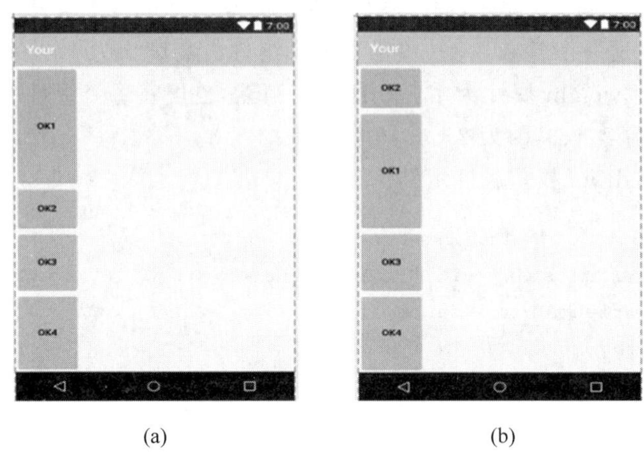

(a) (b)

图 2-4 Android 权重属性示例

5. 对齐

对齐主要设置 android:gravity 及 android:layout_gravity 属性的值。前者含义是子组件相对父容器的相对位置,后者含义是子组件相对父容器某区间范围的相对位置。这两个对齐属性稍显复杂,通过下述实例可加深对它们的理解。

【例 2-5】 android:gravity 属性示例。

```
<?xml version="1.0" encoding="utf-8"?>
<LinearLayout xmlns:android="http://schemas.android.com/apk/res/android"
    android:orientation="vertical"
    android:layout_width="match_parent"
    android:layout_height="match_parent"
    android:gravity="left">
    <Button
        android:layout_width="wrap_content"
        android:layout_height="wrap_content"
        android:text="ok1"/>
    <Button
        android:layout_width="wrap_content"
        android:layout_height="wrap_content"
```

```
        android:text="ok2"/>
</LinearLayout>
```

其界面显示如图 2-5 所示。

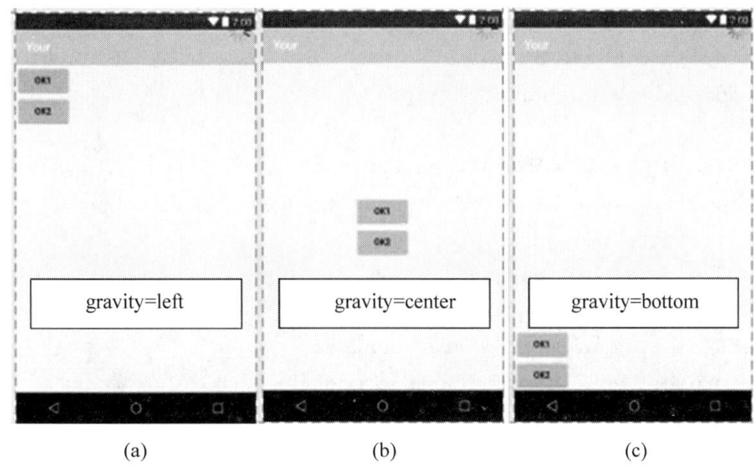

(a)　　(b)　　(c)

图 2-5　android:gravity 属性示例

当代码中的 android:gravity 分别设置为 left、center、bottom 时，界面分别对应图 2-5(a)、(b)、(c)。乍一看，感觉没有规律，其实可以这样理解：由于父容器中对 gravity 进行设置，因此，每个子组件位置都是相对于整个父容器。因此，只理解父容器的上、中、下、左、右的位置及特征标识串即可。常用的特征标识串有 left、center、right、center_horizontal、center_vertical、bottom。父容器位置及特征串对应图如图 2-6 所示。

left	center_horizontal	right
center_vertical	center	
botton		

图 2-6　父容器位置及特征串对应图

【例 2-6】 android:layout_gravity 属性示例。

```
<?xml version="1.0" encoding="utf-8"?>
<LinearLayout xmlns:android="http://schemas.android.com/apk/res/android"
    android:orientation="vertical"
```

```
    android:layout_width="match_parent"
    android:layout_height="match_parent">
    <Button
        android:layout_width="wrap_content"
        android:layout_height="wrap_content"
        android:text="OK1"
        android:layout_gravity="right"/>
    <Button
        android:layout_width="wrap_content"
        android:layout_height="wrap_content"
        android:text="OK2"
        android:layout_gravity="center"/>
    <Button
        android:layout_width="wrap_content"
        android:layout_height="wrap_content"
        android:text="OK3"
        android:layout_gravity="left"/>
</LinearLayout>
```

其界面显示如图 2-7 所示。

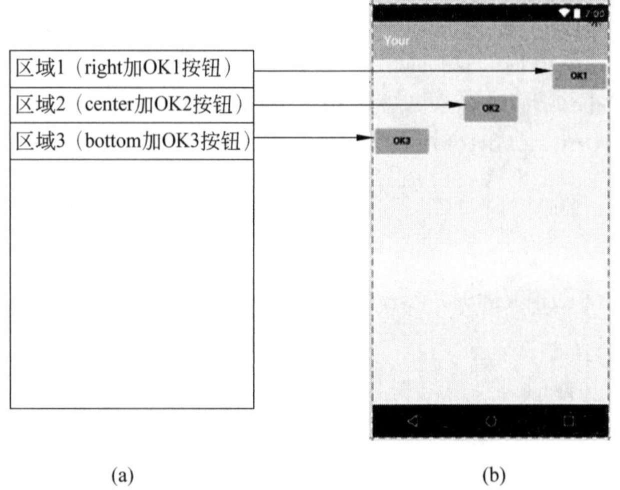

图 2-7 垂直布局 android:layout_gravity 属性示例

示例代码表明是垂直线性布局，在图 2-7(a)的父窗口中的区域 1、区域 2、区域 3 中要加入 3 个按钮。OK1 按钮添加在区域 1 的右侧，OK2 按钮添加在区域 2 的中间，OK3 按钮添加在区域 3 的左侧。因此,界面结果如图 2-7(b)所示。

由上可以得出,当采用垂直线性布局时,子组件中 android:layout_gravity 常用的特征标识字符串为 left、center、right。

若将示例代码改为水平线性布局,OK1、OK2、OK3 按钮的 layout_gravity 分别等于 bottom、center、top,界面如图 2-8 所示。

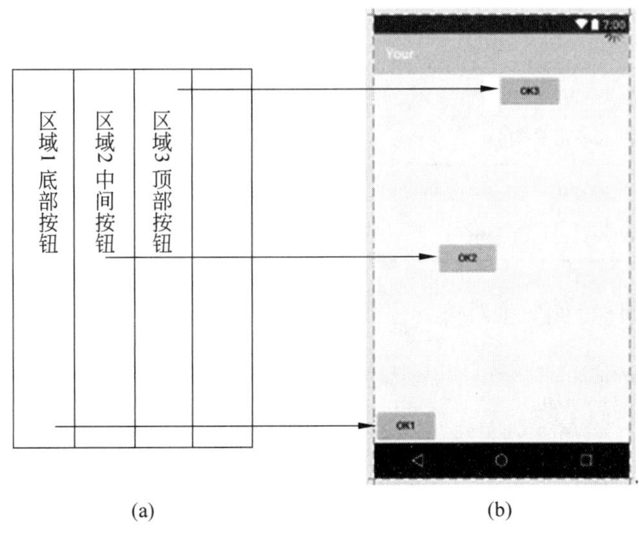

图 2-8　水平布局 android:layout_gravity 属性示例

由上可以得出,当采用水平线性布局时,子组件中 android:layout_gravity 常用的特征标识字符串为 top、center、bottom。

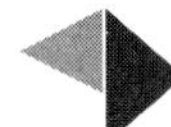

2.3　相对布局

LinearLayout 的功能很强大,能构建出丰富多彩的 UI(用户界面),其缺点是 LinearLayout 嵌套层数多,会大大降低 UI 的渲染速度。而利用 RelativeLayout(相对布局),可能只需一层。RelativeLayout 定义了更为灵活的相对位置,相对性主要体现在以下两个方面:根据父容器定位,根据兄弟组件定位。

2.3.1　根据父容器定位

RelativeLayout 中,可以将子组件添加在父容器特定位置处,在子组件中设置表 2-1

特征标识变量为 true 即可。

表 2-1　相对布局子组件方位特征标识变量表

序号	特征标识变量	含　义	备　注
1	layout_alignParentLeft	水平左对齐	为 true 成立
2	layout_centerHorizontal	水平中间对齐	为 true 成立
3	layout_alignParentRight	水平右对齐	为 true 成立
4	layout_alignParentTop	垂直顶部对齐	为 true 成立
5	layout_centerVerical	垂直中间对齐	为 true 成立
6	layout_alignParentBottom	垂直底部对齐	为 true 成立
7	layout_centerInParent	中心对齐	为 true 成立

相对布局子组件特征变量对应方位图如图 2-9 所示。

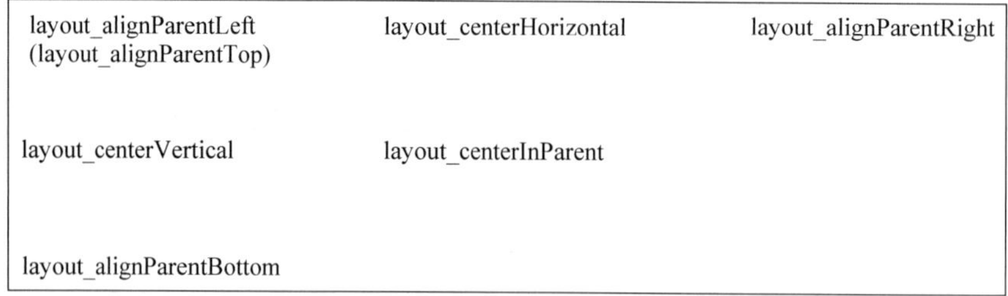

图 2-9　相对布局子组件特征标识变量对应方位图

【例 2-7】 比较线性布局、相对布局的特征方位特点。

```
<?xml version="1.0" encoding="utf-8"?>
<RelativeLayout xmlns:android="http://schemas.android.com/apk/res/android"
    android:layout_width="match_parent"
    android:layout_height="match_parent">
    <Button
        android:layout_width="wrap_content"
        android:layout_height="wrap_content"
        android:text="OK1"
        android:layout_alignParentLeft="true"/>
    <Button
        android:layout_width="wrap_content"
```

```
        android:layout_height="wrap_content"
        android:text="OK2"
        android:layout_alignParentRight="true"/>
</RelativeLayout>
```

其界面显示如图 2-10(a)所示。

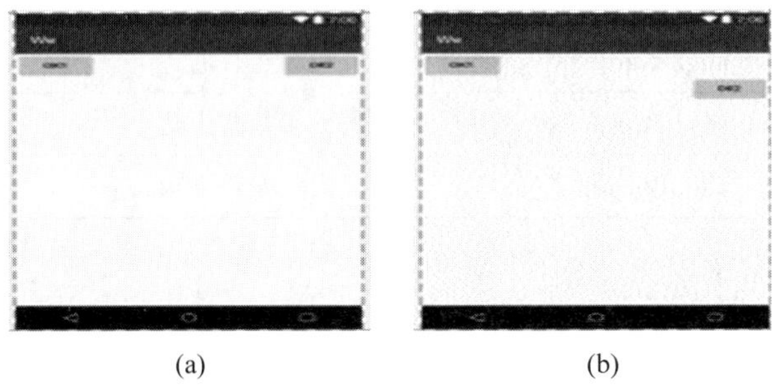

(a)　　(b)

图 2-10　相对布局界面示例

本示例利用相对布局相对父窗口方位分别在 layout_alignParentLeft、layout_alignParentRight 添加了两个按钮。线性布局中也有类似的相对父窗口放置子组件，那么它们有什么不同？假设把示例中的代码 RelativeLayout 改为 LinearLayout，并增加 android:orientation = "vertical" 属性，第 1 个 Button 中的代码 android:layout_alignParentLeft="true"修改为 android:layout_gravity="left"，第 2 个 Button 中的代码 android:layout_alignParentRight="true"修改为 android:layout_gravity="right"，这时界面显示如图 2-10(b)所示。可以看出，OK1 按钮加在了左部，OK2 按钮加在了右部。由于线性布局中所加各组件一定是在不同行(或列)，因此 OK、OK2 按钮是错行的，这也是与相对布局的关键区别之一。

2.3.2 根据兄弟组件定位

RelativeLayout 中，可以将子组件添加在兄弟组件的相对位置处，在子组件中设置表 2-2 所示的特征标识变量值即可。

表 2-2　相对兄弟组件布局特征标识变量表

序号	特征标识变量	含　义	备　注
1	layout_toLeftOf	兄弟组件左边	设置兄弟组件 id 值
2	layout_toRightOf	兄弟组件右边	设置兄弟组件 id 值

续表

序号	特征标识变量	含　义	备　注
3	layout_above	兄弟组件上边	设置兄弟组件 id 值
4	layout_below	兄弟组件下边	设置兄弟组件 id 值
5	layout_alignLeft	对齐兄弟组件左边	设置兄弟组件 id 值
6	layout_alignRight	对齐兄弟组件右边	设置兄弟组件 id 值
7	layout_alignTop	对齐兄弟组件上边	设置兄弟组件 id 值
8	layout_alignBottom	对齐兄弟组件下边	设置兄弟组件 id 值

【例 2-8】 兄弟组件定位示例。

```
<?xml version="1.0" encoding="utf-8"?>
<RelativeLayout xmlns:android="http://schemas.android.com/apk/res/android"
    android:layout_width="match_parent"
    android:layout_height="match_parent">
    <Button
        android:id="@+id/myok"
        android:layout_width="wrap_content"
        android:layout_height="wrap_content"
        android:text="OK"
        android:layout_centerInParent="true"/>
    <Button
        android:layout_width="wrap_content"
        android:layout_height="wrap_content"
        android:text="UP"
        android:layout_above="@id/myok"
        android:layout_centerHorizontal="true"/>
    <Button
        android:layout_width="wrap_content"
        android:layout_height="wrap_content"
        android:text="UP2"
        android:layout_above="@id/myok" />
    <Button
        android:layout_width="wrap_content"
        android:layout_height="wrap_content"
        android:text="DOWN"
```

```
        android:layout_below="@id/myok"
        android:layout_centerHorizontal="true"/>
    <Button
        android:layout_width="wrap_content"
        android:layout_height="wrap_content"
        android:text="LEFT"
        android:layout_toLeftOf="@id/myok"
        android:layout_centerVertical="true"/>
    <Button
        android:layout_width="wrap_content"
        android:layout_height="wrap_content"
        android:text="RIGHT"
        android:layout_toRightOf="@id/myok"
        android:layout_centerVertical="true"/>
</RelativeLayout>
```

其界面显示如图 2-11 所示。

本示例的核心是：以 OK 按钮为中心，上、下、左、右分别加了 4 个子按钮。那么，界面中的 UP2 按钮是什么作用？为什么把它加上去？其实是为了更好地理解兄弟组件的定位。UP2 按钮代码中进行了相对设置：android:layout_above="@id/myok"。从界面看，UP2 按钮确实在 OK 按钮的上方，但是却不是正上方，而是在上方的左侧，若想在正上方，必须在水平位置进行调整，所以，当 UP 按钮代码中加入 android:layout_centerHorizontal="true"后，则 UP 按钮就在 OK 按钮的正上方。同理，DOWN 按钮要进行相应的水平调节，“LEFT、RIGHT”按钮要进行相应的垂直调节。

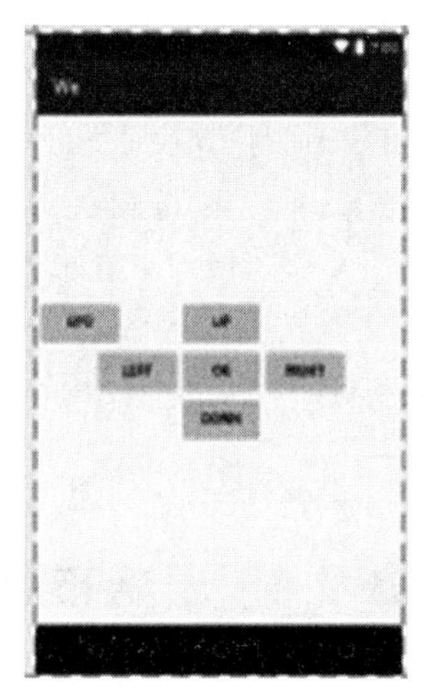

图 2-11　兄弟组件定位示例图

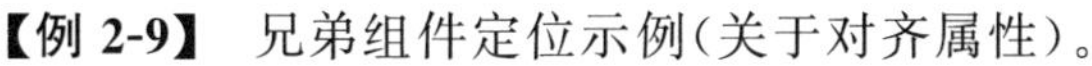

【例 2-9】 兄弟组件定位示例(关于对齐属性)。

```
<?xml version="1.0" encoding="utf-8"?>
<RelativeLayout xmlns:android="http://schemas.android.com/apk/res/android"
    android:layout_width="match_parent"
    android:layout_height="match_parent">
    <Button
        android:id="@+id/myok"
        android:layout_width="wrap_content"
```

```
        android:layout_height="wrap_content"
        android:text="OK"
        android:layout_centerInParent="true"/>
    <Button
        android:layout_width="wrap_content"
        android:layout_height="wrap_content"
        android:text="OKOKOKOKOKOKOKOKOKOK"
        android:layout_alignLeft="@id/myok"/>
    <Button
        android:layout_width="wrap_content"
        android:layout_height="wrap_content"
        android:text="OKOKOKOKOKOKOKOKOKOK2"
        android:layout_above="@id/myok"
        android:layout_alignLeft="@id/myok"/>
</RelativeLayout>
```

其界面显示如图 2-12 所示。

该示例首先定义了中心按钮 OK,然后通过设置 layout_alignLeft 属性,定义了两个与 OK 按钮左对齐的按钮,上方第 1 个 OK……OK 按钮仅定义了 layout_alignLeft 属性,若想调整与 OK 按钮的垂直距离,还要进行其他相关属性的设置。上方第 2 个"OK……OK"按钮进一步设置了 layout_above 属性,保证了不仅与 OK 按钮左对齐,还刚好在 OK 按钮的正上方。

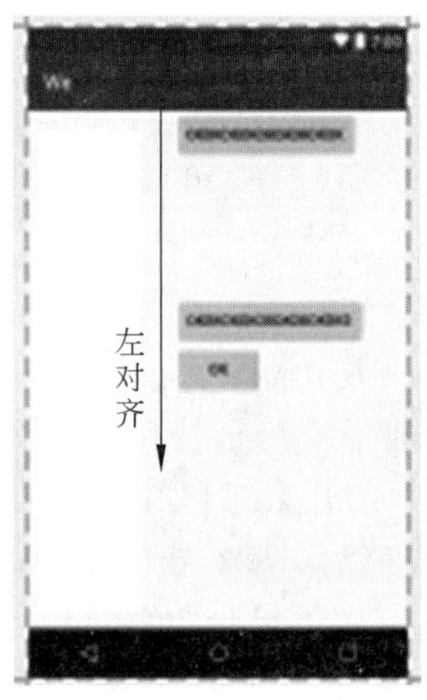

图 2-12 兄弟组件对齐属性示例

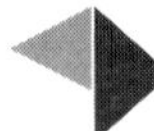

2.4 表格布局

TableLayout 即表格布局,以行、列的形式对每个子控件加以管理。每一行为一个 TableRow 对象,可在其下加所需的各个子控件,每个子控件占据一列。TableRow 中包含 TableRow 对象的个数即表格行数,每个 TableRow 中包含最多的子控件个数即表格列数。例如,TableLayout 有 3 个 TableRow 对象,若每个 TableRow 对象包含的子控件数目为 4、5、6,则该表格的行数为 3,列数为 6。

TableLayout 可设置的属性有全局属性和单元格属性。全局属性直接在 TableLayout 中设定,主要设置表格各列的属性。单元格属性主要在各子组件中设定。具体表述如下所示。

1. 全局属性

- android:stretchColumns：设置可伸展的列。该列可以向行方向伸展，最多占据一整行。
- android:shrinkColumns：设置可收缩的列。如果该列子控件的内容太多，已经挤满所在行，那么该子控件的内容将往列方向显示。
- android:collapseColumns：设置要隐藏的列。

例如，android:stretchColumns="0"表示第 0 列可伸展；android:shrinkColumns="1,2"表示第 1,2 列可收缩，表示多列时中间用逗号分隔；android:collapseColumns="*"表示隐藏所有列。

2. 单元格属性

- android:layout_column：指定该单元格在第几列显示。
- android:layout_span：指定该单元格占据的列数(未指定时，为 1)。

【例 2-10】 表格布局全局属性示例。

```
<?xml version="1.0" encoding="utf-8"?>
<TableLayout xmlns:android="http://schemas.android.com/apk/res/android"
    android:layout_width="match_parent"
    android:layout_height="match_parent"
    android:stretchColumns="0"
    android:shrinkColumns="1"
    android:collapseColumns="2">
<TableRow>
    <Button android:text="1"/>
    <Button android:text="2"/>
    <Button android:text="3"/>
</TableRow>
<TableRow>
    <Button android:text="444444444444444444444444444444444"/>
    <Button android:text="555555555555555"/>
    <Button android:text="6"/>
</TableRow>
</TableLayout>
```

其界面显示如图 2-13 所示。

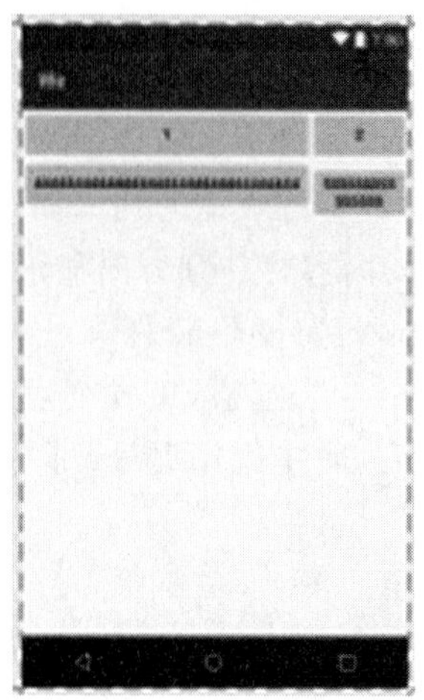

图 2-13　表格布局全局属性示例

由代码可知,①由于定义了全局属性 android:stretchColumns,因此表格所有列的宽度之和等于父容器的总宽度;②由于表格中第 0 列子组件处于伸展模式,因此,若再增加第 2 行第 0 列按钮文本“4”的个数,则会发现表格第 0 列的宽度会继续增加,直至占满该行,而文本“4”会一直在该行显示,而不换行;由于第 1 列子组件处于收缩模式,因此,当增加第 2 行第 1 列按钮文本“5”的个数,会发现当增加到一定程度时,“5”会换行显示;因为第 2 列子组件处于隐藏模式,所以“5、6”按钮会隐藏,不在界面上显示。

【例 2-11】 表格布局单元格属性示例。

```
<?xml version="1.0" encoding="utf-8"?>
<TableLayout xmlns:android="http://schemas.android.com/apk/res/android"
    android:layout_width="match_parent"
    android:layout_height="match_parent"
    android:stretchColumns="0">
<TableRow>
    <Button android:text="1"/>
    <Button android:text="2"/>
    <Button android:text="3"/>
    <Button android:text="4"/>
</TableRow>
<TableRow>
    <Button android:text="5" android:layout_column="1"/>
```

```
    <Button android:text="6" android:layout_span="2"/>
</TableRow>
</TableLayout>
```

其界面显示如图 2-14 所示。

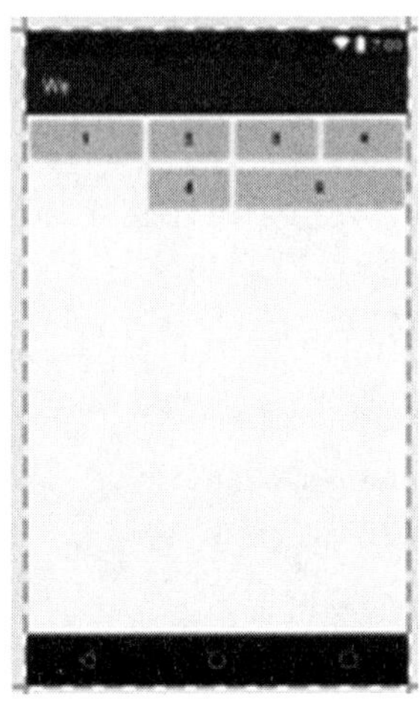

图 2-14　表格布局单元格属性示例

由代码可知，该表格共有 4 列。第 1 行 TableRow 对象加了 4 个按钮，第 2 行 TableRow 对象中仅添加了两个按钮：通过设置“5”按钮的 android:layout_column=1，将“5”按钮放在第 1 列中；通过设置“6”按钮的 android:layout_span=“2”，表明该按钮跨 2 列，即跨第 2、3 列。

2.5　网格布局

GridLayout 即网格布局，与表格布局类似，主要设置以下全局属性和单元格属性。全局属性直接在 GridLayout 中设定，单元格属性主要在各子组件中设定。具体表述如下所示。

1. 全局属性

- android:rowCount：指明表格行数。
- android:columnCount：指明表格列数。
- android:orientation：指明依次水平(或垂直)添加子组件，特征值为 horizontal 及

vertical。

2. 单元格属性

- android:layout_columnSpan：单元格子组件跨几列。
- android:layout_rowSpan：单元格子组件跨几行，此属性是 TableLayout 中不能设置的属性。
- android:layout_row：单元格子组件放置的行数。
- android:layout_column：单元格子组件放置的列数。

【例 2-12】 网格布局单元格属性示例。

```
<GridLayout xmlns:android="http://schemas.android.com/apk/res/android"
    android:id="@+id/GridLayout1"
    android:layout_width="match_parent"
    android:layout_height="wrap_content"
    android:orientation="horizontal"
    android:rowCount="3"
    android:columnCount="4">
    <Button android:layout_gravity="fill" android:text="1" />
    <Button android:layout_gravity="fill" android:text="2" />
    <Button android:layout_gravity="fill" android:text="3" />
    <Button android:layout_gravity="fill" android:text="4" />
    <Button
        android:layout_gravity="fill"
        android:text="5" />
    <Button android:layout_gravity="fill"
        android:text="6"
        android:layout_row="2"
        android:layout_column="0"/>
    <Button android:layout_gravity="fill"
        android:text="7"
        android:layout_row="1"
        android:layout_column="2"
        android:layout_rowSpan="2"
        android:layout_columnSpan="2" />
</GridLayout>
```

其界面显示如图 2-15 所示。

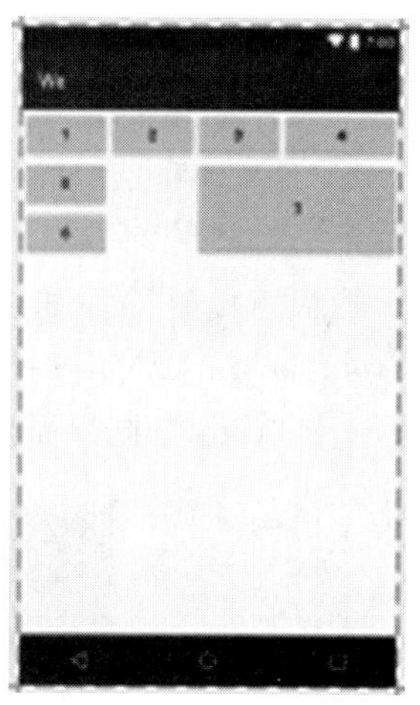

图 2-15 网格布局单元格属性示例

由代码可知，①网格布局不需要 TableRow 对象，但需在网格布局全局属性中定义行列值及子组件添加方式。本例中，android：rowCount ＝ 3，android：columnCount ＝ 4，android：orientation＝“horizontal”，表明是 3 行 4 列，各子组件采取水平添加方式；②添加子组件的过程如下：首先添加“1、2、3、4”按钮，由于第 0 行填了 4 列，因此该行已填满；当添加“5”按钮时，自然而然放在第 1 行第 0 列；当添加“6”按钮时，由于需要放在第 2 行第 0 列（而非自然序第 1 行第 1 列），因此需设置 android：layout_row ＝ 2 及 android：layout_column＝0；当添加“7”按钮时，由于需要放在第 1 行第 2 列（而非自然序第 2 行第 1 列），因此需设置 android：layout_row＝1 及 android：layout_column＝2，又因为跨 2 行 2 列，所以需设置 android：layout_rowSpan＝2 及 android：layout_columnSpan＝2。

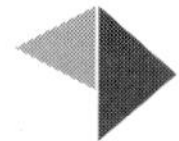

2.6 框架布局

FrameLayout 即框架布局，也称为帧布局，它是 Android 中最简单的一种布局方式。默认情况下，将添加的子组件放在布局的左上角，然后添加的子组件覆盖在之前添加的组件上。当然，也可通过设置 layout_gravity 将子组件添加在相应位置处，但这并不是 FrameLayout 的常用方式。

【例 2-13】 框架布局示例。

```
<FrameLayout xmlns:android="http://schemas.android.com/apk/res/android"
android:layout_width="match_parent"
android:layout_height="match_parent">
    <Button
        android:layout_width="200dp"
        android:layout_height="200dp"
        android:background="#ff0000"/>
   <Button
        android:layout_width="150dp"
        android:layout_height="150dp"
        android:background="#00ff00"/>
```

```
    <Button
        android:layout_width="100dp"
        android:layout_height="100dp"
        android:background="#0000ff"/>
</FrameLayout>
```

本示例代码界面如图 2-16(a)所示，若在 3 个 Button 中依次加入控制位置属性 android:layout_gravity="bottom"、android:layout_gravity="center"、android:layout_gravity="right"，则界面显示如图 2-16(b)所示。

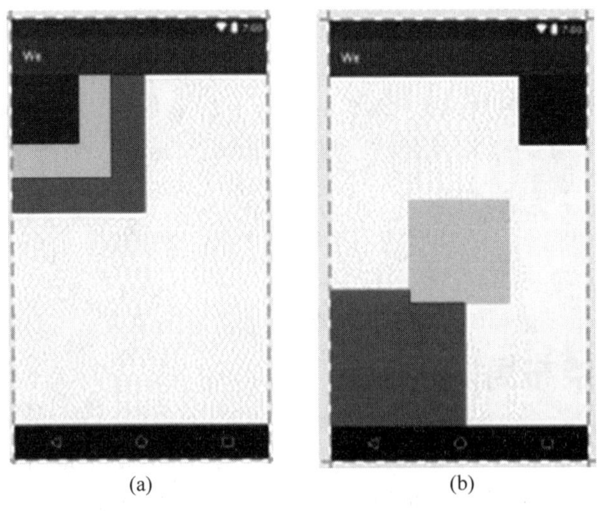

(a) (b)

图 2-16 框架布局示例

图 2-16 所示界面可以很方便地利用 RelativeLayout 实现，那为什么还要有 FrameLayout 呢？框架布局强调的是 Z 轴上的层叠性，如图 2-16(a)所示。更常见的如许多应用软件的"属性页"选项卡页面，都适合利用 FrameLayout 实现；相对布局强调的是 X-Y 平面上的位置性，如图 2-16(b)所示。也就是说，框架布局一般不设置 layout_gravity 属性，默认相对于布局左上角即可，而相对布局一般是设置子组件相对于父容器的相对位置属性的。

2.7 滚动窗口

在 Android 应用过程中，可能会遇到这样的场景：当绘制的 UI 控件超出手机屏幕尺寸的时候，就会导致此 UI 控件无法显示。为了解决这一问题，Android 提供了滚动视图，包含垂直滚动视图 ScrollView、水平滚动视图 HorizontalScrollView。

ScrollView(或 HorizontalScrollView)的子元素只能有一个,可以是一个 View(如 ImageView、TextView 等),也可以是一个 ViewGroup(如 LinearLayout、RelativeLayout 等),其子元素内部则不再限制,否则会出现异常。

【例 2-14】 垂直滚动视图示例。

```
<?xml version="1.0" encoding="utf-8"?>
<ScrollView xmlns:android="http://schemas.android.com/apk/res/android"
    android:layout_width="match_parent"
    android:layout_height="match_parent">
<LinearLayout
     android:layout_width="match_parent"
     android:layout_height="match_parent"
     android:orientation="vertical">
     <Button
            android:layout_width="match_parent"
            android:layout_height="700dp"
            android:text="Button1"
            android:textSize="40sp"/>
      <Button
            android:layout_width="match_parent"
            android:layout_height="700dp"
            android:text="Button2"
            android:textSize="40sp"/>
</LinearLayout>
</ScrollView>
```

本示例在线性布局中定义了两个高为 700dp 的按钮,这个高度已超出手机屏幕高度。这时需将 LinearLayout 容器封装在 ScrollView 中,通过滚动,增大手机屏幕垂直方向的收视高度。

同理,若某布局或视图需要水平滚动,则在外层再加一层 HorizontalScrollView 即可。当然,若某布局或视图既需要水平滚动,又需要垂直滚动,则需按如下格式封装:

```
<?xml version="1.0" encoding="utf-8"?>
<HorizontalScrollView xmlns:android="http://schemas.android.com/apk/res/
android"
    android:layout_width="match_parent"
    android:layout_height="match_parent">
```

```
        <ScrollView xmlns:android="http://schemas.android.com/apk/res/android"
            android:layout_width="match_parent"
            android:layout_height="match_parent">
            //添加某布局或某视图
        </ScrollView>
    </HorizontalScrollView>
```

2.8 综合示例

本章主要讲解了 Android 的五大布局：线性布局、相对布局、表格布局、网格布局、框架布局，Android 界面通常是由上述布局组合或嵌套形成的。那么，做 Android 应用界面从哪里入手比较好？其实，手机就是最好的样本。手机中有许多常用 APK，其对应的界面一般来说都是比较典型的，接下来分析几个示例。

【例 2-15】 手机图标界面示例。

手机中有一类界面，由多行典型的“图标”组成。每个“图标”的结构为：上面是图标，下面是文字说明。由于本书到此，还未讲解图像按钮（ImageButton），因此由 Button 代替，界面如图 2-17(a)所示。

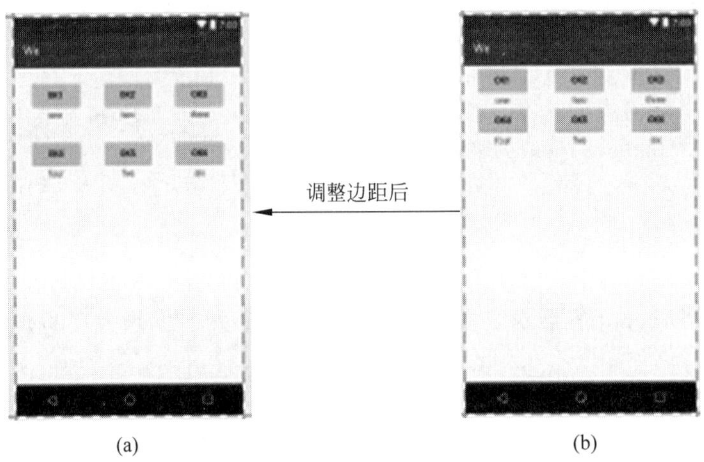

图 2-17 手机图标界面示例

分析：由于图标界面是行列排列，因此选择 TableLayout 或 GridLayout，本题选用 GridLayout；每一列选择 RelativeLayout（因为强调“图标”和“文字说明”的近邻关系）。因此，总体设计如图 2-18 所示。

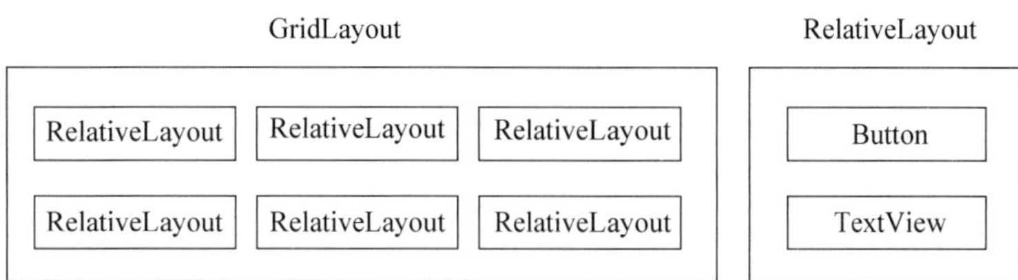

图 2-18　手机图标界面设计思想

代码如下所示。

```
<?xml version="1.0" encoding="utf-8"?>
<GridLayout xmlns:android="http://schemas.android.com/apk/res/android"
    android:layout_width="match_parent"
    android:layout_height="wrap_content"
    android:columnCount="3">
<RelativeLayout
    android:layout_width="wrap_content"
    android:layout_height="wrap_content"
    android:layout_columnWeight="1">
    <Button
        android:id="@+id/myok1"
        android:layout_width="wrap_content"
        android:layout_height="wrap_content"
        android:text="OK1"
        android:layout_centerHorizontal="true"/>
    <TextView
        android:layout_width="wrap_content"
        android:layout_height="wrap_content"
        android:text="one"
        android:layout_below="@id/myok1"
        android:layout_centerHorizontal="true" />
</RelativeLayout>
    <!—
    还有其余 5 个按钮,复制粘贴 5 次上述 RelativeLayout 中的代码
    粘贴后注意修改 Button 代码中的 android:id、android:text 属性值,
    TextView 代码中的 android:text、android:layout_below 属性值。
     -->
</GridLayout>
```

可以看出，上述代码对应图 2-17(b)，与图 2-17(a)有一些细微的差别：①GridLayout 左、右、上边距过小，可通过在 GridLayout 全局属性中设置 paddingLeft、paddingRight、paddingTop 属性值进行修改；②GridLayout 行间距过小，可通过在 TextView 中设置 marginBottom 值进行修改。

在 RelativeLayout 代码中通过设置 layout_columnWeight 权重值，保证了每个单元格的列宽度是一致的。该属性前文中没有论述，这也是正常的，每个布局或组件都有许多属性，不可能面面俱到，希望读者在实践中不断学习，弥补教材的疏漏之处。

【例 2-16】 手机新闻浏览界面示例。

手机新闻浏览界面如图 2-19 所示，左侧是新闻标题及浏览数(或评论数)，右侧是一个新闻图标(示例中为了方便，用按钮代替)。

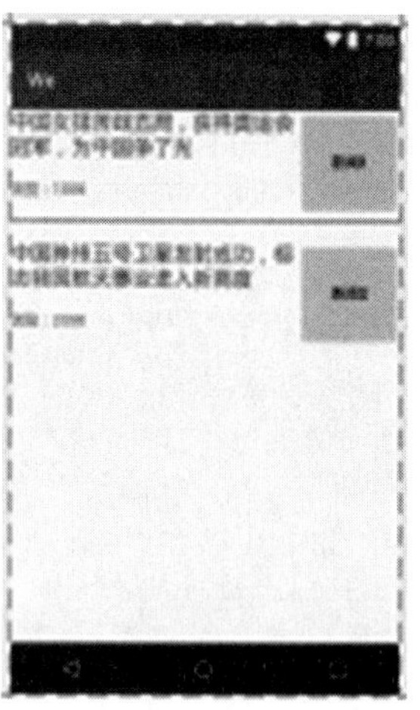

图 2-19　手机新闻浏览界面

分析：该界面主体可采用垂直线性布局 A，每个单元由一个相对布局 B 及一条水平分割线组成，相对布局 B 由一个垂直线性布局 C 及一个按钮组成，线性布局 C 由两个 TextView 组成。其层叠关系如图 2-20 所示。

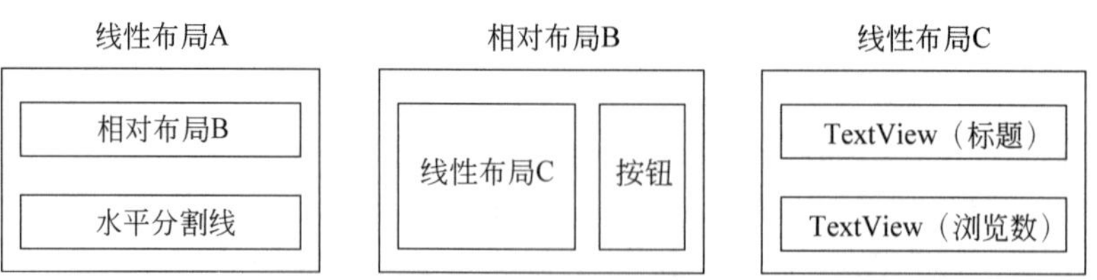

图 2-20　手机新闻浏览界面层次图

代码如下所示。

```
<?xml version="1.0" encoding="utf-8"?>
<LinearLayout xmlns:android="http://schemas.android.com/apk/res/android"
    android:layout_width="match_parent"
    android:layout_height="wrap_content"
    android:orientation="vertical">
    <RelativeLayout
        android:layout_width="match_parent"
        android:layout_height="wrap_content">
        <Button
            android:id="@+id/myicon"
            android:layout_width="100dp"
            android:layout_height="100dp"
            android:text = "新闻 1"
            android:layout_alignParentRight="true"/>
        <LinearLayout
            android:layout_width="match_parent"
            android:layout_height="wrap_content"
            android:orientation="vertical"
            android:layout_toLeftOf="@id/myicon">
            <TextView
                android:layout_width="match_parent"
                android:layout_height="wrap_content"
                android:text="中国女排苦战五局,获得奥运会冠军,为中国争了光"
                android:textSize="20sp"
                android:layout_marginBottom="20dp"/>
            <TextView
                android:layout_width="wrap_content"
                android:layout_height="wrap_content"
                android:text="浏览:1000"
                android:layout_marginBottom="20dp"/>
        </LinearLayout>
    </RelativeLayout>
    <TextView
        android:layout_width="match_parent"
        android:layout_height="1dp"
        android:background="#000000"
        android:layout_marginBottom="20dp"/>
```

```
    <RelativeLayout
        android:layout_width="match_parent"
        android:layout_height="wrap_content">
        <Button
            android:id="@+id/myicon2"
            android:layout_width="100dp"
            android:layout_height="100dp"
            android:text = "新闻 2"
            android:layout_alignParentRight="true"/>
        <LinearLayout
            android:layout_width="match_parent"
            android:layout_height="wrap_content"
            android:orientation="vertical"
            android:layout_toLeftOf="@id/myicon2">
            <TextView
                android:layout_width="match_parent"
                android:layout_height="wrap_content"
                android:text="中国神舟五号卫星发射成功,标志我国航天事业进入新高度"
                android:textSize="20sp"
                android:layout_marginBottom="20dp"/>
            <TextView
                android:layout_width="wrap_content"
                android:layout_height="wrap_content"
                android:text="浏览:2000"/>
        </LinearLayout>
    </RelativeLayout>
</LinearLayout>
```

【例 2-17】 手机登录界面示例。

手机 App 中常用的一种登录界面如图 2-21 所示,主要由以下几部分组成:①应用标题;②用户名、密码输入部分;③登录按钮;④手机底部的工具栏。

分析:形成该图所需的布局及组件关系如图 2-22 所示。

代码如下所示。

```
<?xml version="1.0" encoding="utf-8"?>
<RelativeLayout xmlns:android="http://schemas.android.com/apk/res/android"
    android:layout_width="match_parent"
    android:layout_height="match_parent"
```

图 2-21　手机登录界面

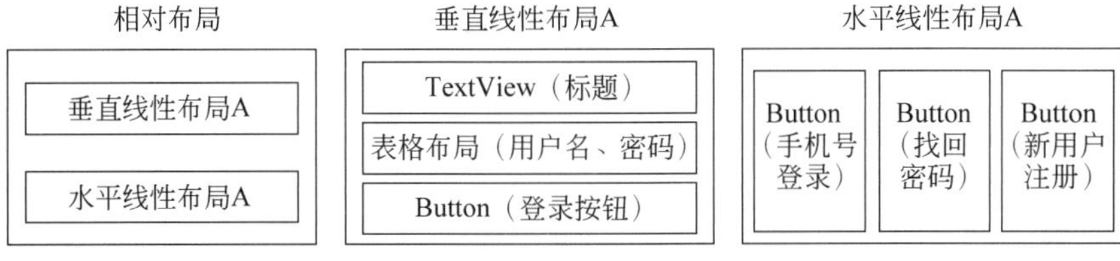

图 2-22　手机登录界面布局层次图

```
android:paddingTop="100dp">
<LinearLayout
    android:id="@+id/mytool"
    android:layout_width="match_parent"
    android:layout_height="wrap_content"
    android:layout_alignParentBottom="true"
    android:orientation="horizontal">
    <Button
        android:layout_width="wrap_content"
        android:layout_height="wrap_content"
        android:text="手机号登录"
        android:layout_weight="1"/>
    <Button
        android:layout_width="wrap_content"
        android:layout_height="wrap_content"
```

```
            android:text="找回密码"
            android:layout_weight="1"/>
        <Button
            android:layout_width="wrap_content"
            android:layout_height="wrap_content"
            android:text="新用户注册"
            android:layout_weight="1"/>
    </LinearLayout>
    <LinearLayout
        android:layout_width="match_parent"
        android:layout_height="match_parent"
        android:orientation="vertical"
        android:layout_above="@id/mytool"
        android:paddingLeft="30dp"
        android:paddingRight="30dp">
        <TextView
            android:layout_width="match_parent"
            android:layout_height="wrap_content"
            android:text="MyApp"
            android:textSize="50sp"
            android:textAlignment="center"
            android:layout_marginBottom="40dp"/>
        <TableLayout
            android:layout_width="match_parent"
            android:layout_height="wrap_content"
            android:stretchColumns="1">
            <TableRow android:layout_marginBottom="30dp">
                <TextView
                    android:text="用户名"
                    android:textSize="25sp"
                    android:layout_marginRight="20dp"/>
                <EditText
                    android:text="admin"/>
            </TableRow>
            <TableRow>
                <TextView
                    android:text="密码"
                    android:textSize="25sp"
```

```
                android:layout_marginRight="20dp"/>
            <EditText
                android:text="......"/>
        </TableRow>
    </TableLayout>
    <Button
        android:layout_width="wrap_content"
        android:layout_height="wrap_content"
        android:text="登录"
        android:textSize = "30sp"
        android:layout_gravity="center"/>
</LinearLayout>
</RelativeLayout>
```

2.9　动态控制布局

前文讲述的线性布局、相对布局、表格布局、网格布局、框架布局形成的界面是由 XML 文件描述的，界面生成是由 Android 系统自动完成的。在实际编程中，有时需要编程人员动态生成界面，因此必须在程序中熟知涉及的各个具体类及函数。一般来说，利用程序生成界面的过程如图 2-23 所示。

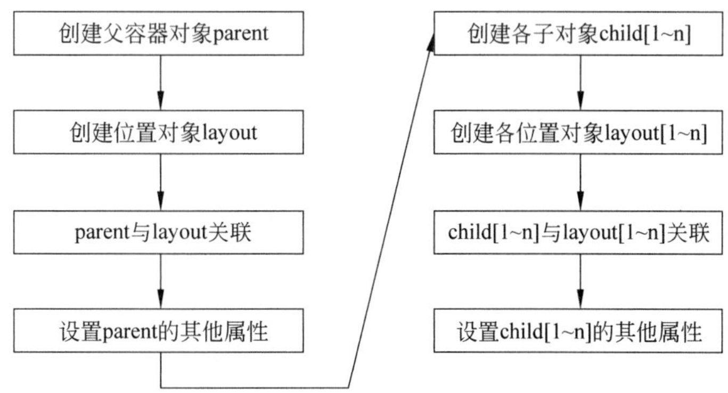

图 2-23　利用程序生成界面的过程

从图 2-23 可以看出，产生父容器、子组件的过程是一致的：产生 UI 对象、产生位置对象、将 UI 对象与位置对象关联，定义其他属性。位置对象对应的系统类是 LayoutParams，对线性布局是 LinearLayout.LayoutParams，对相对布局是 RelativeLayout.LayoutParams，以此类推。每个 UI 对象都有 setLayoutParams(LayoutParams)方法与位置对象相关联。

每个 UI 组件均有若干属性，哪些能用 LayoutParams 对象设置？一个基本原则是：在 XML 界面中配置文件前缀为"layout_"的属性，原则上均可由 Layout 对象设置。

【例 2-18】 编写程序，重新实现例 2-5：android:gravity 属性示例。

```
public class MainActivity extends AppCompatActivity {
protected void onCreate(Bundle savedInstanceState) {
super.onCreate(savedInstanceState);
        LinearLayout parent = new LinearLayout(this);  //产生父容器
        parent.setOrientation(LinearLayout.VERTICAL); //设置方向属性
        parent.setGravity(Gravity.LEFT);               //设置整体对齐属性
        LinearLayout.LayoutParams param = new LinearLayout.LayoutParams(
                                                       //定义位置对象,占满窗口
                LinearLayout. LayoutParams. MATCH _ PARENT, LinearLayout.
LayoutParams.MATCH_PARENT);
        parent.setLayoutParams(param);                 //父容器与位置对象关联

        Button b = new Button(this);                   //产生子组件按钮
        LinearLayout.LayoutParams child = new LinearLayout.LayoutParams
                                                       //定义位置对象,自适应大小
               (LinearLayout. LayoutParams. WRAP _ CONTENT, LinearLayout.
LayoutParams.WRAP_CONTENT);
        b.setText("OK1");                              //设置按钮文本属性
        b.setLayoutParams(child);                      //按钮与位置对象关联

        Button b2 = new Button(this);
        LinearLayout.LayoutParams child2 = new LinearLayout.LayoutParams(
                LinearLayout. LayoutParams. WRAP _ CONTENT, LinearLayout.
LayoutParams.WRAP_CONTENT);
        b2.setText("OK2");
        b2.setLayoutParams(child2);
        parent.addView(b); parent.addView(b2);      //在父窗口中添加两个按钮对象
        setContentView(parent);
    }
}
```

当代码 parent.setGravity(参数)中的"参数"分别用 Gravity.LEFT、Gravity.

CENTER、Gravity.RIGHT 代替时，界面分别如图 2-5(a)、(b)、(c)所示。

Gravity 是 Android 的系统类，其中定义了与方位有关的静态整型常量，而非字符串，如 Gravity. LEFT、Gravity. CENTER _ HORIZONTAL、Gravity. RIGHT、Gravity. CENTER_VERTICAL、Gravity.CENTER、Gravity.TOP、Gravity.BOTTOM 等。

本例主要对父容器 LinearLayout 对象的 gravity 进行设置，由于在 XML 文件中该属性没有前缀“layout_”，因此应用了方法 parent.setGravity(Gravity.LEFT)。那么，若不对父容器 gravity 进行设置，而想让第 1 个 Button 水平居右显示，该如何操作？很明显，在 XML 文件中是通过设置 Button 代码段的 layout_gravity=“right”实现的，由于该属性包含前缀“layout_”，因此一定是通过设置 LayoutParams 对象实现的，关键代码段如下所示。

```
Button b = new Button(this);
LinearLayout.LayoutParams child = new LinearLayout.LayoutParams
        (LinearLayout.LayoutParams.WRAP_CONTENT,
        LinearLayout.LayoutParams.WRAP_CONTENT);
child.gravity = Gravity.RIGHT;                  //设置按钮组件相当于父容器的位置
b.setText("OK1");
b.setLayoutParams(child);
```

由代码可知，LinearLayout.LayoutParams 类中仅有 gravity 属性，而无设置该属性的方法，因此，在代码中设置 child.gradvity = Gravity.RIGHT 即可。

【例 2-19】 利用相对布局动态产生界面：按钮 OK1 位于父容器中心，按钮 OK2 位于按钮 OK1 的正下方。

```
public class MainActivity extends AppCompatActivity {
    protected void onCreate(Bundle savedInstanceState) {
        super.onCreate(savedInstanceState);
        RelativeLayout parent = new RelativeLayout(this);
        RelativeLayout.LayoutParams param = new RelativeLayout.LayoutParams(
          RelativeLayout. Layout. LayoutParams. MATCH _ PARENT, RelativeLayout.
LayoutParams.MATCH_PARENT);
        parent.setLayoutParams(param);
        Button b = new Button(this);            //建立 OK1 按钮对象
        b.setId(100);                           //设置 id 值,方便之后设置相对位置
        RelativeLayout.LayoutParams child = new RelativeLayout.LayoutParams(
              RelativeLayout. LayoutParams. WRAP _ CONTENT, RelativeLayout.
LayoutParams.WRAP_CONTENT);
```

```
        child.addRule(RelativeLayout.CENTER_IN_PARENT,1);
                                                    //按钮位于父容器中心
        b.setText("OK1");
        b.setLayoutParams(child);
        Button b2 = new Button(this);                       //建立 OK2 按钮对象
        RelativeLayout.LayoutParams child2 = new RelativeLayout.LayoutParams(
            RelativeLayout. LayoutParams. WRAP _ CONTENT, RelativeLayout.
LayoutParams.WRAP_CONTENT);
        child2.addRule(RelativeLayout.BELOW, b.getId());      //OK2 位于 OK1 的下方
        child2.addRule(RelativeLayout.CENTER_HORIZONTAL, 1);  //正下方
        b2.setText("OK2");
        b2.setLayoutParams(child2);
        parent.addView(b); parent.addView(b2);
        setContentView(parent);
    }
}
```

由代码可知,需要解决的关键问题如下:①对 OK1 按钮而言,在程序中如何实现 XML 中 layout_centerInParent 属性设置的问题;②对 OK2 按钮而言,在程序中如何实现 XML 中 layout _ below、layout _ centerHorizontal 属性设置的问题。主要通过 LayoutParams 对象的 addRule()方法,其常用原型如下所示。

```
void addRule(int verb, int subject)。
```

该方法的第一个参数 verb 是一个整型数,与表 2-1、表 2-2 对应,可设置的值为 RelativeLayout 类中与父窗口位置相关的静态常量:ALIGN _ PARENT _ LEFT、CENTER_ HORIZONTAL、ALIGN _ PARENT _ RIGHT、ALIGN _ PARENT _ TOP、CENTER_VERTICAL、ALIGN_PARENT_BOTTOM、CENTER_IN_PARENT;与子组件位置相关的静态常量:LEFT_OF、RIGHT_OF、ABOVE、BELOW、ALIGN_LEFT、ALIGN_RIGHT、ALIGN_TOP、ALIGN_BOTTOM。

该方法的第二个参数 subject 是一个整型数:当子组件相对父容器某相对位置为真时,值为整数 1。当子组件相对父容器某相对位置为假时,值为整数 0;当子组件相对同级某子组件某相对位置为真时,值为相对某子组件的 ID 值。例如,上文中 OK1 按钮代码中有"child.addRule(RelativeLayout.CENTER_IN_PARENT,1",表明 OK1 按钮加在父容器的几何中心;上文中 OK2 按钮代码中有"child2.addRule(RelativeLayout.BELOW, b.getId())",表明 OK2 按钮在 OK1 按钮的下方,方法中的第 2 个参数 b.getId()代表 OK1 按钮的 ID 值,因此,在创建 OK1 按钮代码中必须为该按钮设置 ID(一个正整数)值,保证

不重复即可,如示例中的代码“b.setId(100)”。

【例 2-20】 利用表格布局动态产生界面:表格共 3 列,第 0、2 列有伸展属性,第 1 列有收缩属性;第 1 行有 3 个按钮,第 2 行有一个按钮,从第 1 列开始,横跨 2 列。

```
public class MainActivity extends AppCompatActivity {
protected void onCreate(Bundle savedInstanceState) {
super.onCreate(savedInstanceState);
        TableLayout parent = new TableLayout(this);
        parent.setColumnStretchable(0,true);            //第 0 列可伸展
        parent.setColumnShrinkable(1,true);             //第 1 列可收缩
        parent.setColumnStretchable(2,true);            //第 2 列可伸展
        TableLayout.LayoutParams param = new TableLayout.LayoutParams(
            TableLayout.LayoutParams.MATCH_PARENT,TableLayout.LayoutParams.
WRAP_CONTENT);
        parent.setLayoutParams(param);

        TableRow row = new TableRow(this);              //创建第 1 行对象
        Button b = new Button(this); b.setText("OK1");  //创建 3 个子按钮
        Button b2 = new Button(this); b2.setText("OK2");
        Button b3 = new Button(this); b3.setText("OK3");
        row.addView(b); row.addView(b2); row.addView(b3); //在第 1 行添加 3 个按钮
        parent.addView(row);                            //表格布局添加第 1 行

        TableRow row2 = new TableRow(this);             //创建第 2 行对象
        Button b4 = new Button(this); b4.setText("OK4");
        TableRow.LayoutParams params2 = new TableRow.LayoutParams(
            TableRow.LayoutParams.MATCH_PARENT,TableRow.LayoutParams.WRAP_
CONTENT);
        params2.column = 1;                             //单元格位于当前行第 1 列
        params2.span = 2;                               //单元格跨 2 列
        b4.setLayoutParams(params2);
        row2.addView(b4);
        parent.addView(row2);
        setContentView(parent);
    }
}
```

TableLayout 类中需要掌握 3 个重要函数,如下所示。

- void setColumnStretchable(int nColumn,boolean state),当 state 为 true 时,表明

第 nColumn 列是伸展状态。

- void setColumnShrinkable(int nColumn,boolean state),当 state 为 true 时,表明第 nColumn 列是收缩状态。
- void setColumnCollapsed(int nColumn,boolean state),当 state 为 true 时,表明第 nColumn 列是隐藏状态。

TableRow.LayoutParams 类中需要掌握以下关键属性。

- column,设置列号。
- span,设置横跨几列。
- weight,设置列宽权重。

【例 2-21】 利用网格布局动态产生界面:表格共 3 列,第 1 行有 3 个按钮,列宽均为 1;第 2 行有 1 个按钮,从网格(0 基)1 行、1 列开始,跨两列。

```
public class MainActivity extends AppCompatActivity {
    protected void onCreate(Bundle savedInstanceState) {
        super.onCreate(savedInstanceState);
        GridLayout parent = new GridLayout(this);
        parent.setOrientation(GridLayout.HORIZONTAL);     //水平添加子组件
        parent.setColumnCount(3);                         //设置网格为 3 列
        //通过循环添加 3 个按钮
        for(int i=0; i<3; i++) {
            Button b = new Button(this);
            b.setText("OK"+(i+1));
            GridLayout.Spec row = GridLayout.spec(0, 1);
                                                    //按钮加在 0 行,占 1 行
            GridLayout.Spec col = GridLayout.spec(i, 1, 1f);
                                                    //按钮加在 i 列,占 1 列,权重 1.0f
            GridLayout.LayoutParams c = new GridLayout.LayoutParams(row, col);
            b.setLayoutParams(c);
            parent.addView(b);
        }
        //在第 2 行中添加按钮
        Button b4 = new Button(this);
        b4.setText("OK4");
        GridLayout.Spec row4 = GridLayout.spec(1, 1);     //按钮加在 1 行,占 1 行
        GridLayout.Spec col4 = GridLayout.spec(1, 2,1f); //按钮加在 1 列,跨 2 列
        GridLayout.LayoutParams c4 = new GridLayout.LayoutParams(row4, col4);
        b4.setLayoutParams(c4);
```

```
        parent.addView(b4);
        setContentView(parent);
    }
}
```

由代码可知，动态生成网格布局关键要掌握 GridLayout.Spec 类的作用，该类的主要功能是定义子组件在网格中的行、列（均 0 基）位置，权重值等，该类不能直接调用 spec()构造方法，而是通过调用静态方法 GridLayout.spec()获得所需 Spec 类对象的。常用的静态方法 spec()如下所示。

- Spec spec(int pos)：定义行(或列)坐标位置。
- Spec spec(int pos，int size)，pos：定义行(或列)坐标位置；size：跨行(或列)数。
- Spec spec(int pos，float weight)，pos：定义行(或列)坐标位置；weight：行(或列)的权重，必须是单精度浮点数，用于行宽(或列宽)比值。例如，若权重大小是 2，方法参数必须写作 2.0f，而非 2。
- Spec spec(int pos，int size，float weight)，pos：定义行(或列)坐标位置；size：跨行(或列)数目；weight：行(或列)的权重，必须是单精度浮点数，用于行宽(或列宽)比值。

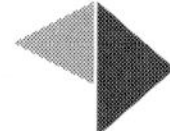

2.10　单位转换

1. 图形基本长度单位 px

px 是 pixel 的简称，指屏幕的像素点，是图形最基本的长度计量单位。例如，手机的分辨率是 1080px×1920px，含义是宽 1080 像素，长 1920 像素。

2. 引入 dp 的原因

假设有 3 款尺寸相同的手机：手机 A，分辨率为 720px * 1600px；手机 B，分辨率为 1080px * 2400px；手机 C，分辨率为 1440px * 3200px。若画一长 360px 的水平直线，则在 A 手机上占屏幕宽度的 1/2，在 B 手机上占屏幕宽度的 1/3，在 C 手机上占屏幕宽度的 1/4。这不是人们看到的效果，人们希望的理想情况是：只要手机尺寸相同，尽管分辨率不一样，但画出的图形对人的视觉比例是一样的。因此，若画某一“定长”直线，在相同尺寸、不同分辨率的手机中，比例相同。也可得出，“定长”直线的单位一定不能是 px，于是 dp 出现了，dp 是 density independent pixels(独立密度像素)的简称。

假设在手机 A、B、C 中画一条 ndp 大小的水平直线，归根结底，要转化为像素 px 来画，如下述公式所示。

N＝n * coef　N：像素数，n：dp 大小，coef：系数

因此，对手机 A、B、C 来说，n 相同，由于系数 coef 不同，因此得到的像素总数 N 也不同，这保证了所画直线的视觉比例是相同的。

那么，1dp 长度等于多少像素长度？先熟悉屏幕分辨率密度概念：即 ppi(pixels per inch)，每英寸有多少像素点。对于 160ppi 的分辨率，1dp = 1px；对于其他分辨率，公式如下所示。

1dp = (真实分辨率 ppi/160)px。

3. 引入 sp 的原因

sp 是 scaled pixels 的简称，主要用于字体显示，与 dp 类似，是与刻度无关的一种单位。区别是 sp 会根据用户调节的字体大小而改变。也就是说，sp 与 dp 的转化关系如下所示。

1sp = 1dp * coff2

通常，系数 coff2＝1 时，1sp＝1dp；当字体大小改变时，coff2 也发生变化，则 1sp 就不等于 1dp。

【例 2-22】 获取手机屏幕分辨率及 1dp 与 1px 的关系，比较 dp、sp 的不同。注：以华为手机 nova 7 为例。

① 界面配置文件 main.xml 如下所示。

```
<?xml version="1.0" encoding="utf-8"?>
<LinearLayout xmlns:android="http://schemas.android.com/apk/res/android"
    android:layout_width="match_parent"
    android:layout_height="match_parent"
    android:orientation="vertical">
    <TextView
        android:id="@+id/mysp"
        android:layout_width="match_parent"
        android:layout_height="wrap_content"
        android:textSize="25sp" />
    <TextView
        android:id="@+id/mydp"
        android:layout_width="match_parent"
        android:layout_height="wrap_content"
        android:textSize="25dp" />
    <TextView
```

```
        android:id="@+id/mydpsp"
        android:layout_width="match_parent"
        android:layout_height="wrap_content"
        android:textSize="25dp" />
</LinearLayout>
```

该界面定义了 3 个 TextView：前两个 TextView 定义的字体大小分别是 25sp、25dp，用于显示相同内容（将获取的分辨率显示在两个 TextView 中）；第三个 TextView 用于显示 dp 与 px 的关系。

② 调用程序的 Java 代码。

```
public class MainActivity extends AppCompatActivity {
    protected void onCreate(Bundle savedInstanceState) {
        super.onCreate(savedInstanceState);
        setContentView(R.layout.main);       //显示界面
        int width = this.getResources().getDisplayMetrics().widthPixels;
                                             //获取 width 分辨率
        int height = this.getResources().getDisplayMetrics().heightPixels;
                                             //获取 height 分辨率
        float scale = this.getResources().getDisplayMetrics().density;
                                             //scale 值表示 1dp 等于多少 px
        String s = "分辨率:" +width+ "px," +height+ "px";
                                             //形成分辨率字符串 s
        TextView tv = (TextView) this.findViewById(R.id.mydp);
                                             //将 s 显示在 R.id.mydp 的 TextView
        tv.setText(s);                       //配置文件中,字体大小是 25dp
        TextView tv2 = (TextView) this.findViewById(R.id.mysp);
                                             //将 s 显示在 R.id.mysp 的 TextView
        tv2.setText(s);                      //配置文件中,字体大小是 25sp
        TextView tv3 = (TextView) this.findViewById(R.id.mydpsp);
                                             //在 R.id.mydpsp 中显示
        tv3.setText("1dp="+scale+"px");      //1dp 等于多少 px
    }
}
```

实验过程如下所示。

① 将华为手机 nova 7 分别设置为低分辨率 720px×1600px、1080px×2400px，分别运行本实验 App。当分辨率为 720px×1600px 时，界面如图 2-24(a)所示；当分辨率为 1080px×2400px 时，界面如图 2-24(b)所示。可以看出，尽管同款手机分辨率设置不同，

但它们的界面是一致的。也就是说,相同的 dp 设置在同款手机中显示比例是一致的,与设备分辨率无关。

② 在分辨率是 1080px×2400px 的基础上,修改手机字体设置,将其设置为超大,再运行本实验 App,界面如图 2-24 右所示。可以看出,定义为 25sp 字体大小对应的 TextView 中的文字明显比定义为 25dp 字体大小对应的 TextView 中的文字大,说明 dp 与分辨率设置一个因素相关,而 dp 与分辨率及字体大小设置两个因素相关。

③ 不同分辨率下 1dp 等于多少 px 也是不一致的。本例中,当分辨率为 720px×1600px 时,1dp=2px;当分辨率为 1080px×2400px 时,1dp=3px。

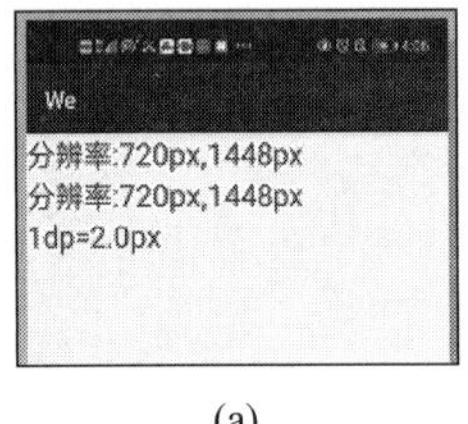

(a)

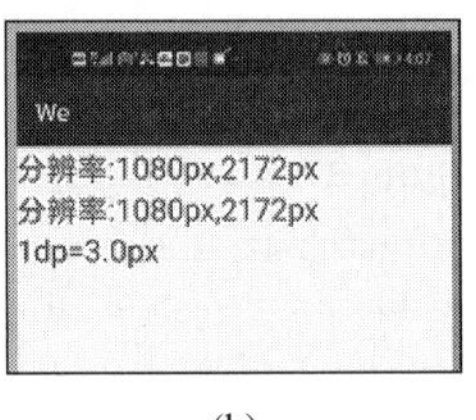

(b)

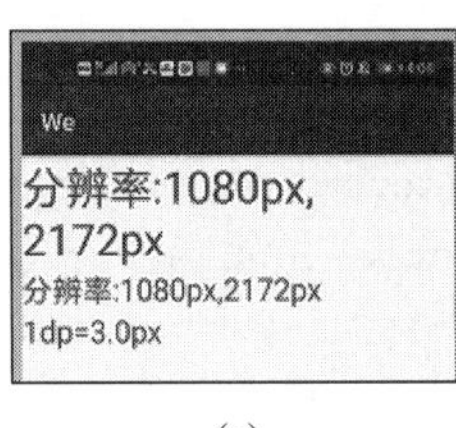

(c)

图 2-24 dp、sp 对比界面

习题 2

一、单选题

1. 下列属性中,用于设置线性布局方向的是(　　)。

   A. orientation　　B. margin　　C. layout_gravity　　D. padding

2. 下列选项中,不属于 Android 布局的是(　　)。

   A. FrameLayout　　B. LinearLayout

   C. FlowLayout　　D. RelativeLayout

3. 下列关于 RelativeLayout 的描述,正确的是(　　)。

   A. RelativeLayout 表示绝对布局,能够自定义控件的 x、y 的位置

   B. RelativeLayout 表示帧布局,能够实现标签切换的功能

   C. RelativeLayout 表示相对布局,其中控件的位置都是相对位置

   D. RelativeLayout 表示表格布局,须配合 TableRow 一块使用

4. Android 布局文件一般存放的目录是(　　)。

   A. layout　　B. drawable　　C. values　　D. anim

二、简答题

1. 线性布局中，属性 gravity、layout_gravity 的作用是什么？
2. 滚动窗口中 ScrollView 作用是什么？

三、程序题

1. 实现下列程序界面，3 个按钮水平居中。

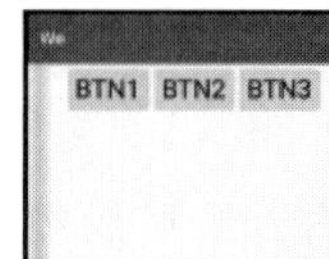

2. 实现下列程序界面。

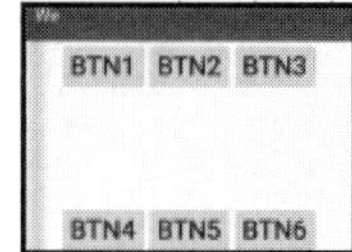

3. 实现如下界面。

4. 实现如下界面。

第 3 章

Android 控件

Android 控件是 UI 的重要组成部分。常用的控件有：按钮控件 Button、ImageButton，状态开关控件 ToggleButton、Switch，单选多选按钮 RadioButton、CheckBox，图片控件 ImageView，列表控件 ListView，文本控件 TextView、EditText，下拉控件 Spinner，日期与时间选择控件 DatePicker、TimePicker。主要掌握这些控件的创建方法、类中的主要函数及事件处理机制。

3.1 类层次关系

在讲述具体控件内容之前，熟悉一下 Android 中类层次关系是非常有必要的，主要包括 View 与 ViewGroup、ViewGroup 与布局的关系等。

在 Android 知识体系中，View 是所有 UI 视图的基类，ViewGroup 是 View 的子类，所以它也具有 View 的特性，但主要用来充当 View 的容器，将其中的 View 视作自己的孩子，对它的子 View 进行管理。当然，它的孩子也可以是 ViewGroup 类型。ViewGroup（树根）和它的孩子（View 和 ViewGroup）以树形结构形成一个层次结构。View 类有接受和处理消息的功能，Android 系统所产生的消息会在这些 ViewGroup 和 View 之间传递。

所有布局类 LinearLayout、RelativeLayout、TableLayout、GridLayout、FrameLayout 等都是 ViewGroup 的子类。

总之，View、ViewGroup、部分控件类、部分布局类的 UML 类图如图 3-1 所示。

View 中常用的方法如下所示。

- View findViewById(int id)，这是 View 中最常用的方法，根据控件 id 值返回 View 对象。一般来说，若进一步使用，需要将 View 强制转换成所需对象，形如 Button obj=(Button)findViewById(id)；

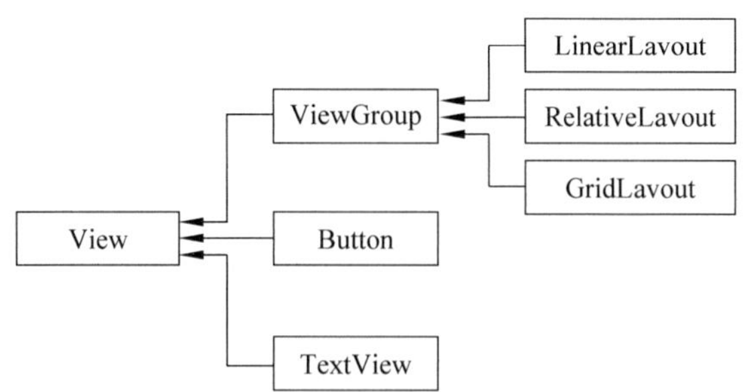

图 3-1　Android 基本类层次 UML 框图

ViewGroup 中常用的方法如下所示。

- void addView(View v)，父容器添加一个子组件对象。
- void removeAllViews()，父容器删除所有子组件。
- void removeView(View v)，父容器删除子组件 v。
- int getChildCount()，返回子组件数目。
- Resource getResources()，获取系统资源对象。

3.2　按钮控件

3.2.1　基本按钮 Button

Button 是 Android UI 的重要元素，其常用方法如下所示。

- Button(Context ctx)，构造方法，ctx 是上下文对象。
- void setText(CharSequence cs)，对应 android:text 属性，cs 是字符串序列，设置按钮文本内容。
- void setText(int resid)，resid 是字符串资源 id，即原字符串定义在资源文件中。
- void setTextSize(float size)，对应 android:textSize 属性，设置文本字体大小，浮点数，单位是 px。
- void setTextSize(int unit,float size)，设置文本字体大小，浮点数，单位由 unit 决定。unit 是系统类 TypedValue 中定义的静态常量：COMPLEX_UNIT_PX，整数 0，代表单位是 px；COMPLEX_UNIT_DIP，整数 1，代表单位是 dp；COMPLEX_UNIT_SP，整数 2，代表单位是 sp。
- void setBackgroundColor(Color c)，对应 android:background 属性，设置背景颜色。

- void setBackgroundResource(int resid),设置背景图片,resid 是图片资源 id。
- void setEnabled(boolean mark),对应 android:enabled 属性,按钮使能标志。

Button 按钮常用的动作有: click,短时单击事件;long click,长时单击事件;touch,触摸事件。若对按钮事件进行响应,必须完成以下步骤: ①注册事件; ②实现事件接口,编制响应函数。按钮事件常用的注册方法如下所示。

- void setOnClickListener(OnClickListener l),注册 click 事件,响应方法写在实现 OnClickListener 接口类中定义的方法中。
- void setOnLongClickListener(OnLongClickListener l),注册 long click 事件,响应方法写在实现 OnLongClickListener 接口类中定义的方法中。
- void setOnTouchListener(OnTouchListener l),注册 click 事件,响应方法写在实现 OnTouchListener 接口类中定义的方法中。

【例 3-1】 按钮事件响应示例:定义线性布局 UI 文件,一个"开始"按钮,一个 TextView 组件。当单击"开始"按钮时,在 TextView 组件中显示"Hello"字符串。

① UI 配置文件 main.xml。

```
<?xml version="1.0" encoding="utf-8"?>
<LinearLayout xmlns:android="http://schemas.android.com/apk/res/android"
    android:layout_width="match_parent"
    android:layout_height="match_parent">
    <Button
        android:id="@+id/mystart"
        android:layout_width="wrap_content"
        android:layout_height="wrap_content"
        android:text="开始"
        android:textSize="25sp"/>
    <TextView
        android:id="@+id/mytext"
        android:layout_width="match_parent"
        android:layout_height="wrap_content"
        android:textSize="25sp"/>
</LinearLayout>
```

② 事件响应:常用的方法有 4 种,如下所示。

方法 1:匿名内部类。

```
public class MainActivity extends AppCompatActivity {
    protected void onCreate(Bundle savedInstanceState) {
```

```
        super.onCreate(savedInstanceState);
        setContentView(R.layout.main);
        Button btn = (Button)findViewById(R.id.mystart);
        btn.setOnClickListener(new View.OnClickListener() {
                                        //注册按钮 click 事件,定义匿名内部类
            public void onClick(View v) {      //实现响应函数
                Button srcBtn = (Button)v;     //强制转换成 button
                srcBtn.setEnabled(false);      //禁止按钮操作
                TextView tv = (TextView)MainActivity.this.findViewById(R.id
                    .mytext);                  //获得 TextView 对象
                tv.setText("Hello");           //显示 Hello
            }
        });
    }
}
```

由以上代码可知，匿名内部类一般用于函数参数中，见 btn.setOnClickListener()函数内部的参数定义。OnClickListener 是一个接口，内部仅定义了一个函数 onClick(View v)，参数 v 是事件源对象，因此 v 可强制转换成 Button 对象，通过 setEnabled(false)禁止其使能状态。

由于在匿名内部类中函数获得外部类 MainActivity 中的 TextView 对象，因此一定要用如下代码："TextView tv = (TextView) MainActivity. this. findViewById(R. id. mytext)"。注意，一定是 MainActivity.this，而不是 this。

方法 2：内部类实现。

```
public class MainActivity extends AppCompatActivity {
    class innerOper implements View.OnClickListener{
        public void onClick(View v) {
            Button srcBtn = (Button)v;
            srcBtn.setEnabled(false);
            TextView tv = (TextView)findViewById(R.id.mytext);
            tv.setText("Hello");
        }
    }
    protected void onCreate(Bundle savedInstanceState) {
        super.onCreate(savedInstanceState);
        setContentView(R.layout.main);
        Button btn = (Button)findViewById(R.id.mystart);
```

```
            btn.setOnClickListener(new innerOper());
        }
    }
```

innerOper 是内部类，且实现了 OnClickListener 接口，重写了 onClick()函数，与界面中“开始”按钮的 click 事件对应。

方法 3：外部类实现。

把方法 2 中的类 innerOper 移出 MainActivity 类，作为外部类进行编译，会出现编译错误：“findViewById()函数没定义”。这是因为当 innerOper 作为内部类时，它可以共享外部类 MainActivity 的一切变量和方法，当然就能用 findViewById()函数。有读者说把 MainActivity 对象作为参数传到外部类中，不就解决问题了吗？当然可以，但有更专业的方法，那就是应用系统 Context 上下文对象，查看 Android 帮助文档，可以看出：对许多系统类，假设抽象为 XXX 类，都有 XXX(Context ct)构造方法。本示例中，入口类 MainActivity 是 Activity 的派生类，Activity 是 Context 的派生类，因此 MainActivity 对象也是 Context 对象。基于此，编制的按钮外部响应事件类及相关类代码如下所示。

```
//外部事件响应类:OuterOper.java
public class OuterOper implements View.OnClickListener {
    Context ct;
    OuterOper(Context ct){
        this.ct = ct;
    }
    public void onClick(View v) {
        Button srcBtn = (Button)v;
        srcBtn.setEnabled(false);
        MainActivity obj = (MainActivity)ct;
                                    //将 Context 对象转化为 MainActivity 对象
        TextView tv = (TextView)obj.findViewById(R.id.mytext);
        tv.setText("Hello");
    }
}
//入口类:MainActivity.java
public class MainActivity extends AppCompatActivity {
    protected void onCreate(Bundle savedInstanceState) {
        super.onCreate(savedInstanceState);
        setContentView(R.layout.main);
        Button btn = (Button) findViewById(R.id.mystart);
```

```
        btn.setOnClickListener(new OuterOper(this));
    }
}
```

方法 4：配置文件方法。

在 main.xml 配置文件中的 Button 配置部分增加 onClick 属性，如下所示（仅列出 Button 配置部分）。

```
<Button
    android:id="@+id/mystart"
    android:layout_width="wrap_content"
    android:layout_height="wrap_content"
    android:text="开始"
    android:textSize="25sp"
    android:onClick="myClick" />
```

然后，在 MainActivity 类中增加 myClick(View v)函数，代码如下所示。

```
public class MainActivity extends AppCompatActivity {
    public void onClick(View v) {
        Button srcBtn = (Button)v;
        srcBtn.setEnabled(false);
        TextView tv = (TextView)findViewById(R.id.mytext);
        tv.setText("Hello");
    }
    protected void onCreate(Bundle savedInstanceState) {
        super.onCreate(savedInstanceState);
        setContentView(R.layout.main);
    }
}
```

3.2.2　图像按钮 ImageButton

为丰富 Android UI，有时需要图像按钮。图像按钮可以给设计者更广阔的想象空间。ImageButton 与 Button 最大的区别是，前者不需要文字，android：text 属性对 ImageButton 是不起作用的。

ImageButton 类常用的函数如下所示。

- ImageButton(Context ct)，构造方法，ct 是上下文对象。
- void setImageResource(int resid)，对应 android：src 属性，设置按钮图像，resid 是

图像资源号,图像一般保存在 res\drawable\目录下。

【例 3-2】 图像按钮事件响应示例:定制线性布局页面,有一个图像按钮,初始时显示图像 a.png。当手按下后显示图像 b.png、手抬起后显示图像 a.png。

一般来说,将图像 a.png、b.png 保存在 res\drawable\目录下。

① 界面配置文件 main.xml。

```
<?xml version="1.0" encoding="utf-8"?>
<LinearLayout xmlns:android="http://schemas.android.com/apk/res/android"
    android:layout_width="match_parent"
    android:layout_height="match_parent">
    <ImageButton
        android:id="@+id/mybtn"
        android:layout_width="wrap_content"
        android:layout_height="wrap_content"
        android:src="@drawable/a"/>
</LinearLayout>
```

由代码可知,默认初始图像按钮显示图像 res\drawable\a.png,其路径用@drawable/a 描述,不带文件名后缀,不要写成 a.png。

② MainActivity.java。

```
public class MainActivity extends AppCompatActivity {
    protected void onCreate(Bundle savedInstanceState) {
        super.onCreate(savedInstanceState);
        setContentView(R.layout.main);
        final ImageButton btn = (ImageButton)findViewById(R.id.mybtn);
        btn.setOnTouchListener(new View.OnTouchListener() {  //注册触摸事件
            public boolean onTouch(View v, MotionEvent event) {
                int r = event.getAction();                   //获得特征值
                if(r==MotionEvent.ACTION_DOWN){              //动作是按下
                    btn.setImageResource(R.drawable.b);      //设置 b 图像
                }
                if(r==MotionEvent.ACTION_UP){                //动作是抬起
                    btn.setImageResource(R.drawable.a);      //设置 a 图像
                }
                return true;
            }
        });
```

```
    }
}
```

很明显,需对图像按钮注册 touch 触摸事件,通过 setOnTouchListener()函数完成,响应函数是匿名内部类中的 onTouch()。touch 触摸事件一般包括 3 部分:按住、移动、松开,它们均属于 MotionEvent 对象,具体哪一个是由 MotionEvent 类中的 getAction()函数的返回值决定。返回值等于 MotionEvent.ACTION_DOWN,表明是按住事件;返回值等于 MotionEvent.ACTION_MOVE,表明是移动事件;返回值等于 MotionEvent.ACTION_UP,表明是松开事件。

3.3 状态开关

3.3.1 ToggleButton 开关

ToggleButton 是一个具有选中和未选中双状态的按钮,并且需要为不同的状态设置不同的显示文本。其常用函数如下所示。

- ToggleButton(Context ct),构造方法,ct 是上下文对象。
- void setTextOn(CharSequence ontxt),对应 android:textOn 属性,设置开关打开时的文字内容。
- void setTextOff(CharSequence offtxt),对应 android:textOff 属性,设置开关关闭时的文字内容。
- boolean isChecked(),对应 android:checked 属性,返回开关状态布尔值。

ToggleButton 按钮的常用事件注册函数如下所示。

- setOnCheckedChangeListener(OnCheckedChangeListener l),注册 check 变化侦听事件,OnCheckedChangeListener 是系统接口,响应函数是该接口中定义的 onCheckedChange()函数。

【例 3-3】 ToggleButton 示例:定制线性布局页面,有一个 ToggleButton 双态按钮(打开、关闭),初始时处于"打开"状态。当不断地"按下-松开"该按钮,在手机屏幕上利用 Toast 交替显示"关闭状态""打开状态"。

① 在 UI 配置文件 main.xml。

```
<?xml version="1.0" encoding="utf-8"?>
<LinearLayout xmlns:android="http://schemas.android.com/apk/res/android"
    android:layout_width="match_parent"
    android:layout_height="match_parent">
```

```
    <ToggleButton
        android:id="@+id/mytoggle"
        android:layout_width="wrap_content"
        android:layout_height="wrap_content"
        android:textSize="30sp"
        android:checked="true"
        android:textOff="关闭"
        android:textOn="打开" />
</LinearLayout>
```

② MainActivity.java。

```
public class MainActivity extends AppCompatActivity {
    protected void onCreate(Bundle savedInstanceState) {
        super.onCreate(savedInstanceState);
        setContentView(R.layout.main);
        ToggleButton btn = (ToggleButton)findViewById(R.id.mytoggle);
        btn.setOnCheckedChangeListener(new CompoundButton.OnCheckedChange-
Listener() {
            public void onCheckedChanged(CompoundButton buttonView, boolean
isChecked) {
                if(isChecked)
                    Toast.makeText(MainActivity.this,"开关打开",Toast.LENGTH_
LONG).show();
                else
                    Toast.makeText(MainActivity.this,"开关关闭",Toast.LENGTH_
LONG).show();
            }
        });
    }
}
```

其界面如图 3-2 所示。

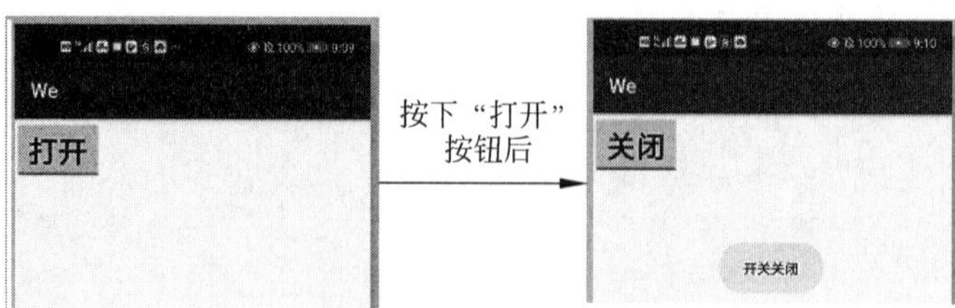

图 3-2 ToggleButton 示例

3.3.2 Switch 开关

Switch 开关也是一个双态开关，与 ToggleButton 相比，从界面来说，它更美观。其常用函数及事件注册函数与 ToggleButton 几乎是一致的，增量的知识点如下所示。

- Switch(Context ct)，构造方法，ct 是上下文对象。
- void setShowText(boolean mark)，对应 android:showText。当 mark 为 true 时，android:textOn 及 android:textOff 设置的字符串内容才能显示在屏幕上。
- void setTrackResource(int resid)，设置 Track 滑道图片资源，对应 android:track 属性，resid 是资源号，图片资源一般保存在 res\drawable\目录下。
- void setThumbResource(int resid)，设置 Thumb 滑块图片资源，对应 android:thumb 属性，resid 是资源号，图片资源一般保存在 res\drawable\目录下。

Switch 开关也可叫作滑动开关，当“关闭”时，一般滑块(Thumb)位于滑道(Track)的左侧；当“打开”时，一般滑块位于滑道的右侧，同时将其“高亮”显示。

【例 3-4】 系统 Switch 开关示例：定制线性布局页面，有一个 Switch 双态按钮(打开、关闭)，初始时处于“关闭”状态。当不断地“按下-松开”该按钮，在手机屏幕上利用 Toast 交替显示“打开状态”“关闭状态”。

① 在 UI 配置文件 main.xml。

```
<?xml version="1.0" encoding="utf-8"?>
<LinearLayout xmlns:android="http://schemas.android.com/apk/res/android"
    android:layout_width="match_parent"
    android:layout_height="match_parent">
    <Switch
        android:id="@+id/myswitch"
        android:layout_width="wrap_content"
        android:layout_height="wrap_content"
        android:textSize="30sp"
        android:text="开关状态"
        android:textOn="打开"
        android:textOff="关闭"
        android:showText="true" />
</LinearLayout>
```

② MainActivity.java。

```
public class MainActivity extends AppCompatActivity {
```

```
    protected void onCreate(Bundle savedInstanceState) {
        super.onCreate(savedInstanceState);
        setContentView(R.layout.main);
        Switch btn = (Switch)findViewById(R.id.myswitch);
        btn.setOnCheckedChangeListener(new CompoundButton
.OnCheckedChangeListener() {
            public void onCheckedChanged(CompoundButton buttonView, boolean
isChecked) {
                if(isChecked)
                    Toast.makeText(MainActivity.this,"开关打开",Toast.LENGTH_
LONG).show();
                else
                    Toast.makeText(MainActivity.this,"开关关闭",Toast.LENGTH_
LONG).show();
            }
        });
    }
}
```

其界面如图 3-3 所示。

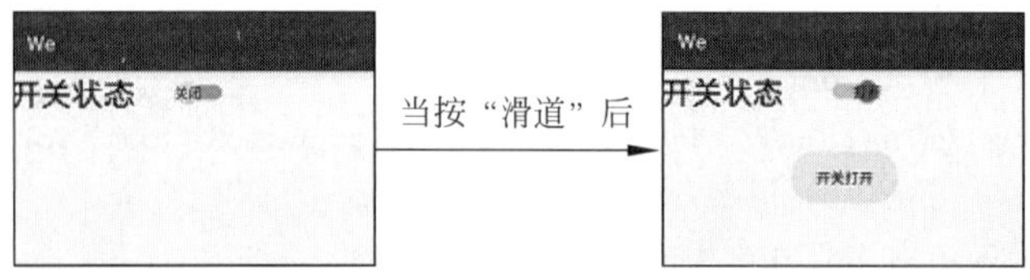

图 3-3 系统 Switch 开关示例

可以看出，Android 系统默认的滑道图片形似长方形，滑块图片是一个圆形。当 Switch 开关关闭时，圆形滑块位于左侧，且是白色；当 Switch 开关打开时，圆形滑块位于右侧，且是红色(表示高亮)。

实际上，从此例可推出：ToggleButton 开关必须在按钮上显示文字，用以区分打开、关闭状态，而 Switch 开关完全可以用滑道、滑块颜色的变化表示开关的打开、关闭。读者可以看看自己手机的设置部分，许多双态按钮都类似 Switch 开关，完全是由颜色变化区分状态的。

【例 3-5】 自定义 Switch 开关示例：主要对 track 及 thumb 参数进行设置。track 有打开、关闭状态；thumb 也有打开、关闭状态，因此每个 track、thumb 图片资源最多定义 2 个。为了方便，令 track 图片在打开、关闭两个状态下不变化，定义为一个资源即可，假设为 track.jpg。thumb 图片定义为 2 个：一个为 off.jpg(对应关闭状态)；一个为 on.jpg(对应打开状态)。

注意：一般来说，滑块、滑道图片的高度应一致，滑道图片的宽度要大于或等于 2 倍滑块图片的宽度。

① 在 res\drawable\目录下，按题意建立滑道图片 track.jpg 和滑块图片 off.jpg、on.jpg。

② 在 UI 配置文件 main.xml。

```
<?xml version="1.0" encoding="utf-8"?>
<LinearLayout xmlns:android="http://schemas.android.com/apk/res/android"
    android:layout_width="match_parent"
    android:layout_height="match_parent">
    <Switch
        android:id="@+id/myswitch"
        android:layout_width="wrap_content"
        android:layout_height="wrap_content"
        android:text="开关状态" />
</LinearLayout>
```

配置文件中仅定义了 Switch 组件，其滑道、滑块图片动态设置是在程序中实现的，勿需在此设置。

③ MainActivity.java。

```
public class MainActivity extends AppCompatActivity {
    protected void onCreate(Bundle savedInstanceState) {
        super.onCreate(savedInstanceState);
        setContentView(R.layout.main);
        final Switch btn = (Switch)findViewById(R.id.myswitch);
        btn.setTrackResource(R.drawable.track);//设置滑道图片
        btn.setThumbResource(R.drawable.off);  //设置滑块图片(初始为关闭状态)
        btn.setOnCheckedChangeListener(new CompoundButton
.OnCheckedChangeListener() {
            public void onCheckedChanged(CompoundButton buttonView, boolean
isChecked) {
                if(isChecked) {
                    btn.setThumbResource(R.drawable.on);
                                                        //若开关打开,则设置 on 图片
```

```
                Toast.makeText(MainActivity.this, "开关打开", Toast.LENGTH_
LONG).show();
            }
            else {
                btn.setThumbResource(R.drawable.off);
                                //若开关关闭,则设置 off 图片
                Toast.makeText(MainActivity.this, "开关关闭", Toast.LENGTH_
LONG).show();
            }
        }
    });
  }
}
```

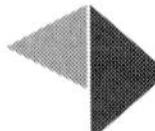

3.4 单选按钮和多选按钮

3.4.1 RadioButton 单选按钮

RadioButton 是常用的单选按钮。为方便实现单选功能,一般将 RadioGroup 作为父容器,内部放置多个 RadioButton 子控件,这些子控件要么没有选中,要么同时只能有一个 RadioButton 子控件被选中。

RadioGroup 类常用的函数如下所示。

- void setOrientation(int mark),设置添加子组件方式,对应 android:orientation 属性。当 mark = RadioGroup.VERTICAL 时,垂直添加子组件;当 mark = RadioGroup.HORIZONTAL 时,水平添加子组件。
- void getChildCount(),获得子组件数量。
- int getCheckedRadioButtonId(),获得选中的 RadioButton 对象 ID 值,若无选中的单选按钮对象,则返回-1。

RadioButton 类常用的函数如下所示。

- boolean isChecked(),获得布尔值,若为 true,则表示选中;若为 false,则表示未选中。

【例 3-6】 RadioButton 示例:定义一组单选按钮选项,分别是语文、数学、英语;再定义一个 OK 按钮,当单击 OK 按钮时,利用 Toast 显示选中单选按钮的内容,若没有选中的按钮,则显示"无"。

① 在 UI 布局文件 main.xml。

```
<?xml version="1.0" encoding="utf-8"?>
<LinearLayout xmlns:android="http://schemas.android.com/apk/res/android"
    android:layout_width="match_parent"
    android:layout_height="match_parent"
    android:orientation="vertical">
<RadioGroup
        android:id="@+id/mygrp"
        android:layout_width="wrap_content"
        android:layout_height="wrap_content"
        android:orientation="vertical">
    <RadioButton
        android:layout_width="wrap_content"
        android:layout_height="wrap_content"
        android:text="语文"/>
    <RadioButton
        android:layout_width="wrap_content"
        android:layout_height="wrap_content"
        android:text="数学"/>
    <RadioButton
        android:layout_width="wrap_content"
        android:layout_height="wrap_content"
        android:text="英语"/>
</RadioGroup>
<Button
        android:id="@+id/myok"
        android:layout_width="wrap_content"
        android:layout_height="wrap_content"
        android:text="OK"/>
</LinearLayout>
```

代码中仅对 RadioGroup 节点定义了 android:id 属性，各单选 RadioButton 子节点没有定义 android:id 属性。这是因为根据 id 属性值可获取父节点 RadioGroup 对象。当然，可遍历其子节点对象，因此可不对各 RadioButton 子节点设置 android:id 值。

② MainActivity.java。

```
public class MainActivity extends AppCompatActivity {
    protected void onCreate(Bundle savedInstanceState) {
```

```
            super.onCreate(savedInstanceState);
            setContentView(R.layout.main);
            Button btn = (Button)findViewById(R.id.myok);
            btn.setOnClickListener(new View.OnClickListener() {
                public void onClick(View v) {
                    String s = "";
                    RadioGroup rg = (RadioGroup)MainActivity.this.findViewById
(R.id.mygrp);
                    int id = rg.getCheckedRadioButtonId();
                    if(id==-1)
                        s = "你没有选中任何科目!";
                    else{
                        Button sel = (Button)MainActivity.this.findViewById(id);
                        s = "你选中的科目是:" +sel.getText();
                    }
                    Toast.makeText(MainActivity.this, s, Toast.LENGTH_LONG).show();
                }
            });
        }
    }
```

示例中获取选中单选按钮的步骤是：①根据 id 值获取 RadioGroup 对象 rg；②通过调用 RadioGroup 类中的 getCheckedRadioButtonId()，获取选中的单选按钮 id 值；③根据 id 值，获得选中的单选按钮对象 sel；④根据 sel，调用 getText()，获取选中单选按钮的文本字符串值。

当然，也可用第二种方法确定选中的单选对象，步骤是：①根据 id 值，获取 RadioGroup 对象 rg；②遍历 rg 的各 RadioButton 子对象，若某子对象的 isChecked()函数返回 true，则该单选按钮被选中。部分关键代码如下所示。

```
String s = "你没有选中任何科目!";
RadioGroup rg = (RadioGroup)MainActivity.this.findViewById(R.id.mygrp);
for(int i=0; i<rg.getChildCount(); i++) {                //遍历各子节点
    RadioButton rb = (RadioButton)rg.getChildAt(i);      //获取各具体子节点
if(rb.isChecked()){                                      //若选中
        s = "你选中的科目是:" + rb.getText(); break;
    }
}
Toast.makeText(MainActivity.this, s, Toast.LENGTH_LONG).show();
```

该示例界面如图 3-4 所示。

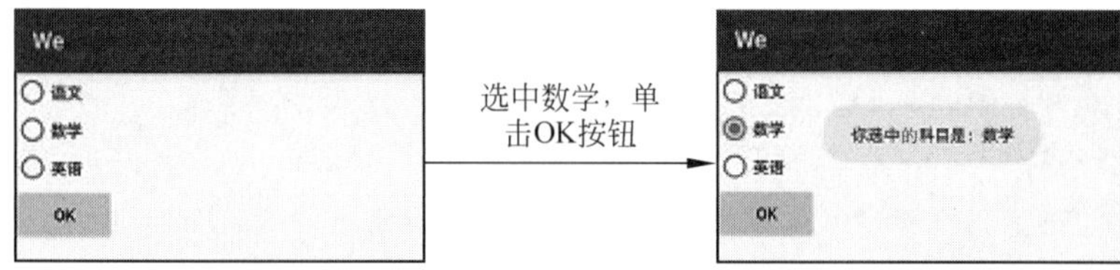

图 3-4　单选按钮示例

3.4.2　深入探究

1. 动态生成界面

人们经常利用手机做单选的调查问卷，有的读者说利用布局配置文件很容易实现这样的界面，有 N 个调查问卷，就定制 N 个单选布局文件。诚然是可以的，但有无其他方法？当然有！即主要由程序动态生成界面，分析如下所示。

- 每个单选题目由题目内容＋选项组成，需求分析是稳定的，完全可以利用循环生成 N 个单选题目的界面。
- 为了方便，可以借助 UI 布局文件，仅定制一道题目（包括题目内容及选项）的样式，包括字体、边距等，作为生成其他题目的样式模板。

仿真单选调查问卷功能的关键代码如下所示。

① 在 UI 布局文件 main.xml。

```
<?xml version="1.0" encoding="utf-8"?>
<LinearLayout xmlns:android="http://schemas.android.com/apk/res/android"
    android:id="@+id/myline"
    android:layout_width="match_parent"
    android:layout_height="match_parent"
    android:orientation="vertical">
    <TextView
        android:id="@+id/mytext"
        android:layout_width="match_parent"
        android:layout_height="wrap_content"
        android:visibility="gone"
        android:layout_marginBottom="10dp"/>
    <RadioGroup
        android:id="@+id/mygrp"
```

```
        android:layout_width="wrap_content"
        android:layout_height="wrap_content"
        android:orientation="horizontal"
        android:visibility="gone"
        android:layout_marginBottom="15dp">
        <RadioButton
            android:id="@+id/mybtn"
            android:layout_width="wrap_content"
            android:layout_height="wrap_content"
            android:text-"语文"/>
    </RadioGroup>
</LinearLayout>
```

该文件主要定义了题目内容与选项、相邻两道题目之间的位置关系模板。模板主要涉及以 layout_为前缀的所有属性内容。TextView 节点是题目内容节点模板，由 android:layout_marginBottom="10dp"得出：题目内容与选项之间的间距为 10dp；RadioGroup 节点定义了该题与下一道题目的间距，由 android:layout_marginBottom="15dp"得出：两道相邻题目之间的间距是 15dp。

由于 TextView、RadioGroup 节点都不是真实的调查问卷所需内容，因此把这两个节点的 android:visibility 均设置为 gone，表示此两个节点在实际界面中是隐藏的，且不占空间，若设置成 invisible，也是不显示，但占用实际空间。

本例仅对 margin_bottom 属性做了设置，实际中可对 padding、margin 等所有需要的以 layout_为前缀的属性进行设置。这样，在程序中读一次模板参数，为所需要的 UI 控件进行设置，程序会非常简洁。看下述的实际代码。

② MainActivity.java。

```
public class MainActivity extends AppCompatActivity {
    protected void onCreate(Bundle savedInstanceState) {
        super.onCreate(savedInstanceState);
        setContentView(R.layout.main);
        TextView tv = (TextView)findViewById(R.id.mytext);
                                                        //获得题目内容模板对象
        RadioGroup rg = (RadioGroup)findViewById(R.id.mygrp);
                                                        //获得题目选项模板对象
        RadioButton rb = (RadioButton)findViewById(R.id.mybtn);
                                                        //获得单选按钮模板对象
        LinearLayout parent = (LinearLayout)findViewById(R.id.myline);
                                                        //获得总的父容器对象
```

```
        //二维字符串数组模拟题目内容和选项内容
        String s[][] = {{"你的性别","男","女"},
                {"你最喜欢的科目?","语文","数学","外语"},
                {"你最喜欢的乐器","钢琴","笛子","手风琴","二胡"}};
        for(int i=0; i<s.length; i++){
            TextView text = new TextView(this);           //产生题目内容对象
            text.setLayoutParams(tv.getLayoutParams());   //设置位置对象
            text.setText(s[i][0]);                        //设置题目内容
            RadioGroup grp = new RadioGroup(this);        //产生选项组对象
            grp.setLayoutParams(rg.getLayoutParams());    //设置位置对象
            for(int j=1; j<s[i].length; j++){
                RadioButton btn = new RadioButton(this);  //产生第 j 个单选按钮对象
                RadioGroup.LayoutParams param2 = new RadioGroup.LayoutParams
    (rb.getLayoutParams());
                btn.setLayoutParams(param2);              //设置位置对象
                btn.setText(s[i][j]);                     //设置单选按钮文本内容
                grp.addView(btn);                         //单选按钮加入选项组
            }
            parent.addView(text);                        //加入第 i 道题目内容对象
            parent.addView(grp);                         //加入第 i 道题目选项组对象
        }
    }
}
```

示例中利用不规则二维数组 s 模仿了题目及选项内容，每行中第 0 项表示题目内容，第 1～n 项表示选项文本内容。例如，第 1 道题目的内容是“你的性别?”，选项有“男”和“女”。改变字符串内容，其他不用修改，即可生成新的调查问卷。或者从数据源(文件或数据库)按规则动态读入，即可形成新的所需界面。

代码中，核心思想是：利用循环添加题目内容及选项组对象，产生每个 UI 组件时，都要调用“组件.setLayoutParams(模板对象.getLayoutParams())”，确定组件的安放位置后，代码非常简洁。反之，知道每个 UI 组件以 layout_为前缀的属性非常多，如果一个个设置，代码会显得非常臃肿。从中也可体会出：每个组件都由“位置＋其他”属性组成，把“位置”单独提出来很有必要，可减少代码规模，而这一点单纯从布局配置文件中是体会不出来的，只有深入动态编程时，才可体会其中的精髓。

因此，一定不要被布局配置文件所左右，要灵活应用配置文件，让静态的配置文件动起来。

2. 属性增加问题

RadioButton 有属性 text，即显示在 UI 上的文本，但缺少一个属性 value。如何为 RadioButton 增加属性 value？如何在 UI 布局 XML 文件中应用 value 属性？步骤如下所示。

① 定义 RadioButton 派生类 MyRadioButton，value 成员变量（字符串类型）以及 setter()和 getter()方法。

② 为了能在布局配置文件中应用 MyRadioButton，要在工程 res\values\attrs.xml 中增加如下定义。当然，也可命名其他名字的文件，增加相同的内容。

```
<?xml version="1.0" encoding="utf-8"?>
<resources>
<declare-styleable name="MyRadioButton"><!-- 控件名称-->
<attr name="value" format="string"/><!-- 属性名称,类型-->
</declare-styleable>
</resources>
```

此文件中定义了名为 MyRadioButton 的按钮，并定义了一个属性，名字为 value，类型是字符串 string。名字 value 必须与 MyRadioButton 定义的成员变量 value 的名称一致。若再增加其他属性，只增加相应的<attr />节点即可。

③ 在②的基础上，完善的 MyRadioButton 类关键代码如下所示。

```
public class MyRadioButton extends RadioButton {
    String value;
    public MyRadioButton(Context context, AttributeSet attrs) {
        super(context, attrs);
        try {
            //获得 MyRadioButton 的属性-值集合
             TypedArray a = context.obtainStyledAttributes(attrs,R.styleable.
MyRadioButton);
            this.value = a.getString(R.styleable.MyRadioButton_value);
                                                                      //获得 value 值
            a.recycle();
        } catch (Exception e) {
            e.printStackTrace();
        }
    }
    public String getValue() {return value;}
```

```
    public void setValue(String value) {this.value = value;}
}
```

构造方法 MyRadioButton()不是主动调用的，而是 Android 系统自动调用的。一些关键代码的解释如下所示。

- super(context, attrs)，调用基类 RadioButton 构造方法，将 RadioButton 定义的“属性-值”集合保存在内存中。
- TypedArray a = context.obtainStyledAttributes(attrs, R.styleable.MyRadioButton);

一方面将 MyRadioButton 自定义的“属性-值”集合添加到 attrs 对象中，也就是说，attrs 包含了 RadioButton 及派生类 MyRadioButton 的“属性-值”集合；另一方面获得 MyRadioButton 自定义“属性-值”集合对象（由 TypedArray a 描述）。代码中，obtainStyleAttributes()函数的第 2 个参数前缀 R.styleable 是固定的，末尾的 MyRadioButton 必须与上文 attrs.xml 中<declare-styleable name="MyRadioButton">定义的 name 属性值相一致。

- this.value = a.getString(R.styleable.MyRadioButton_value)

获取 MyRadioButton 自定义属性 value 值。代码中，getString()函数的第 2 个参数前缀 R.styleable 是固定的，MyRadioButton 是控件名称，末尾的 value 必须与上文 attrs.xml 中<attr name="value" format="string"/>定义的 name 属性值相一致。

④ 简单测试。

前提条件是上文论述的 MyRadioButton 类，attrs.xml 已存在。测试功能是：定义布局文件 main.xml，仅包含 MyRadioButton 节点，并定义 value 属性值，在程序中获得该属性值并利用 Toast 输出该值。

//布局文件 main.xml。

```
<?xml version="1.0" encoding="utf-8"?>
<LinearLayout
    xmlns:android="http://schemas.android.com/apk/res/android"
    xmlns:myattr="http://schemas.android.com/apk/res/com.example.dqjbd.we"
    android:id="@+id/myline"
    android:layout_width="match_parent"
    android:layout_height="match_parent">
    <com.example.dqjbd.we.MyRadioButton
        android:id="@+id/myradio"
        android:layout_width="wrap_content"
        android:layout_height="wrap_content"
        android:text="Test"
```

```
        android:textSize="30sp"
        myattr:value="1000"/><!--定义自定义属性值 value="1000" -->
</LinearLayout>
```

注意：有两行 xmlns 节点，xmlns 是 XML namespace 命名空间的意思，命名空间是避免命名冲突的解决方法。例如，A、B 学校都有名叫小张的学生，如何进行区分？这时命名空间就起作用了，A、B 可以作为命名空间，A.小张就区分了 B.小张。由此可得，命名空间里存放的是特定属性的集合。

第一个 xmlns:android=http://schemas.android.com/apk/res/android，各组成部分的含义如下。

- xmlns：即 xml namespace，表明要定义一个命名空间。
- android：命名空间名称，若有 XXX 属性，则写作 android:XXX。
- http://schemas.android.com/apk/res/android：看起来是一个 URL，但是却不可访问，实际上是一个 URI，所以它的值是固定不变的，相当于一个常量，末尾的 Android 串表明是系统固有资源，起到了标识作用。

第二个 xmlns：myattr=http://schemas.android.com/apk/res/com.example.dqjbd.we 定义了 myattr 命名空间，作用于 com.example.dqjbd.we 包下定义的各派生 UI 控件类。

```
//MainActivity.java
public class MainActivity extends AppCompatActivity {
    protected void onCreate(Bundle savedInstanceState) {
        super.onCreate(savedInstanceState);
        setContentView(R.layout.main);
        MyRadioButton obj = (MyRadioButton)findViewById(R.id.myradio);
        Toast.makeText(this,obj.getValue(),Toast.LENGTH_LONG).show();
                                                    //显示自定义属性 value 的值
    }
}
```

3.4.3 CheckBox 多选按钮

CheckBox 是 Android 中常用的多选按钮，即对一组多选按钮来说，可以同时选中多个选项。其常用函数及事件响应与单选按钮几乎是一致的。对单选按钮来说，一般是由 RadioGroup+RadioButton 组合完成；对多选按钮来说，其实也可以是由 RadioGroup+CheckBox 组合完成，这样的好处是获得父节点 RadioGroup 对象，就能循环遍历各

CheckBox 子节点,获取所需功能。

【例 3-7】 CheckBox 示例:定义一组多选按钮选项,分别是语文、数学、英语;再定义一个 OK 按钮,当单击 OK 按钮时,利用 Toast 显示选中多选按钮的内容。

① 在 UI 布局文件 main.xml。

```
<?xml version="1.0" encoding="utf-8"?>
<LinearLayout
    xmlns:android="http://schemas.android.com/apk/res/android"
    android:layout_width="match_parent"
    android:layout_height="match_parent"
    android:orientation="vertical">
    <RadioGroup
        android:id="@+id/mygrp"
        android:layout_width="match_parent"
        android:layout_height="wrap_content">
        <CheckBox
            android:layout_width="wrap_content"
            android:layout_height="wrap_content"
            android:text="语文"
            android:textSize="30sp"/>
        <CheckBox
            android:layout_width="wrap_content"
            android:layout_height="wrap_content"
            android:text="数学"
            android:textSize="30sp"/>
        <CheckBox
            android:layout_width="wrap_content"
            android:layout_height="wrap_content"
            android:text="英语"
            android:textSize="30sp"/>
    </RadioGroup>
    <Button
        android:id="@+id/myok"
        android:layout_width="wrap_content"
        android:layout_height="wrap_content"
        android:text="OK"
        android:textSize="30sp"/>
</LinearLayout>
```

代码中仅对 RadioGroup 节点定义了 android:id 属性,各多选 CheckBox 子节点没有定义 android:id 属性。这是因为根据 id 属性值可获取父节点 RadioGroup 对象,当然可遍历其子节点对象,因此可不对各 CheckBox 子节点设置 android:id 值。

② MainActivity.java。

```
public class MainActivity extends AppCompatActivity {
    protected void onCreate(Bundle savedInstanceState) {
        super.onCreate(savedInstanceState);
        setContentView(R.layout.main);
        Button btn = (Button)findViewById(R.id.myok);
        btn.setOnClickListener(new View.OnClickListener() {
            public void onClick(View v) {
                String s = "";
                RadioGroup rg = (RadioGroup)MainActivity.this.findViewById
(R.id.mygrp);                                                   //获取父节点对象
                for(int i=0; i<rg.getChildCount(); i++){        //遍历各子节点
                    CheckBox cb = (CheckBox)rg.getChildAt(i);   //获取子节点对象
                    if(cb.isChecked())                          //若选中
                        s += cb.getText()+",";                  //加上文字标识
                }
                Toast.makeText(MainActivity.this,"选中科目:"+s, Toast.LENGTH_
LONG).show();
            }
        });
    }
}
```

该示例界面如图 3-5 所示。

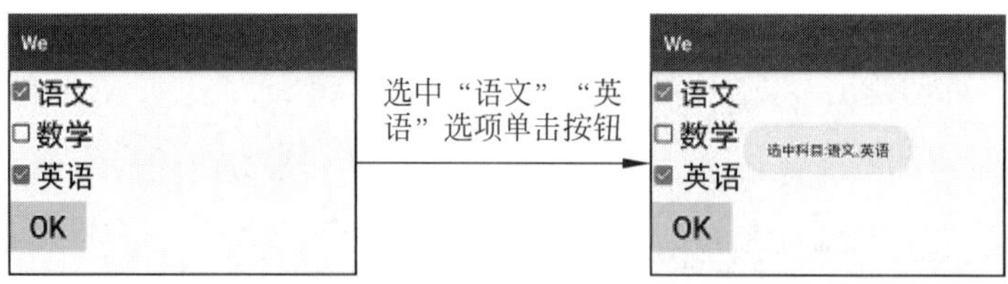

图 3-5 CheckBox 多选按钮示例

从表意形式来说,用 RadioGroup 作为 Checkbox 父节点也许不恰当,当然也可用 LinearLayout,或者定义 LinearLayout 的自定义派生类 CheckGroup 作为多选题目的父

节点，读者可以自行完成。

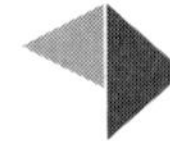

3.5 图片控件 ImageView

3.5.1 基本函数

ImageView 是用来显示图像的控件，图像是 UI 的重要组成部分。一般来说，图像尺寸是固定的，但是必须能适应不同情况：原大小显示，等比缩放显示，不等比缩放显示等。达到这些目的，ImageView 图像控件只需设置相关参数即可，其中用到的主要函数如下所示。

- void setImageResource(int resid)，设置图像文件，对应 android:src 属性，一般在 res\drawable\目录下，resid 是图像文件资源号。
- void setAdjustViewBounds(boolean mark)，对应 android:adjustViewBounds 属性，用于缩放时是否保持图像原长宽比。当 mark 为 true 时，缩放后图像长宽比与原图一致。
- void setMaxWidth(int width)，对应 android:maxWidth，设置图像的最大宽度。
- void setMaxHeight(int height)，对应 android:maxHeight，设置图像的最大高度。
- void setScaleType(ScaleType st)，对应 android:scaleType，用于设置显示的图片如何缩放或者移动，以适应 ImageView 的大小，函数的参数为 st，形如 ImageView.ScaleType.XXX，XXX 的取值见表 3-1。

表 3-1 ScaleType 取值表

序号	程序参数值	含　义
1	CENTER	按图片原大小居中显示，若图片长/宽超过 View 的长/宽，则截取图片居中部分显示
2	CENTER_CROP	按比例扩大图片居中显示，使得图片长（或宽）等于或大于 View 的长（或宽）
3	CENTER_INSIDE	将图片完整居中显示，通过按比例缩小使得图片的长（或宽）等于或小于 View 的长（或宽）
4	FIT_CENTER	把图片按比例扩大（或缩小）到 View 的大小，居中显示
5	FIT_END	把图片按比例扩大（或缩小）到 View 的宽度，显示在 View 的下部
6	FIT_START	把图片按比例扩大（或缩小）到 View 的宽度，显示在 View 的上部

续表

序号	程序参数值	含　　义
7	MATRIX	运用图片进行绘制时的矩阵进行缩放。默认情况下,展现形式为:不进行任何缩放,从 ImageView 的左上角开始摆放,对原图超过 ImageView 的部分作裁剪处理

3.5.2 数学基础

ImageView 进行了大量的图像缩放操作,因此,了解其中简单的缩放原理是必要的,这样会懂得如何设置 ImageView 的参数,而不是死记硬背。下面分情况加以描述。

1. 不等比缩放

首先要明确什么情况下能发生不等比缩放,那一定是目的视图的大小已经确定(这句话非常关键)。已知原始图像宽 a、高 b,目的视图宽 c、高 d,如图 3-6 所示。

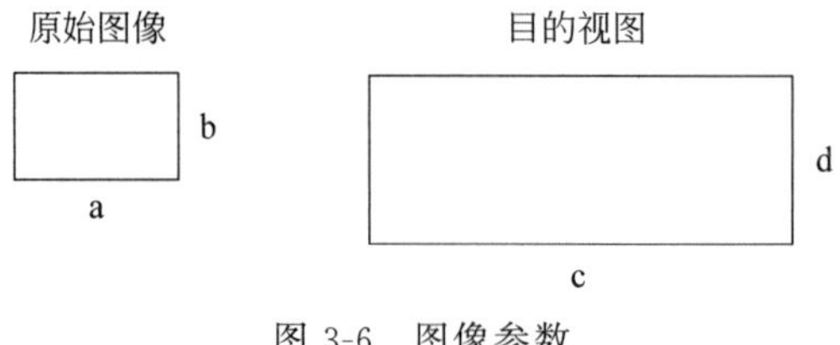

图 3-6　图像参数

则纵向比 r0=d/b,横向比 r1=c/a。一般来说,r0≠r1,为不等比缩放,当然也有 r0=r1 的特殊情况。

2. 等比缩放

原始图像等比缩放置于目的视图中,目的视图有 3 种情况:大小确定;水平宽度确定、垂直高度未知;垂直高度确定,水平宽度未知。

- 目的视图大小确定。

已知原始图像宽 a、高 b,目的视图宽 c、高 d,那么原始图像在目的视图中进行等比缩放后,宽、高分别为多少?

由纵向比 r0=d/b,横向比 r1=c/a,可知真实缩放比例为 r。很明显,r 一定是 r0、r1 中的最小值,由此可计算等比缩放后的图像大小。

- 目的视图水平宽度确定、垂直高度未知。

已知原始图像宽 a、高 b,目的视图宽 c、高未知,那么原始图像在目的视图中进行等比缩放后,宽、高分别为多少?

根据已知条件只能确定横向比 r1=c/a,因此真实缩放比例 r 只能等于 r1。缩放后图像的水平宽度等于目的视图的宽度 c,垂直高度等于原始图像的高度与缩放比例的乘积,即 b * r1。

- 目的视图的垂直高度确定、水平宽度未知。

已知原始图像宽 a、高 b,目的视图高 d、宽未知,那么原始图像在目的视图中进行等比缩放后,宽、高分别为多少?

根据已知条件只能确定纵向比 r0=d/b,因此真实缩放比例 r 只能等于 r0。缩放后图像的垂直高度等于原始图像的高度 b,水平宽度等于原始图像的宽度与缩放比例的乘积,即 a * r0。

3.5.3 典型事例

【例 3-8】 ImageView 基本属性中仅包含 android:src 的情况分析。

布局 UI 文件 main.xml 如下所示。

```
<?xml version="1.0" encoding="utf-8"?>
<LinearLayout
    xmlns:android="http://schemas.android.com/apk/res/android"
    android:layout_width="match_parent"
    android:layout_height="match_parent">
    <ImageView
        android:layout_width="wrap_content"
        android:layout_height="wrap_content"
        android:src="@drawable/my"/>
</LinearLayout>
```

该配置文件的含义是将图像文件 my.jpg 显示在 ImageView 中,但是,由于 ImageView 的宽、高都设置为 wrap_content,是不确定的,因此该 ImageView 的大小是原图显示。

当 ImageView 的宽、高发生变化,ImageView 上的 my.jpg 图像会如何变化?结论是均等比例缩放,但比例系数稍有不同,详细描述如下所示。

- layout_width="**match_parent**",**layout_height**="**wrap_content**"

若 **ImageView** 的宽度固定,高度不确定,则等比例系数为 1,显示的图像与原始图像大小一致。

- layout_width="**wrap_content**",**layout_height="200dp"**

若 **ImageView** 的高度固定,宽度不确定,则等比例系数为 1,显示的图像与原始图像大小一致。

- layout_width="**300dp**",**layout_height="200dp"**

若 ImageView 的高度、宽度都固定,则等比例系数等于横向比与纵向比中的最小值,图像放大或缩小,或与原图一致。

无论上述哪种情况,发现图像都显示在 ImageView 的中心。

【例 3-9】 ImageView 基本属性中 adjustViewBounds 对应情况分析。

adjustViewBounds 是 ImageView 类中的一个主要属性,也是关于等比缩放的内容,那么它与例 3-8 中的等比缩放有什么不同?该属性也时常与 maxWidth、maxHeight 属性联合使用,它们之间又是如何关联的?为此,定义一幅图在 3 个 ImageView 中,布局 UI 配置文件 main.xml 代码如下所示。

```
<?xml version="1.0" encoding="utf-8"?>
<LinearLayout
    xmlns:android="http://schemas.android.com/apk/res/android"
    android:layout_width="match_parent"
    android:layout_height="match_parent"
    android:orientation="vertical">
    <ImageView
        android:layout_width="200px"
        android:layout_height="wrap_content"
        android:src="@drawable/my"
        android:layout_marginBottom="10dp"/>
    <ImageView
        android:layout_width="200px"
        android:layout_height="wrap_content"
        android:src="@drawable/my"
        android:adjustViewBounds="true"
        android:layout_marginBottom="10dp"/>
    <ImageView
        android:layout_width="200px"
        android:layout_height="wrap_content"
        android:src="@drawable/my"
        android:adjustViewBounds="true"
        android:maxHeight="150px"/>
</LinearLayout>
```

对应界面如图 3-7 所示。

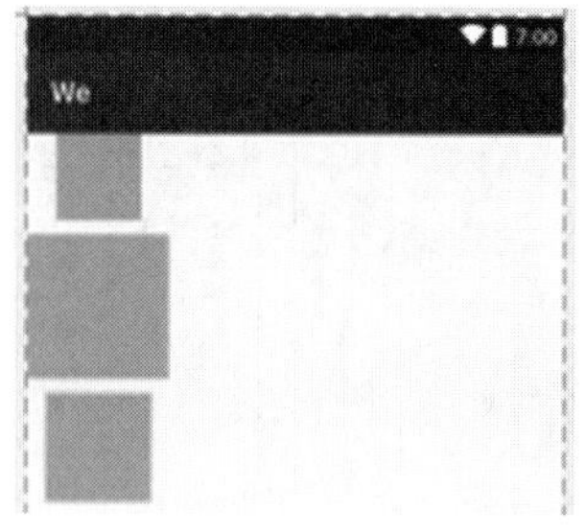

图 3-7 3 个 ImageView 显示图

3 个 ImageView 定义的视窗大小都是宽 200px，高自适应。第一个 ImageView 代码对应图 3-7 上图，与原图大小(100px×100px)一致。第二个 ImageView 代码在第一个 ImageView 代码的基础上增加了 android:adjustViewBounds="true"，表示进行等比缩放，由于宽 200px，高不确定，因此缩放比等于横向比，为 200px/100px=2，图像横、纵方向都扩大了 2 倍，对应图 3-7 中图。第三个 ImageView 代码在第二个 ImageView 代码的基础上增加了 android:maxHeight="150px"，含义是图像等比缩放后，高度不能超过 150px，已知该图经等比缩放后，图像理论大小为 200px×200px，由于 maxHeight=150px，200px>150px，因此图像实际高度为 150px，又由于等比缩放，因此图像实际宽度为 150px。

有一个知识点需要加深理解：仅当 ImageView 视图宽、高有一个大小确定、一个大小不确定的时候，adjustViewBounds 属性设置为 true 时，才能实现等比缩放功能。maxWidth、maxHeight 属性是对缩放后的图像进行进一步限制。当 ImageView 视图宽大小确定，高不确定的时候，maxHeight 起到约束作用，maxWidth 不起作用；当 ImageView 视图宽不确定，高大小确定的时候，maxWidth 起到约束作用，maxHeight 不起作用。

【例 3-10】 ImageView 基本属性中 ScaleType 对应情况分析。

ScaleType 是 ImageView 类中的又一重要属性，可以取表 3-1 中的 7 个常用值，这些值设置后，图形有哪些变化是理解的重点。由于涉及图形的放大和缩小，因此一般设置 3 个不同大小的 ImageView：第 1 个 ImageView 与原图大小相等；第 2 个 ImageView 的尺寸要大于原图(可能放大)；第 3 个 ImageView 的尺寸要小于原图(可能缩小)。对第 2、3 个 ImageView 设置相同的 ScaleType 值，根据这 3 幅图的变化特征，就能大致分析出相应 ScaleType 值对应的图像变化特点。

本例也是按上述思路理解 ScaleType 中值为 CENTER、CENTER_CROP、CENTER_INSIDE 的情况，但用动态程序实现，程序仿真界面如图 3-8 所示。

主面板包含左、右两部分。左面是功能按钮，用 ScaleType 属性字符串作为按钮名称；右面是图像显示区，当按左侧某一功能按钮时，右侧显示 3 幅图像。所编制的程序代码文件如下所示。

① 布局配置文件 main.xml。

```
<?xml version="1.0" encoding="utf-8"?>
```

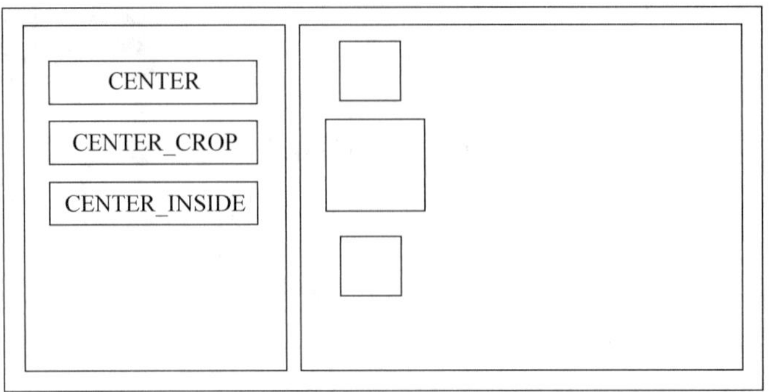

图 3-8 动态程序 ScaleType 属性仿真界面

```
<LinearLayout
    xmlns:android="http://schemas.android.com/apk/res/android"
    android:layout_width="match_parent"
    android:layout_height="match_parent">
    <LinearLayout
        android:id="@+id/myleft"
        android:layout_width="150dp"
        android:layout_height="match_parent"
        android:orientation="vertical">
    </LinearLayout>
    <LinearLayout
        android:layout_width="2px"
        android:layout_height="match_parent"
        android:background="#000">
    </LinearLayout>
    <LinearLayout
        android:id="@+id/myright"myleft
        android:layout_width="match_parent"
        android:layout_height="match_parent"
        android:orientation="vertical">
    </LinearLayout>
</LinearLayout>
```

可以看出，主布局采用水平线性布局，加了 3 个线性布局，左侧线性布局(id=myleft)用于添加按钮，右侧线性布局(id=myright)用于添加图像，中间线性布局用于添加竖直分割线，将左、右线性布局区域加以区分。

该文件本质上来说仅是一个界面框架，所有子组件都是由下面的程序动态添加的。

② 程序文件 MainActivity.java。

该类稍复杂，先将某些函数方法体略去，代码如下所示。

```
public class MainActivity extends AppCompatActivity {
    String text[] ={"center","centercrop","centerinside"};
     ImageView. ScaleType  st [ ] = { ImageView. ScaleType. CENTER,  ImageView.
ScaleType.CENTER_CROP,
    ImageView.ScaleType.CENTER_INSIDE};
    void fillLeft(){……}                //动态填充左侧按钮
    void process(View v){}             //按钮响应函数
    protected void onCreate(Bundle savedInstanceState) {
    super.onCreate(savedInstanceState);
            setContentView(R.layout.main);
            fillLeft();
        }
    }
```

fillLeft()函数的功能是动态生成左侧布局功能按钮及添加事件响应，按钮文字由 text 数组定义，由程序将 text 数组与 st 数组关联，st 数组元素是 ScaleType 类型的数据。例如，当单击文字为 CENTER 的按钮时，已知现在要为 ImageView 设置 ScaleType.CENTER 参数，看其对应的图像显示情况。

process()是按钮响应函数，负责动态设置 ScaleType 参数，动态创建 3 个 ImageView 对象，并将图像显示出来。

下面将进一步介绍两个主要函数 fillLeft()、process()。

```
//fillLeft()函数代码
void fillLeft(){
    LinearLayout left = (LinearLayout)findViewById(R.id.myleft);//获得左侧布局
父容器对象
    for(int i=0; i<text.length; i++){
        Button b = new Button(this);
        ViewGroup.LayoutParams lay = new ViewGroup.LayoutParams(
            ViewGroup.LayoutParams.WRAP_CONTENT,ViewGroup.LayoutParams.WRAP_
CONTENT);
        b.setLayoutParams(lay);
        b.setText(text[i]);
        left.addView(b);
```

```
            b.setOnClickListener(new View.OnClickListener() {
                public void onClick(View v) {
                    process(v);
                }
            });
        }
    }
```

首先获得左侧布局父容器对象，然后依据 text 数组，利用 for 循环动态产生 Button，并通过匿名类添加响应函数，直接将处理函数转到 MainActivity 类中的 process()函数。

当然，通过修改 text 及 st 数组，可把表 3-1 中 ScaleType 的 7 个值都加进来，动态产生 7 个按钮。

```
//process()函数代码
void process(View v){
    //根据按钮文字,确定待测的 st 数组元素位置
    int pos = -1 ;
    String s = ((Button)v).getText().toString();
    if(s.equals("center")) pos = 0;
    else if(s.equals("centercrop")) pos = 1;
    else if(s.equals("centerinside")) pos = 2;
    //删除右侧布局中的所有子组件
    ViewGroup.LayoutParams lay = new ViewGroup.LayoutParams(100,100);
    LinearLayout right = (LinearLayout)findViewById(R.id.myright);
    right.removeAllViews();
    //添加第 1 个 ImageView,大小与原图一致
    ImageView img = new ImageView(this); img.setLayoutParams(lay);
    img.setImageResource(R.drawable.my2);
        right.addView(img);
    //添加第 2 个 ImageView,大小大于原图
    lay.width = 200; lay.height=300;
    ImageView img2 = new ImageView(this); img2.setLayoutParams(lay);
    img2.setImageResource(R.drawable.my2); img2.setScaleType(st[pos]);
        right.addView(img2);
    //添加第 3 个 ImageView,大小小于原图
    lay.width = 80; lay.height=60;
    ImageView img3 = new ImageView(this); img3.setLayoutParams(lay);
    img3.setImageResource(R.drawable.my2); img3.setScaleType(st[pos]);
    right.addView(img3);
}
```

本示例中，原图像的大小为 100px×100px，添加的 3 个 ImageView 的大小分别为：100px×200px，等于原图；200px×300px，大于原图；80px×60px，小于原图。

代码中，首先根据按钮文本内容确定要测试的 ScaleType 值在 st 数组中的位置 pos；然后删除当前右线性布局中的所有子 ImageView 对象(此功能是必需的，避免右布局对象子 ImageView 对象越加越多)；最后动态添加 3 个子 ImageView 对象，对第 2、3 个 ImageView 设置 ScaleType 值。

实际界面如图 3-9 所示。

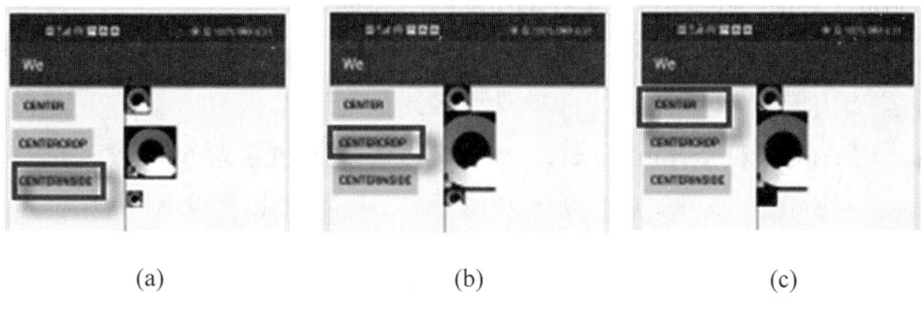

(a)　　(b)　　(c)

图 3-9　ScaleType 测试界面

通过展示界面可得，最好理解的是 CENTER_INSIDE 属性(图 3-9(a))，原图均是等比缩放，完整地存放在 ImageView 视图中；对 CENTER 属性而言(图 3-9(c))：当 ImageView 视图大小小于或等于原图时，ImageView 视图显示的是相当于把该图几何中心放在 ImageView 视图几何中心中，同时两图的水平轴、垂直轴平行时的图像截取部分；当 ImageView 视图大小大于原图时，则原图进行等比放大，由图看出，放大比选择的是横向比、纵向比中的最大值；对 CENTER_CROP 而言(见图 3-9(b))，原图进行等比缩放，当 ImageView 视图大小大于原图时，放大比选择的是横向比、纵向比中的最大值；当 ImageView 视图大小小于或等于原图时，放大比选择的是横向比、纵向比中的最小值。

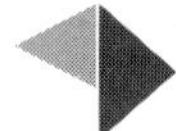

3.6　文本控件

3.6.1　TextView

TextView 控件是文本表示控件，主要功能是向用户展示文本的内容，它是不可编辑的。文本控件的主要操作有颜色、大小、字体等的设置，其相关主要函数如下所示。

- void setTextColor(int value)，对应 android：textColor，设置字体颜色。value 可为 Color.RED、Color.GREEN、Color.BLUE 等，也可由 Color.rgb(int r,int g,int b)

定义，r、g、b 为三基色的分量值，范围均为[0,255]。

- void setTextSize(float size)，对应 android:textSize，设置字体大小，且是以 sp 为单位，并不是以 px 为单位。
- void setTextSize(int unit, float size)，对应 android:textSize，设置字体大小，unit 整型数是单位标识，可为下述值。

TypedValue.COMPLEX_UNIT_PX，表明单位是 px。

TypedValue.COMPLEX_UNIT_SP，表明单位是 sp。

TypedValue.COMPLEX_UNIT_DIP，表明单位是 dp。

- void setTypeFace(Typeface tf)，对应 android:typeface 属性，设置字体，Typeface 对象 tf 代表字体。
- void setTypeFace(Typeface tf,int style)，该函数包含两个参数：第一个参数对应 android:typeface 属性，用于设置字体；第二个参数用于设置字体风格，如粗体、斜体等。

手机中常用的字体有 4 种，见表 3-2。

表 3-2　4 种常用字体 Typeface 标识表

序号	名　　称	说　　明
1	Typeface.NORMAL	普通字体，系统默认使用的字体
2	Typeface.SERIF	非衬线字体
3	Typeface.SANS_SERIF	衬线字体
4	Typeface.MONOSPACE	等宽字体

手机中常用的字体风格有 4 种，也由 Typeface 标识，见表 3-3。

表 3-3　字体风格 Typeface 标识表

序号	名　　称	说　　明
1	Typeface.NORMAL	系统默认字体风格
2	Typeface.BOLD	粗体
3	Typeface.ITALIC	斜体
4	Typeface. BOLD_ITALIC	粗体＋斜体

- void setLineSpacing(float add,float multi)，定义行间距的加因子和乘因子，加因子对应 android:lineSpacingExtra 属性，乘因子对应 android:lineSpacingMultiplier 属性。

【例 3-11】 利用程序设置字体大小。

主要是应用 setTextSize(int unit,float size)双参数函数,在 size 相同的情况下,测试不同单位(unit)下的字体大小变化情况。本示例不用布局文件,完全用程序动态实现。

```
public class MainActivity extends AppCompatActivity {
    protected void onCreate(Bundle savedInstanceState) {
        super.onCreate(savedInstanceState);
        LinearLayout parent = new LinearLayout(this);
        parent.setOrientation(LinearLayout.VERTICAL);
        LinearLayout.LayoutParams param = new LinearLayout.LayoutParams(
        ViewGroup.LayoutParams.MATCH_PARENT,ViewGroup.LayoutParams.MATCH_
PARENT);
        int d[] = {TypedValue.COMPLEX_UNIT_PX,TypedValue.COMPLEX_UNIT_SP,
            TypedValue.COMPLEX_UNIT_DIP};
        String s[] = {"px","sp","dp"};
        for(int i=0; i<d.length; i++){
            TextView tv = new TextView(this);
            tv.setText("Hello:" + s[i]);
            tv.setTextSize(d[i], 50);parent.addView(tv);
        }
        setContentView(parent);
    }
}
```

由代码可知,主界面采用垂直线性布局,d 数组与 s 数组是相关的,s 数组是单位的文字描述,d 数组是 Android 系统单位的标识常量。动态产生 3 个 TextView 子组件,大小都设置为 50,只是相应的单位设置不同,显示的内容是"Hello:单位"。实际界面如图 3-10 所示。

(a)　　　　(b)

图 3-10　动态设置字体单位界面

程序运行后,初始显示界面如图 3-10(a),可以看出,相同 size 下,sp、dp 单位对应的显示一致，与单位 px 对应的显示不一致。进一步实验，打开手机设置功能，将显示字体调为最大，这时界面与图 3-10(b)一致，可以看出，在 size 相同的情况下，单位为 sp 对应的字体随手机字体大小设置的变化而变化，而单位为 dp 对应的字体大小不变化。

3.6.2 深入探究

TextView 通常用来显示普通文本，但 TextView 的功能并不局限于此，它还有一些高级特性，如 html 嵌入、插入图片等，也就是 TextView 可以进行富文本操作，直接显示结果。TextView 可操作的常用 Html 标签见表 3-4。

表 3-4 TextView 可操作的常用 Html 标签

名　称	说　明	名　称	说　明
<a>	定义超链接	<font>	定义字体、字体大小及颜色
<b>	定义粗体文字效果	<h1> - <h6>	定义标题
<i>	定义斜体文字效果	<p>	定义段落
<u>	定义下画线文本	<div>	定义分区
<big>	定义大号字体效果	<img>	定义图像
<small>	定义小号字体效果	<sup>	定义上标
 	定义换行	<sub>	定义下标

Android 系统通过 Html 类中的静态方法 fromHtml()将 HTML 富文本字符串格式化，函数说明如下所示。

- CharSequence fromHTML(String source, int flags)，source 是待格式化的 HTML 字符串，flags 是整数标识，常用以下两个 Html 类中的静态常量。

FROM_HTML_MODE_COMPACT：HTML 块元素之间使用一个换行符分隔。

FROM_HTML_MODE_LEGACY：HTML 块元素之间使用两个换行符分隔。

- CharSequence fromHTML(String source, int flags, ImageGetter imgproc, TagHandler tagproc)，source 是待格式化的 HTML 字符串，flags 是整数标识，解释见上文，imgproc 是图像处理接口对象，tagproc 是标签接口处理对象。

【例 3-12】 在 TextView 中显示简单的 HTML 语句。

定义线性布局文件，包含一个 TextView 子控件，利用程序在 TextView 中显示 HTML 内容。

① main.xml。

```
<?xml version="1.0" encoding="utf-8"?>
<LinearLayout
    xmlns:android="http://schemas.android.com/apk/res/android"
    android:layout_width="match_parent"
    android:layout_height="match_parent">
    <TextView
        android:id="@+id/mytext"
        android:layout_width="match_parent"
        android:layout_height="wrap_content" />
</LinearLayout>
```

定义 TextView 子组件 id 为 mytext,方便程序获得该对象。

② MainActivity.java。

```
public class MainActivity extends AppCompatActivity {
    protected void onCreate(Bundle savedInstanceState) {
    super.onCreate(savedInstanceState);
    setContentView(R.layout.main);
    String s = "<font>Html 格式文本</font><br>";          //黑色文本
    String s2 = "<font color='red'>Html 格式文本</font><br>";
                                                         //红色文本
    String s3 = "<h1>Html 格式文本</h1><br>";              //h1 大小的文本
    TextView tv = (TextView) findViewById(R.id.mytext);
    tv.setText(Html.fromHtml(s+s2+s3, Html.FROM_HTML_MODE_COMPACT));
    }
}
```

本示例较简单,用的是前文介绍的两个参数的 fromHtml()函数,界面如图 3-11 所示。

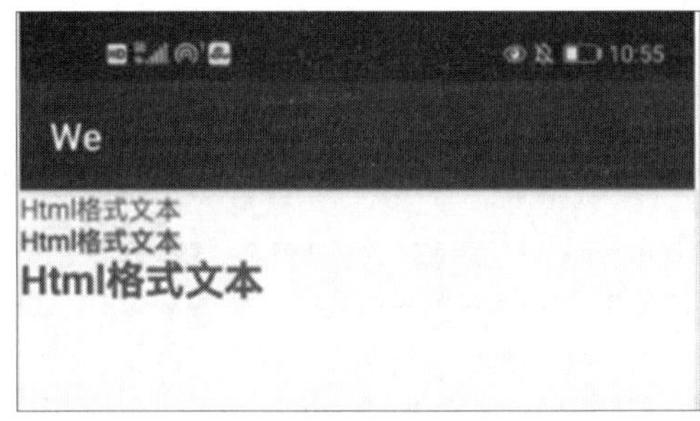

图 3-11 TextView 显示 HTML 语句图

【例 3-13】 在 TextView 中显示图像标签<img>。

在 HTML 语句中，<img>标签与其他标签不同，其他标签形如"<XXX>……</XXX>"均包含所需的全部内容，因此易于解析；而<img>标签还包含图像文件，必须将图像文件解析，才能完成对<img>标签的解析，Html.fromHtml()函数提供了处理图像的接口，但需自行完成。先看具体代码，如下所示。

```
//main.xml
<?xml version="1.0" encoding="utf-8"?>
<LinearLayout
    xmlns:android="http://schemas.android.com/apk/res/android"
    android:layout_width="match_parent"
    android:layout_height="match_parent">
    <TextView
        android:id="@+id/mytext"
        android:layout_width="match_parent"
        android:layout_height="wrap_content" />
</LinearLayout>
```

定义 TextView 子组件 id 为 mytext，显示<img>标签图像内容。当然，需在 res\drawable\目录下建一个图像文件，假设为 my.jpg。

```
//MainActivity.java
public class MainActivity extends AppCompatActivity {
class A implements Html.ImageGetter{
    public Drawable getDrawable(String source) {
        int id = Integer.parseInt(source);                          //获得图像 id
        Drawable drawable = getResources().getDrawable(id, null);
                                                                    //获得图像对象
        drawable.setBounds(0, 0, drawable.getIntrinsicWidth(),
        drawable.getIntrinsicHeight());
        return drawable;                                            //返回图像对象
    }
}
protected void onCreate(Bundle savedInstanceState) {
    super.onCreate(savedInstanceState);
    setContentView(R.layout.main);
    String s = "图像 1<img src='" +R.drawable.my+ "' />";
    TextView tv = (TextView)findViewById(R.id.mytext);
```

```
    tv.setText(Html.fromHtml(s, Html.FROM_HTML_MODE_COMPACT,new A(), null));
  }
  }
```

本示例用的是前文介绍的 4 个参数的 fromHtml()函数，第 3 个参数，形参必须是 ImageGetter 接口对象，是专门处理<img>标签的，因此编制了该接口的派生内部类 A。

对<img>标签而言，src 属性是标识图像位置的，而测试的图像为 res\drawable\my.jpg，那怎么写 src，能这样写(<img src='res\drawable\my.jpg' />)？很明显不行。由于 Android 系统中每个资源都有唯一的整型 id，因此应该由 R.drawable.my 决定，但由于该值是变量，因此应如代码中形成特征字符串：s = "图像 1<img src='" + R.drawable.my+ "' />"，若写成：s = "图像 1<img src=R.drawable.my />"，则是错误的。

内部类 A 实现了接口定义的图像处理函数 getDrawable(String source)，source 与<img src='XXX'>中的'XXX'是一致的。本例中，'XXX'代表的是 R.drawable.my 的数字字符串，因此必须将数字字符串'XXX'转化为数字 XXX，即图像 my.jpg 的 id 号，根据 id 号就能获得 Drawable 对象，将其返回即可。

【例 3-14】　自定义标签的应用。

假设 HTML 语句：<font size="60">Hello</font><font size="100">Hello</font>，该语句的含义是按不同字体大小输出 Hello，但应用 Html.fromHtml()函数将其显示在 TextView 中却发现 size 属性根本没有起作用，两个 Hello 显示大小是一致的。经查 fromHtml()函数源码，发现系统根本没有 font 标签中的 size 属性，怎么办？自定义标签！例如，<mysize>Hello</mysize>用到的 fromHtml(参数 1、参数 2、参数 3、TagHandle obj)中的第 4 个参数 TagHandle 技术。TagHandle 是标签处理接口，简单来说，当 fromHtml()函数处理 HTML 语句中各标签的时候，会调用 TagHandle 子类对象，通过截获所需的标签，从而实现相应的功能。

本示例针对 HTML 语句<mysize>Hello</mysize>，设置字体大小为 100，相关代码如下所示。

```
//布局配置文件 main.xml
<?xml version="1.0" encoding="utf-8"?>
<LinearLayout
    xmlns:android="http://schemas.android.com/apk/res/android"
    android:layout_width="match_parent"
    android:layout_height="match_parent">
    <TextView
```

```
        android:id="@+id/mytext"
        android:layout_width="match_parent"
        android:layout_height="wrap_content" />
</LinearLayout>
```

定义了 id 为 mytest 的 TextView 节点,用于自定义标签显示效果。

```
//MainActivity.java
public class MainActivity extends AppCompatActivity {
    int start ;                                                   //文本起始位置
    int end;                                                      //文本结束位置
    class A implements Html.TagHandler{
        int size;
        A(int size){
            this.size = size;
        }
         public void handleTag(boolean opening, String tag, Editable output,
XMLReader xmlReader) {
            try {
                if (tag.toLowerCase().equals("mysize") ) {
                    if(opening)                                   //解析标签开始
                        start = output.length();                  //获得起始位置
                    else{                                         //解析标签结束
                        end = output.length();                    //获得结束位置
                        //将[start,end)间的文本字体大小设置成 size
                        output.setSpan(new AbsoluteSizeSpan(size), start, end,
                            Spanned.SPAN_EXCLUSIVE_EXCLUSIVE);
                    }
                }
            }catch(Exception e){   }
        }
    }
    protected void onCreate(Bundle savedInstanceState) {
        super.onCreate(savedInstanceState);
        setContentView(R.layout.main);
        String s = "<mysize>Hello</mysize>";
        TextView tv = (TextView) findViewById(R.id.mytext);
         tv.setText(Html.fromHtml(s, Html.FROM_HTML_MODE_COMPACT,null, new A
```

```
(100)));
    }
}
```

为了操作自定义标签，定义了内部类 A，派生于 TagHandler 接口，重写了 handleTag()函数，当示例中应用 Html.fromHtml 函数的时候，会不断地调用 handleTag()函数，该函数是理解的重点。

函数定义：void handleTag(boolean open, String tag, Editable out, XMLReader reader)

- open：为 true 时，表示标签开始解析；为 false 时，表示解析完。
- tag：当前解析的标签。
- out：文本内容对象。
- reader：xml 解析器。

为了更好地说明 handleTag()函数的作用，通过解析 HTML 串<p>aaa</p><mysize>bbb</mysize>的流程加以论述，见表 3-5。

表 3-5 HTML 串解析流程表

源　　串	<p>aaa</p><mysize>bbb</mysize>
解析流程	第 1 次调用 handleTag 时(遇到<p>)：tag=p,open=true,out 为空串 第 2 次调用 handleTag 时(遇到</p>)：tag=p,open=false,out 为 aaa 第 3 次调用 handleTag 时(遇到<mysize>)：tag=mysize,open=true,out 为 aaa 第 2 次调用 handleTag 时(遇到</mysize>)：tag=mysize,open=false,out 为 aaabbb

由此得本例中 handleTag 的功能思路是：当解析标签是自定义标签 mysize 时，根据 opening 布尔值，获得输出文本在内存中的起止位置 start、end，最后将[start,end)间的内存字符串设置成所需的大小并输出。

本例中所设字体大小是通过构造方法 A(int size)传进去的，这样再增加几个<mysize>标签，则字体大小显示是一致的，根本问题还没有完全解决。希望<mysize size=60>aaa</aaa><mysize size=100>bbb</mysize>能以不同大小的字体显示，也就是说，必须能获得 size 属性值大小，才能解决字体设置的根本问题，但涉及底层源码，笔者仅给出简要的代码说明，有能力的读者可深入研究。

```
public class MainActivity extends AppCompatActivity {
    int start ;int end;
    final HashMap<String, String>attributes = new HashMap<String, String>();
                                                        //定义属性集合
```

```
    private void processAttributes(final XMLReader xmlReader) {
        try {
            attributes.clear() ;                                    //清空属性集合
              Field elementField = xmlReader. getClass ( ) . getDeclaredField ( "
theNewElement");
            elementField.setAccessible(true);
            Object element = elementField.get(xmlReader);
            Field attsField = element.getClass().getDeclaredField("theAtts");
            attsField.setAccessible(true);
            Object atts = attsField.get(element);
            Field dataField = atts.getClass().getDeclaredField("data");
            dataField.setAccessible(true);
            String[] data = (String[]) dataField.get(atts);
            Field lengthField = atts.getClass().getDeclaredField("length");
            lengthField.setAccessible(true);
            int len = (Integer) lengthField.get(atts);
            for (int i = 0; i < len; i++) {
                attributes.put(data[i * 5 + 1], data[i * 5 + 4]);
            }
        } catch (Exception e) {     }
    }
    class A implements Html.TagHandler{
         public void handleTag (boolean opening, String tag, Editable output,
XMLReader xmlReader) {
            try {
                if (tag.toLowerCase().equals("mysize") ) {
                    if(opening){                          //标签解析开始
                        processAttributes(xmlReader); //解析属性函数
                        start = output.length();  //获得输出文本在内存中的起始位置
                    }
                    else{                                 //标签解析结束
                        end = output.length();    //获得输出文本在内存中的结束位置
                        int value = Integer.parseInt(attributes.get("size"));
                                                          //获取 size 属性值
                        output.setSpan(new AbsoluteSizeSpan(value), start, end,
                                Spanned.SPAN_EXCLUSIVE_EXCLUSIVE);
                    }
                }
```

```
            }catch(Exception e){   }
        }
    }
    protected void onCreate(Bundle savedInstanceState) {
        super.onCreate(savedInstanceState);
        setContentView(R.layout.main);
        String s = "<p><mysize size='100'>Hello</mysize></p><mysize size='
200'>Hello</mysize>";
        TextView tv = (TextView)findViewById(R.id.mytext);
        tv.setText(Html.fromHtml(s, Html.FROM_HTML_MODE_COMPACT,null, new A()));
    }
}
```

以上代码定义了标签键-值映射成员变量 attributes；processAttributes()函数的功能是将解析的自定义标签的键-值保存到映射 attributes。由此得本例中 handleTag 的流程是：当解析标签是自定义标签 mysize，opening 为 true 时，一方面获得输出文本在内存中的起始位置 start，另一方面调用 processAttributes()函数，将键-值保存到 attributes 成员变量中；当 opening 为 false 时，一方面获得输出文本在内存中的结束位置 end，另一方面通过 attributes 获得自定义字体大小 value，最后将[start，end)间的内存字符串设置成 value 大小并输出。

processAttributes()函数涉及系统底层，一般情况下是不变的，将其作为一个系统函数即可。

3.6.3　EditText

1. 基本函数

EditText 是一个输入框，是获取用户数据的一种方式。EditText 是 TextView 的子类，它继承了 TextView 的所有属性。其常用的函数（与 TextView 相同的略去）如下所示。

- void setInputType(int type)，对应 android:inputType 属性，设置输入类型，type 为系统 InputType 类中的静态常量，见表 3-6。

表 3-6　InputType 常用静态常量表

android:XXX XXX 可为	InputType.XXX XXX 可为	说　　明
Text	TYPE_CLASS_TEXT	普通文本

续表

android:XXX XXX 可为	InputType.XXX XXX 可为	说　明
Number	TYPE_CLASS_NUMBER	数字文本
Phone	TYPE_CLASS_PHONE	电话号码
Datetime	TYPE_CLASS_DATETIME	日期时间
Date	TYPE_DATETIME_VARIATION_DATE	日期
Time	TYPE_DATETIME_VARIATION_TIME	时间
numberDecimal	TYPE_NUMBER_FLAG_DECIMAL	小数数字
numberSigned	TYPE_NUMBER_FLAG_SIGNED	带正负号的数字
textEmailAddress	TYPE_TEXT_VARIATION_EMAIL_ADDRESS	电子邮件地址
textPassword	TYPE_TEXT_VARIATION_PASSWORD	密码

- void setFocusable(boolean mark)，对应 ansroid: focusable，设置是否为当前输入焦点。
- void setHint(CharSequence s)，设置提示文本。
- void setSingleLine(boolean mark)，对应 android:singleLine，设置是否单行显示，若 mark 为 true，则单行显示；若 mark 为 false，则多行显示。当该行写满时，自动转入下一行。
- void setMaxLines(int size)，对应 android:maxLines，设置最大显示行数 size，超过的内容部分利用滚动实现。
- void setMinLines(int size)，对应 android:minLines，设置最小显示行数 size，即使内容为空，也要显示 size 行。

2. 事件响应

TextWatcher 是一个用来监听文本实时变化的接口，使用该接口可以很方便地对 EditText 控件中的文字进行监听和修改，消息注册函数是 addTextChangedListener (TextWatcher obj)。

TextWatcher 接口定义了 3 个接口函数：beforeTextChanged()、onTextChanged()、afterTextChanged()。详细解释如下所示。

① void beforeTextChanged(CharSequence s, int start, int count, int after)

该方法在文本改变之前调用，传入了 4 个参数，如下所示。

CharSequence s：文本改变之前的内容。

int start：文本开始改变时的起点位置，从 0 开始计算。

int count：要被改变的文本字数，即将要被替代的选中文本字数。

int after：改变后添加的文本字数，即替代选中文本后的文本字数。

该方法是在文本没有被改变，但将要被改变的时候调用，把 4 个参数组成一句话就是：在当前文本 s 中，从 start 位置开始之后的 count 个字符即将被 after 个字符替换掉。

② void onTextChanged(CharSequence s, int start, int before, int count)

该方法是在当文本改变时被调用，同样传入了 4 个参数，如下所示。

CharSequence s：文本改变之后的内容。

int start：文本开始改变时的起点位置，从 0 开始计算。

int before：要被改变的文本字数，即已经被替代的选中文本字数。

int count：改变后添加的文本字数，即替代选中文本后的文本字数。

该方法是在文本被改变，并且改变的结果已经可以显示时调用，把 4 个参数组成一句话就是：在当前文本 s 中，从 start 位置开始之后的 before 个字符(已经)被 count 个字符替换掉。

③ void afterTextChanged(Editable s)

该方法是在文本改变结束后调用，传入了一个参数 s，即改变后的最终文本。

【例 3-15】 登录输入信息校验。

功能是：输入用户名、密码，要求两者不能为空，用户名只能是数字，密码长度必须大于 6。若输入不满足条件，则在界面上给出错误信息：用户名或密码不能为空，或者密码长度必须大于 6。

① 布局配置文件 main.xml。

```
<?xml version="1.0" encoding="utf-8"?>
<LinearLayout
    xmlns:android="http://schemas.android.com/apk/res/android"
    android:layout_width="match_parent"
    android:layout_height="match_parent"
    android:orientation="vertical">
    <EditText
        android:id="@+id/myuser"
        android:layout_width="match_parent"
        android:layout_height="wrap_content"
        android:inputType="number"
        android:hint="user"/>
```

```
    <EditText
        android:id="@+id/mypwd"
        android:layout_width="match_parent"
        android:layout_height="wrap_content"
        android:inputType="textPassword"
        android:hint="password"/>
    <TextView
        android:id="@+id/myinfo"
        android:layout_width="wrap_content"
        android:layout_height="wrap_content"
        android:textColor="#f00"
        android:textSize="30sp"/>
    <Button
        android:id="@+id/myok"
        android:layout_width="wrap_content"
        android:layout_height="wrap_content"
        android:text="OK"/>
</LinearLayout>
```

以上代码定义了两个 EditText 节点：id 为 myuser 的节点用于用户名输入，由于要求用户名必须是数字，因此定义了 inputtype="number"；id 为 mypwd 的节点用于密码输入，由于其与密码有关，因此定义了 inputtype="textPassword"。

定义了 id=myinfo 的 TextView 节点，用于显示对用户名、密码输入后的校验结果。id 为 myok 的按钮用于获取用户名、密码的值，并加以校验，将结果填入 id 为 myinfo 的 TextView 节点中。

② MainActivity.java。

```
public class MainActivity extends AppCompatActivity {
    protected void onCreate(Bundle savedInstanceState) {
        super.onCreate(savedInstanceState);
        setContentView(R.layout.main);
        Button btn = (Button)findViewById(R.id.myok);
        btn.setOnClickListener(new View.OnClickListener() {
            public void onClick(View v) {
                EditText user = (EditText)MainActivity.this.findViewById
(R.id.myuser);
                EditText pwd = (EditText)MainActivity.this.findViewById
(R.id.mypwd);
```

```
                TextView tv = (TextView)MainActivity.this.findViewById
(R.id.myinfo);
                String strUser = user.getText().toString().trim();
                String strPwd = pwd.getText().toString().trim();
                if(strUser.length()==0 ||strPwd.length()==0)
                    tv.setText("用户名或密码不能为空!");
                else if(strPwd.length()<6)
                    tv.setText("密码长度必须大于或等于 6!");
                else
                    tv.setText("输入正确!");
            }
        });
    }
}
```

该示例的运行界面如图 3-12 所示。

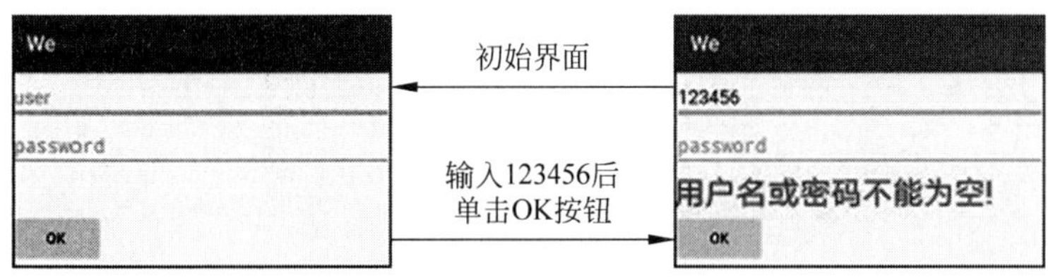

图 3-12 登录输入信息校验界面

当程序运行后发现：对 EditText 子组件设置输入类型 inputtype＝number，仅能保证它输入数字，其他字符不能输入。它不会检查输入为空的情况，而在界面中给出相应的提示信息。对表 3-6 中的其他常用类型亦是如此：仅能保证输入符合条件的字符，不会判断结果字符串是否符合要求。例如，假设 EditText 子组件设置 inputtype＝textEmailAddress，表示输入 E-mail 地址，它能保证输入@字符，但输入几个@字符，它是不会判定的。因此，数据校验仍是编程中非常重要的环节。再如，本例中密码输入长度要求必须大于或等于 6，这也需要编程加以实现。

【例 3-16】 深入理解 TextWatcher，仿真界面如图 3-13 所示。

功能是：开发文本同步跟踪器，在"文本输入"处输入文本（假设某段文字），在"文本同步"处同步输出文本，若出错（与原文比较），则在"删除次数"处显示删除字符的次数。

很明显，输入部分应用的是 EditText 组件，由于实现同步跟踪功能，因此一定应用到 TextWatcher 接口，相关代码如下所示。

删除次数：	2
文本同步：	abcde
文本输入：	abcde

图 3-13　文本同步跟踪器仿真界面

① 布局配置文件 main.xml。

```
<?xml version="1.0" encoding="utf-8"?>
<LinearLayout
    xmlns:android="http://schemas.android.com/apk/res/android"
    android:layout_width="match_parent"
    android:layout_height="match_parent"
    android:orientation="vertical">
    <TextView
        android:id="@+id/mysize"
        android:layout_width="wrap_content"
        android:layout_height="wrap_content"
        android:textSize="30sp"/>
    <TextView
        android:id="@+id/mydes"
        android:layout_width="wrap_content"
        android:layout_height="wrap_content"
        android:textSize="30sp"
        android:minLines="3"
        android:maxLines="3"/>
    <EditText
        android:id="@+id/mysrc"
        android:layout_width="match_parent"
        android:layout_height="wrap_content"
        android:textSize = "30sp"
        android:minLines="3"
        android:maxLines="3"/>
</LinearLayout>
```

采用垂直线性布局，id 为 mysize 的 TextView 组件用于显示删除次数，id 为 mysrc

的 EditText 组件用于输入文本内容，id 为 mydes 的 TextView 组件用于同步显示 EditText 组件的输入内容。mysrc、mydes 对应的子组件的 minLines、maxLines 属性都设置为 3，将这两个子组件的高固定为对应 3 行文本的高度。

② MainActivity.java。

```
public class MainActivity extends AppCompatActivity {
    int size = 0;                                              //累积的删除次数
    protected void onCreate(Bundle savedInstanceState) {
        super.onCreate(savedInstanceState);
        setContentView(R.layout.main);
        final TextView tv = (TextView)findViewById(R.id.mysize);
        final TextView des = (TextView)findViewById(R.id.mydes);
        EditText src = (EditText)findViewById(R.id.mysrc);
        src.addTextChangedListener(new TextWatcher() {
    *       public void beforeTextChanged(CharSequence s, int start, int count,
int after) {
                if(after==0) {
                    size += count;                             //累积删除次数
                    tv.setText("删除次数:" + size);
                }
            }
            public void onTextChanged(CharSequence s, int start, int before, int
count) {   }
            public void afterTextChanged(Editable s) {
                des.setText(s.toString());                     //同步输出
            }
        });
    }
}
```

id 为 mysrc 的 EditText 子组件内容同步输出到 id 为 mydes 的 TextView 中较易实现，只在实现 TextWatcher 的匿名内部类中重写 afterTextChanged()函数，直接将 mysrc 的内容设置到 mydes 中即可。读者可以思考：能在 onTextChanged()函数中实现同步输出吗？

如何实现“删除次数”功能？本例是在 beforeTextChanged()函数中实现的，关键是要理解其 4 个函数参数的变化情况。虽然前文简述过它们的含义，但从实践中加以理解是必然的。表 3-7 列举了输入动态变化对应 beforeTextChanged()函数 4 个参数的变化情况。

表 3-7 输入动态变化对应 before TextChanged()函数 4 个参数的变化情况

序号	操　　作	s	start	count	after
1	输入 a	空	0	0	1
2	输入 b	a	0	1	2
3	输入 c	ab	0	2	3
4	输入 d	abc	0	3	4
5	输入 e	abcd	0	4	5
6	将光标移到 c 后，按删除键	abcde	2	1	0
7	再输入 f	abce	2	0	1

通过表 3-7 可进一步总结(对 beforeTextChanged()函数 4 个参数)，如下所示。

输入操作有两种情况：在序列尾插入，start 永远为零，count 是插入前字符序列的总个数，after 是插入后字符序列的总个数；在序列中间插入，start 为插入位置，count 永远是 0，after 表示插入字符的个数。

对于删除操作而言，start 是删除位置，count 是删除的字符个数，after 永远是 0。

明白上述道理，很容易写出示例中 beforeTextChanged()的代码。读者可进一步思考：该功能在 onTextChanged()函数中能实现吗？如何写代码？

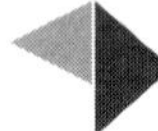

3.7 列表控件

3.7.1 基本函数与事件响应

ListView 是最常用的列表控件。由于手机屏幕空间比较有限，能一次性在屏幕上显示的内容并不多，当程序中有大量数据需要展示的时候，就可以利用 ListView 实现内置滚动功能。其常用的主要函数如下所示。

- void setAdapter(ListAdapter ad)，设置数据适配器。
- int getChildCount()，返回列表项总数目。
- View getChildAt(int index)，返回第 index 列表项视窗对象。
- void addHeaderView(View v)，添加列表头对象。
- void removeAllViews()，清空列表所有子视图对象。

ListView 常用事件有以下两个：短按事件 OnItemClick；长按事件 OnItemLongClick。详细说明如下所示。

1. 短按事件 OnItemClick

消息注册函数是 void setOnItemClickListener（OnItemClickListener obj），OnItemClickListener 接口定义了事件响应函数原型，如下所示。

函数原型：void onItemClick(AdapterView<?> parent, View view, int position, long id)

parent：相当于 listview 适配器的一个指针。

view：轻触对应条目的子视窗 view 对象。

position：是在 ListView 中的顺序位置(0 基)。

id：一般与 position 一致。

2. 长按事件 OnItemLongClick

消息注册函数是 voidsetOnItemLongClickListener(OnItemLongClickListener obj)，OnItemLongClickListener 接口定义了事件响应函数原型，如下所示。

```
boolean onItemLongClick(AdapterView<? > parent, View view, int position, long
id)
```

该函数的参数与 onItemClick()函数的参数是一致的，但返回值不一样，它返回布尔值。若对某 ListView 添加了长按事件、短按事件侦听，当短按的时候，onItemClick()函数运行；当长按的时候，onItemLongClick()函数先运行。当手抬起的时候，若 onItemLongClick()函数返回 true，则 onItemClick()函数不运行；若 onItemLongClick()函数返回 false，则 onItemClick()函数运行。

【例 3-17】 静态显示 ListView。

若某些条目在应用中不变，可以用静态 ListView 显示，例如，静态显示学习科目语文、数学、英语。当短按某条目时，利用 Toast 显示："你选中了:XX"。按如下步骤操作即可。

① 将静态选项添加在 res\values\strings.xml 文件中。当然，也可添加在其他命名的文件中，系统会统一处理的。

```
<resources>
  <string-array name="subject">
    <item>语文</item>
```

```
    <item>数学</item>
    <item>英语</item>
  </string-array>
</resources>
```

在<resources>标签下增加新的<string-array>子标签，将 name 属性命名为 subject，在<string-array>标签下增加 3 个<item>子标签，值分别为语文、数学、英语。

② 定义布局文件 main.xml。

```
<?xml version="1.0" encoding="utf-8"?>
<LinearLayout
    xmlns:android="http://schemas.android.com/apk/res/android"
    android:layout_width="match_parent"
    android:layout_height="match_parent"
    android:orientation="vertical">
    <ListView
        android:id="@+id/mylist"
        android:layout_width="match_parent"
        android:layout_height="wrap_content"
        android:entries="@array/subject"></ListView>
</LinearLayout>
```

最关键的一行 android:entries="@array/subject"，其中 entries 表示资源路径，"@array/subject"的含义为资源是字符串数组，数组从资源文件中 name 为 subject 的<string-array>标签中获取。

③ MainActivity.java。

```
public class MainActivity extends AppCompatActivity {
    protected void onCreate(Bundle savedInstanceState) {
        super.onCreate(savedInstanceState);
        setContentView(R.layout.main);
        final ListView lv = (ListView)findViewById(R.id.mylist);
        lv.setOnItemClickListener(new AdapterView.OnItemClickListener() {
            public void onItemClick(AdapterView<?> parent, View view, int
position, long id) {
                TextView tv = (TextView)view;
                Toast.makeText(MainActivity.this,"你选中了:"+tv.getText(),
Toast.LENGTH_LONG).show();
            }
```

```
        });
    }
}
```

最终运行界面如图 3-14 所示。

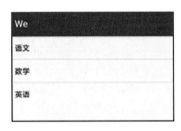

图 3-14　ListView 静态显示图

ListView 对象是父窗口，每行都是 TextView 子视图对象，因此 onItemClick()函数的第 3 个形参 view 就是 TextView 对象。将 view 强制转化成 TextView 对象，再用 getText()函数就能获得选中的科目字符串名称。进一步思考利用 onItemClick()函数的第 4 个形参 position 能获得选中的科目名称吗？可以，position 表示该条目在 ListView 中的顺序位置，调用 ListView 中的 getItemAtPosition(position)函数，把返回结果强制转化成字符串，即选中科目名称字符串。

3.7.2　数据适配器

1. Adapter

Adapter 是适配器最基本的接口，定义了将数据绑定到 UI 上的接口函数。Adapter 负责创建显示每个项目的子 View 和提供对下层数据的访问。在多数情况下，你不需要创建自己的 Adapter。Android 提供了一系列 Adapter 将数据绑定到 UI 上。Adapter 接口中定义的主要接口函数如下所示。

- int getCount()，返回条目总数目。
- Object getItem(int i)，返回第 i 个数据项对象。
- long getItemId(int i)，返回第 i 个数据项 ID。
- View getView(int i,View view,ViewGroup viewGroup)，为第 i 个数据项返回相应的视图。

2. ArrayAdapter

ArrayAdapter 是最常用的数组适配器，读者只要掌握好构造方法即可。常用的构造方法有 2 个，如下所示。

- ArrayAdapter(Context ctx,int layoutID,Object[] ary)；
- ArrayAdapter(Context ctx,int layoutID,List list)。

两个构造方法中前两个参数是一致的：ctx 是上下文对象，layoutID 是每行子视图布局文件资源号；最后一个参数稍有不同，ary 是数据数组对象，list 是线性数组集合对象。

【例 3-18】 利用 ArrayAdapter 重新实现例 3-17 的功能。

① 界面布局文件 main.xml。

```
<?xml version="1.0" encoding="utf-8"?>
<LinearLayout
    xmlns:android="http://schemas.android.com/apk/res/android"
    android:layout_width="match_parent"
    android:layout_height="match_parent">
    <ListView
        android:id="@+id/mylist"
        android:layout_width="match_parent"
        android:layout_height="wrap_content"></ListView>
</LinearLayout>
```

以上代码仅包含<ListView>空节点，没有设置 android:entries 属性。

② MainActivity.java。

```
public class MainActivity extends AppCompatActivity {
    protected void onCreate(Bundle savedInstanceState) {
        super.onCreate(savedInstanceState);
        setContentView(R.layout.main);
        String s[] = {"语文","数学","英语"};
        ArrayAdapter ada = new ArrayAdapter(this, android.R.layout.simple_list
_item_multiple_choice, s);
        ListView lv = (ListView) findViewById(R.id.mylist);
        lv.setAdapter(ada);
    }
}
```

注意 ArrayAdapter 构造方法中，第二个子布局资源号形如 android.R.layout_XXX，前缀是 android.R，表明 XXX 是系统已有资源，又由于包含 layout，所以 XXX 是子布局配置文件。Android 系统提供了多个子布局模板，例中的 simple_list_item_multiple_choice.xml 只是其中之一，还可取如下值，见表 3-8。

表 3-8 Android 系统提供的 ListView 子视图布局

序号	标　识	子视图模板说明
1	simple_list_item_1	单行显示子布局
2	simple_list_item_multiple_choice	带有 CheckBox 多选按钮单行显示布局
3	simple_list_item_single_choice	带有 RadioButton 单选按钮单行显示布局

在已安装的 Android SDK 中找到表 3-8 中序号 1、2 对应的布局文件，内容见表 3-9。

表 3-9 两个子布局文件内容表

文件名称标识	文 件 内 容
simple_list_item_1	<TextView xmlns:android="http://schemas.android.com/apk/res/android" android:id="@android:id/text1" android:layout_width="match_parent" android:layout_height="wrap_content" android:textAppearance="? android:attr/textAppearanceListItemSmall" android:gravity="center_vertical" android:paddingStart="? android:attr/listPreferredItemPaddingStart" android:paddingEnd="? android:attr/listPreferredItemPaddingEnd" android:minHeight="? android:attr/listPreferredItemHeightSmall" />
simple_list_item_multiple_choice	<CheckedTextView xmlns:android="http://schemas.android.com/apk/res/android" android:id="@android:id/text1" android:layout_width="match_parent" android:layout_height="? android:attr/listPreferredItemHeightSmall" android:textAppearance="? android:attr/textAppearanceListItemSmall" android:gravity="center_vertical" android:checkMark="? android:attr/listChoiceIndicatorMultiple" android:paddingStart="? android:attr/listPreferredItemPaddingStart" android:paddingEnd="? android:attr/listPreferredItemPaddingEnd"/>

可以看出，系统提供的 ListView 子布局文件模板均是 TextView 或其子类节点，不包含其他节点。这也为自定义子布局文件提供了方向，少走弯路，所以经常看系统源码是一个很好的习惯，后文还有论述。

示例中，当 ArrayAdapter 构造方法分别选中表 3-9 中的两个资源号时，程序运行界面如图 3-15 所示。

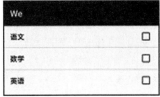

图 3-15 ListView 适配不同模板显示图

3. SimpleAdapter

SimpleAdapter 叫作简单适配器，与 ArrayAdapter 相比，它可以形成更复杂的界面，如多列 ListView、带图像的多列 ListView 等。List 常用的构造方法只有一个，如下所示。

```
SimpleAdapter(Context ctx, List<Map<String,?>> data, int layoutId, String[]
from, int []to);
```

ctx：上下文对象。

data：数据集合，内层 Map 代表每一行数据映射，外层 List 是线性结构，每行数据依次存放在 List 中。每个 Map 中的键都定义在 from 字符串数组中。

layoutId：每行子布局模板资源号。

from：Map 映射中键值组成的字符串数组。

to："from 到 to"的映射，to 数组元素是模板视图各子视图的 ID，将绑定数据视图的 ID 与 from 参数对应。详细解释如下所示：根据某行数据 map<key，value>，由 key 可推出数据项 value，在 from 数组中查找 key 的位置，假设为 pos，则将数据 value 绑定到 to[pos]所对应的模板子视图。

【例 3-19】 在 ListView 中应用系统模板 simple_list_2.xml 完成显示。

在 Android SDK 中找到 simple_list_2.xml，其主要内容如下所示。

```
<TwoLineListItem xmlns:android="http://schemas.android.com/apk/res/android"
    android:layout_width="match_parent"
    android:layout_height="wrap_content"
    android:minHeight="?attr/listPreferredItemHeight"
    android:mode="twoLine"
    android:paddingStart="?attr/listPreferredItemPaddingStart"
    android:paddingEnd="?attr/listPreferredItemPaddingEnd">
    <TextView android:id="@id/text1"
        android:layout_width="match_parent"
        android:layout_height="wrap_content"
        android:layout_marginTop="8dp"
        android:textAppearance="?attr/textAppearanceListItem" />
    <TextView android:id="@id/text2"
        android:layout_width="match_parent"
        android:layout_height="wrap_content"
        android:layout_below="@id/text1"
        android:layout_alignStart="@id/text1"
```

```
        android:textAppearance="?attr/textAppearanceListItemSecondary" />
</TwoLineListItem>
```

可以看出，该模板采用的是相对布局，有两个 TextView 子控件，着重注意它们的 id 值(text1、text2)，调用下文的 SimpleAdapter 构造方法时用到。text1 组件在上方，text2 组件在下方。根据该文件，就能看出 ListView 的显示风格。主要编程代码如下所示。

① 布局配置文件 main.xml：同例 3-18 中一致。

② MainActivity.java。

```
public class MainActivity extends AppCompatActivity {
    protected void onCreate(Bundle savedInstanceState) {
        super.onCreate(savedInstanceState);
        setContentView(R.layout.main);
        List l = new ArrayList();
        //形成仿真数据
        Map m = new HashMap();
        m.put("name","zhang"); m.put("age", 25);l.add(m);
        Map m2 = new HashMap();
        m2.put("name","sun"); m2.put("age", 30);l.add(m2);
        String from[] = {"name", "age"};
        int to[] = {android.R.id.text1, android.R.id.text2};
        SimpleAdapter ad = new SimpleAdapter(this,l,android.R.layout.simple_
list_item_2, from,to);
        ListView lv = (ListView)findViewById(R.id.mylist);
        lv.setAdapter(ad);
    }
}
```

若在 ListView 中应用系统模板，一定要看它的源文件，否则代码中的 to[]数组是写不出来的，且不可死记硬背。

讨论 1：若显示 n 列列表，如 n=3，该如何实现？采用自定义模板形式。按照 simple_list_item_2.xml 文件内容，它的根节点是<TwoLineListItem>，以此类推，难道写<ThreeLineListItem>？其实根据 simple_list_item_2.xml 内容，可看出 TwoLineListItem 一定是 RelativeLayout 相对布局的派生类。根节点写成 RelativeLayout，ListView 显示效果是一样的，只不过 TwoLineListItem 更表意罢了。对于读者而言，一般不需要定义派生类，做模板视图与做常规视图一样。对 SimpleAdapter 而言，做模板视图时，一定要对需要绑定的子 View 定义好 id。

若列表需要3列，则其中的一个自定义模板如下所示(假设文件名为threelist.xml)。

```
<?xml version="1.0" encoding="utf-8"?>
<LinearLayout xmlns:android="http://schemas.android.com/apk/res/android"
    android:layout_width="match_parent"
    android:layout_height="match_parent"
    android:orientation="horizontal">
    <TextView
        android:id="@+id/one"
        android:layout_width="wrap_content"
        android:layout_height="wrap_content"
        android:layout_weight="1"/>
    <TextView
        android:id="@+id/two"
        android:layout_width="wrap_content"
        android:layout_height="wrap_content"
        android:layout_weight="1"/>
    <TextView
        android:id="@+id/three"
        android:layout_width="wrap_content"
        android:layout_height="wrap_content"
        android:layout_weight="1"/>
</LinearLayout>
```

该模板采用的是水平线性布局，有3个id为one、two、three且等宽的TextView节点，这也是ListView中每行的显示效果。简单修改MainActivity.java中的onCreate()函数，代码如下所示，即可正常显示3列ListView。

```
protected void onCreate(Bundle savedInstanceState) {
    super.onCreate(savedInstanceState);setContentView(R.layout.main);
    List l = new ArrayList();
    //形成仿真数据
    Map m = new HashMap();
    m.put("name","zhang"); m.put("age", 25);m.put("school","lnnu");l.add(m);
    Map m2 = new HashMap();
    m2.put("name","sun"); m2.put("age", 30);m2.put("school","dlut");l.add(m2);
    Map m3 = new HashMap();
    m3.put("name","Li"); m3.put("age", 32);m3.put("school","neu");l.add(m3);
    String from[] = {"name", "age","school"};
```

```
        int to[] = {R.id.one,R.id.two,R.id.three};
        SimpleAdapter ad = new SimpleAdapter(this,l,R.layout.threeitem, from,to);
        ListView lv = (ListView)findViewById(R.id.mylist);
        lv.setAdapter(ad);
    }
```

若把模板 orientation 的属性值改为 vertical，去掉 3 个 Text 节点的权重 weight 属性，则 ListView 每行变为 3 个子行显示。

有多少种 3 个 TextView 的自定义模板？其实是无穷尽的。因为只要定义的模板布局文件中 3 个 TextView 与 ListView 绑定，至于 3 个 TextView 如何摆放，模板文件中还有无其他节点，这些都是无关的。

讨论 2：SimpleAdapter 支持图像显示吗？可以。对文本来说，在模板文件中对应 TextView 节点，传入适配器的是字符串数据类型；对图像文件来说，在模板文件中对应 ImageView 节点，传入适配器的是整数类型，每个整数代表图像的资源号。步骤如下所示。

① 主布局文件 main.xml：同例 3-18 中一致。

② 在 res\drawable\ 目录下复制两个图像文件 a.jpg、b.jpg，然后定义模板文件 imglist.xml。

```
<?xml version="1.0" encoding="utf-8"?>
<LinearLayout xmlns:android="http://schemas.android.com/apk/res/android"
    android:layout_width="match_parent"
    android:layout_height="match_parent"
    android:orientation="horizontal">
    <ImageView
        android:id="@+id/one"
        android:layout_width="wrap_content"
        android:layout_height="wrap_content" />
    <TextView
        android:id="@+id/two"
        android:layout_width="wrap_content"
        android:layout_height="wrap_content" />
</LinearLayout>
```

以上代码采用的是水平线性布局，一个 ImageView 显示图像，一个 TextView 显示文本，一定要为 ImageView、TextView 节点定义 id 值，用于与 ListView 绑定。

③ MainActivity.Java。

```
public class MainActivity extends AppCompatActivity {
```

```
    protected void onCreate(Bundle savedInstanceState) {
        super.onCreate(savedInstanceState);
        setContentView(R.layout.main);
        List l = new ArrayList();
        //形成仿真数据
        Map m = new HashMap();
        m.put("img",R.drawable.a);m.put("name","zhang");l.add(m);
        Map m2 = new HashMap();
        m2.put("img",R.drawable.b);m2.put("name","sun");l.add(m2);
        String from[] = {"img", "name"};
        int to[] = {R.id.one,R.id.two};
        SimpleAdapter ad = new SimpleAdapter(this,l,R.layout.main2, from,to);
        ListView lv = (ListView)findViewById(R.id.mylist);
        lv.setAdapter(ad);
    }
}
```

可以看出，Map 映射中存的是不同类型的值。键 img 对应的值是整型数 R.drawable.XXX，即图像资源号；键 name 对应的值是字符串，即具体姓名。

4. BaseAdapter

SimpleAdapter 的功能已经很强，但有时涉及细节，例如，若对 ListView 中的单选、按钮添加事件响应，操作就不方便了。因此，只有操作适配器的底层，才容易使 ListView 功能适用更复杂的情况。应用 BaseAdapter 基本适配器是较好的解决方法。BaseAdapter 是一个抽象类，实现了 Adapter 接口。在 ListView 中应用 BaseAdapter 的基本步骤如下所示。

- 定义 ListView 数据项类，包括所需各种数据。
- 定义 ListView 行视图模板文件。
- 定义抽象类 BaseAdapter 的派生类 MyAdapter，成员变量必须包含上下文对象、数据项集合类对象，必须重写 getCount()、getItem()、getItemId()、getView()4 个函数。
- 在应用程序中建立 ListView 与 BaseAdapter 的关联。

【例 3-20】 ListView 显示学生信息，每行显示学号、姓名及一个多选框。

① 主布局配置文件 main.xml：同例 3-18 中一致。

② 定义数据项源数据。

```
public class ItemBean {
    String no;                                            //学号
    String name;                                          //姓名
    ItemBean(String no,String name){
        this.no=no;this.name=name;
    }
}
```

③ 定义数据项视图模板。

```
<?xml version="1.0" encoding="utf-8"?>
<RelativeLayout xmlns:android="http://schemas.android.com/apk/res/android"
    android:layout_width="match_parent"
    android:layout_height="match_parent">
    <TextView
        android:id="@+id/myno"
        android:layout_width="wrap_content"
        android:layout_height="wrap_content"
        android:textSize="30sp"
        android:layout_marginRight="15dp"
        android:focusable="false"/>
    <TextView
        android:id="@+id/myname"
        android:layout_width="wrap_content"
        android:layout_height="wrap_content"
        android:layout_toRightOf="@id/myno"
        android:textSize="30sp"
        android:focusable="false"/>
    <CheckBox
        android:layout_width="wrap_content"
        android:layout_height="wrap_content"
        android:layout_alignParentRight="true"
        android:focusable="false"/>
</RelativeLayout>
```

以上代码采用的是相对布局,第 1 个 TextView 节点 id 为 myno,用于绑定学号;第 2 个 TextView 节点 id 为 myname,用于绑定姓名;第 3 个节点是多选按钮 CheckBox,用于显示,没有定义 id。所有相对布局的子节点 android:focusable 均设置为 false,为 ListView 添加事件响应作准备。若 android:focusable 不设置或设置为 true,则 ListView

不响应事件,读者记住即可。

④ 自定义适配器类 MyAdapter。

```
public class MyAdapter extends BaseAdapter {
    List<ItemBean>list;
    Context ctx;
    MyAdapter (Context ctx, List< ItemBean> list) {this.ctx=ctx; this.list=
list;}
    public int getCount() { return list.size();}
    public Object getItem(int position) {return list.get(position);}
    public long getItemId(int position) {return position;}
    public View getView(int position, View convertView, ViewGroup parent) {
        LayoutInflater inflater = LayoutInflater.from(ctx);
                                                        //获得布局装载器对象
        View view = inflater.inflate(R.layout.mylist,null);    //装载模板布局
        TextView noView = (TextView) view.findViewById(R.id.myno);
        TextView nameView = (TextView) view.findViewById(R.id.myname);
        ItemBean bean = list.get(position);        //获取相应索引的 ItemBean 对象
        noView.setText(bean.no);                                //数据项与模板绑定
        nameView.setText(bean.name);
        return view;
    }
}
```

着重理解 getView()函数,读者会发现:代码中根本没用到两个入口参数 convertView 和 parent。这和 ListView 系统缓存有关,超出本书研究范围。只要知道在该函数内加载 ListView 行视图模板,获得行数据项,将数据与子视图绑定即可。

另外,getItemId(position)函数返回的是数据项的 ID,而不是行视图 View 的 ID,只要保证每个数据项的 ID 唯一即可,示例中返回的是 position,改为返回 position * 2 或其他有效的表达式均可。

这 4 个重写函数一般不是主动调用的,而是系统根据需要调用的,只写好要求的内容即可。

⑤ 主应用 MainActivity.java。

```
public class MainActivity extends AppCompatActivity {
    protected void onCreate(Bundle savedInstanceState) {
        super.onCreate(savedInstanceState);
        setContentView(R.layout.main);
```

```
        List l = new ArrayList();
        //形成仿真数据
        ItemBean item=new ItemBean("1000","zhang");
        ItemBean item2=new ItemBean("1001","li");
        List<ItemBean> list = new ArrayList();
        list.add(item); list.add(item2);
        MyAdapter ma = new MyAdapter(this,list);
        ListView lv = (ListView)findViewById(R.id.mylist);
        lv.setAdapter(ma);
    }
}
```

可以看出，主类代码应用自定义适配器与应用 SimpleAdapter、ArrayAdapter 代码几乎是一致的，实际界面如图 3-16 所示。

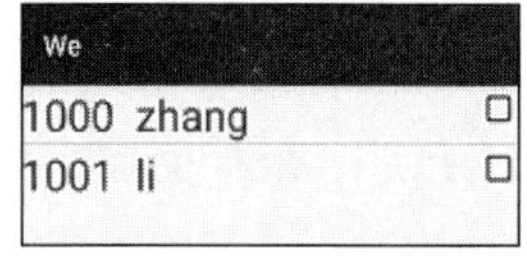

图 3-16　自定义适配器结果图

讨论 1：为 ListView 增加 OnItemClick 事件。当单击该条目时，使对应的多选按钮处于选中状态，再单击该条目时，多选按钮处于未选中状态。onCreate()函数的实现代码如下所示。

```
protected void onCreate(Bundle savedInstanceState) {
    //其余所有的代码同例 3-20 中的⑤
    lv.setOnItemClickListener(new AdapterView.OnItemClickListener() {
        public void onItemClick(AdapterView<?> parent, View view, int
position, long id) {
            RelativeLayout r = (RelativeLayout)view;
            CheckBox cb = (CheckBox)r.getChildAt(2);
            boolean state = cb.isChecked();
            cb.setChecked(!state);
        }
    });
}
```

OnItemClick()函数中的第 2 个参数 view 是当前数据项所在的子视图，由于该子视图从模板视图来，在模板视图中首层节点时 RelativeLayout，因此该 view 即 RelativeLayout。因此，代码中将 view 强制转换成 RelativeLayout 对象 r。在模板视图中，CheckBox 节点是 RelativeLyout 的第 2(0 基)个节点，利用 getChildAt()函数可获得 CheckBox 对象 cb，当然对多选按钮的操作也就轻而易举了。

很明显，参数 view 是进一步进行各子窗口操作的关键，它和模板视图息息相关。提醒读者：若应用到 Android 系统本身适配器模板，一定要查看源文件，这是进行深层次控制的基础。

讨论 2：如何让 ListView 中的数据项多选按钮全选中？就像手机中的批量删除、批量移动功能一样。对本示例而言，若想实现类似功能，只需在主布局文件中增加一个按钮，增加其消息响应函数（读者自行完成这两步工作），其函数内部关键代码如下所示。

```
ListView lv = (ListView)MainActivity.this.findViewById(R.id.mylist);
int count = lv.getChildCount();
for(int i=0; i<count; i++){
    RelativeLayout obj = (RelativeLayout)lv.getChildAt(i);
    CheckBox cb = (CheckBox)obj.getChildAt(2);
    cb.setChecked(true);
}
```

由以上代码可知，主要思想是获得 ListView 对象 lv，对其子视图遍历，每个子视图即 RelativeLayout 对象 obj，obj 中的第 2(0 基)个子视图即 CheckBox 对象 cb，获得 cb 后，就可对其进行选中与否的操作。

因此，已知父视图，级联遍历子视图，或者已知子视图反向遍历到父视图，这些是编制一些实用功能时必须掌握的技术。

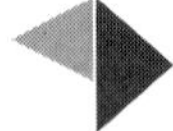

3.8 下拉控件

Spinner 是常用的下拉列表控件，其使用方法几乎与 ListView 完全相同，也支持 ArrayAdapter、SimpleAdapter、BaseAdapter 适配器，因此本节主要讲述它们不同的地方。

Spinner 常用的函数如下所示。

- View getSelectedView()，返回选中的列表项视窗对象。
- Object getSelectedItem()，返回选中的列表项视窗对象绑定的数据。

Spinner 常用的事件是数据项选中事件 OnItemSelectedClick，消息注册函数是 void setOnItemSelectedClickListener (OnItemSelectedClickListener obj)，OnItemSelectedClickListener 接口定义了两个函数原型，如下所示。

- void onItemSelected(AdapterView<?> parent, View view, int position, long id)

parent：相当于 ListView 适配器的一个指针。

view：轻触对应条目的子视窗 view 对象。

position：是在 ListView 中的顺序位置(0 基)。

id：一般与 position 一致。

- void onNothingSelected(AdapterView<?> parent)，该函数用得较少，仅当 spinner 绑定的数据集为空时才响应。

【例 3-21】 Spinner 显示学生科目信息。主布局定义一个“获得数据”按钮及一个 Spinner 下拉控件，Spinner 用于显示科目信息，当选中某项时，利用 Toast 显示“你正在选择：XXX”。当单击“获得数据”按钮时，利用 Toast 显示“你最终选择：XXX”。

① 主布局文件 main.xml。

```
<?xml version="1.0" encoding="utf-8"?>
<LinearLayout
    xmlns:android="http://schemas.android.com/apk/res/android"
    android:layout_width="match_parent"
    android:layout_height="match_parent"
    android:orientation="vertical">
    <Spinner
        android:id="@+id/myspan"
        android:layout_width="wrap_content"
        android:layout_height="wrap_content">
    </Spinner>
    <Button
        android:id="@+id/myok"
        android:layout_width="wrap_content"
        android:layout_height="wrap_content"
        android:text="获得数据" />
</LinearLayout>
//MainActivity.java
public class MainActivity extends AppCompatActivity {
    protected void onCreate(Bundle savedInstanceState) {
        super.onCreate(savedInstanceState);
        setContentView(R.layout.main);
        Spinner sp = (Spinner) findViewById(R.id.myspan);
        String str[]={"语文","数学","英语"};
         ArrayAdapter ad = new ArrayAdapter(this, android.R.layout.simple_
spinner_dropdown_item,str);
        sp.setAdapter(ad);
        sp.setOnItemSelectedListener(new AdapterView.OnItemSelectedListener() {
            public void onItemSelected(AdapterView<?> parent, View view, int
position, long id) {
```

```
                String s = "你正在选择:"+ ((TextView) view).getText();
                Toast.makeText(MainActivity.this,s,Toast.LENGTH_LONG).show();
            }
            public void onNothingSelected(AdapterView<? > parent) { }
        });
        Button b = (Button) findViewById(R.id.myok);
        b.setOnClickListener(new View.OnClickListener() {
            public void onClick(View v) {
                Spinner sp = (Spinner) findViewById(R.id.myspan);
                String s = "你最终选择:"+ (String) sp.getSelectedItem();
                Toast.makeText(MainActivity.this,s,Toast.LENGTH_LONG).show();
            }
        });
    }
}
```

实际界面如图 3-17 所示。

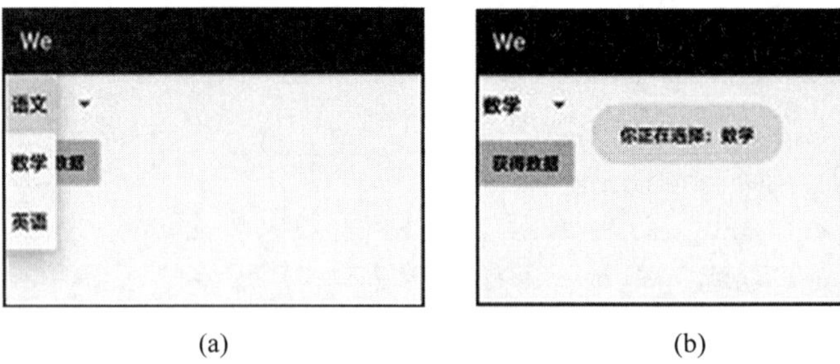

(a) (b)

图 3-17 Spinner 简单示例图

初始单击下拉控件,界面如图 3-17(a)所示;单击“获取数据”按钮后,界面如图 3-17(b)所示。示例中采用系统模板文件 simple_spinner_dropdown_item.xml,读者可自行查看。

【例 3-22】 级联 Spinner 显示省-市信息。第一个 Spinner 用于显示省名称,当选中某省时,利用第二个 Spinner 显示市名称。很明显,第一个 Spinner 绑定的数据是固定的,第二个 Spinner 绑定的数据是动态变化的。关键代码如下所示。

① 主布局文件 mian.xml。

```
<? xml version="1.0" encoding="utf-8"? >
<LinearLayout
    xmlns:android="http://schemas.android.com/apk/res/android"
    android:layout_width="match_parent"
```

```
    android:layout_height="match_parent"
    android:orientation="vertical">
    <Spinner
        android:id="@+id/myprov"
        android:layout_width="wrap_content"
        android:layout_height="wrap_content"
        android:dropDownWidth="100dp">
    </Spinner>
    <Spinner
        android:id="@+id/mycity"
        android:layout_width="wrap_content"
        android:layout_height="wrap_content"
        android:dropDownWidth="100dp">
    </Spinner>
</LinearLayout>
```

id 为 myprov 的 Spinner 用于显示省名称，id 为 mycity 的 Spinner 用于显示市名称。

② MainActivity.java。

```
public class MainActivity extends AppCompatActivity {
    String one[]={"辽宁","吉林"};
    String two[][] = {{"沈阳","大连"},{"长春","四平"}};
    protected void onCreate(Bundle savedInstanceState) {
        super.onCreate(savedInstanceState);
        setContentView(R.layout.main);
        Spinner prov = (Spinner) findViewById(R.id.myprov);
        ArrayAdapter ad = new ArrayAdapter(this,android.R.layout.simple_
spinner_dropdown_item,one);
        prov.setAdapter(ad);
        prov.setOnItemSelectedListener(new AdapterView
.OnItemSelectedListener() {
            public void onItemSelected(AdapterView<?> parent, View view, int
position, long id) {
                Spinner city = (Spinner)MainActivity.this.findViewById(R.id
.mycity);
                ArrayAdapter ad = new ArrayAdapter(MainActivity.this,
                    android.R.layout.simple_spinner_dropdown_item,two[position]);
                city.setAdapter(ad);
            }
```

```
            public void onNothingSelected(AdapterView<?> parent) {    }
        });
    }
}
```

利用一维字符串数组 one[]仿真省名称,二维字符串数组 two[][]仿真城市数据,one[i]省包含 two[i]对应的城市数据。

由于省的下拉菜单内容是固定的,因此在 onCreate()函数部分将下拉菜单对象 prov 通过 ArrayAdapter 直接绑定到 one 上;由于市的下拉菜单 city 对象内容随选择的省变化而变化,因此在 prov 的 OnItemSelectedClick 选中省份事件对应的响应事件中通过 ArrayAdapter 动态绑定到 two 数组中。

总之,处理级联(N 级)Spinner,在对应的响应事件中动态绑定数据即可。

3.9 进度条控件

进度条控件即 ProgressBar 控件,分为确定模式和不确定模式。确定模式下(任务长度已知),ProgressBar 作为可视化进度指示器,为用户呈现操作的进度。不确定模式下(任务长度未知),ProgressBar 显示循环动画。

ProgressBar 常用的函数如下所示。

- void setMax(int value);设置进度条最大值。进度条的范围为[0,value]。
- int getMax();返回这个进度条范围的上限。
- void setProgress(int value);设置当前进度值。
- int getProgress();返回当前进度值。
- void incrementProgressBy(int value);指定增加的进度值,value 可正可负。
- void setIndeterminate(boolean mark);若 mark 为 true,则为不确定模式;否则为确定模式。

利用 XML 文件定义 ProgressBar 组件有一个重要的属性 style,它定义了 ProgressBar 的显示风格,其取值如下所示。

- Widget.ProgressBar.Horizontal;
- Widget.ProgressBar.Small;
- Widget.ProgressBar.Large;
- Widget.ProgressBar.Inverse;
- Widget.ProgressBar.Small.Inverse;
- Widget.ProgressBar.Large.Inverse;

应用的时候不要带前缀 android，直接写：style＝"@android：style/Widget.ProgressBar.Small"。

【例 3-23】 利用按钮演示增加进度、减少对应的进度条显示情况。本示例涉及的文件如下所示。

① 主界面文件 main.xml。

```
<?xml version="1.0" encoding="utf-8"?>
<LinearLayout xmlns:android="http://schemas.android.com/apk/res/android"
xmlns:app="http://schemas.android.com/apk/res-auto"
android:layout_width="match_parent"
android:layout_height="match_parent"
android:orientation="vertical">
    <Button
        android:id="@+id/myinc"
        android:layout_width="wrap_content"
        android:layout_height="wrap_content"
        android:text="增加"/>
    <Button
        android:id="@+id/mydec"
        android:layout_width="wrap_content"
        android:layout_height="wrap_content"
        android:text="减少"/>
    <ProgressBar
        android:id="@+id/mybar"
        android:layout_width="match_parent"
        android:layout_height="wrap_content"
        android:max="100"
        android:progress="50"
        style="@android:style/Widget.ProgressBar.Horizontal" />
</LinearLayout>
```

以上代码定义了两个按钮组件：一个 id 为 myinc，用于增加进度值；一个 id 为 mydec，用于减少进度值。而且定义了一个进度条 ProgressBar 组件，id 为 mybar，最大显示进度值为 100，初始进度值为 50。

② 主应用文件 MianActivity.java。

```
public class MainActivity extends AppCompatActivity{
```

```
    protected void onCreate(Bundle savedInstanceState) {
        super.onCreate(savedInstanceState);
        setContentView(R.layout.main);
        final ProgressBar p = (ProgressBar)findViewById(R.id.mybar);
        Button b = (Button)findViewById(R.id.myinc);
        b.setOnClickListener(new View.OnClickListener() {
            public void onClick(View v) {
               int value = p.getProgress();  p.setProgress(value+5);
            }
        });
        Button b2 = (Button)findViewById(R.id.mydec);
        b2.setOnClickListener(new View.OnClickListener() {
            public void onClick(View v) {
                int value = p.getProgress();  p.setProgress(value-5);
            }
        });
    }
}
```

在增加按钮响应事件中，首先利用 getProgress()获得当前进度值 value，然后利用 setProgress()函数设置进度值为 value＋5，进度条显示也随之发生变化。减少按钮事件响应与增加按钮事件响应大同小异，略。

也可利用 incrementProgressBy()函数实现本示例功能。在增加按钮响应事件中用 p.incrementProgressBy(5)代替原来函数中的所有代码，在减少按钮响应事件中用 p.incrementProgressBy(－5)代替原来函数中的所有代码即可。

本示例运行界面如图 3-18 所示。

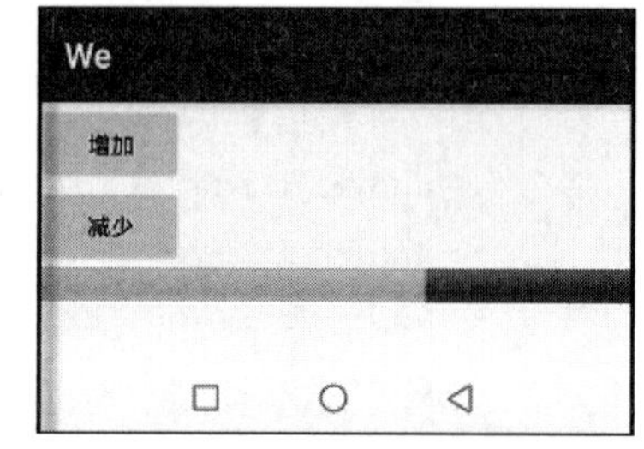

图 3-18　进度条界面图

3.10　形状文件

学习默认 EditText 控件时，发现其显示感觉像缺少边界似的，学习 TableLayout、GridLayout 时各单元格也缺少边界。如何增加边界？如何在一定程度上美化控件？利用 Android 形状文件是重要的方法之一，它是 XML 文件，以 shape 为根节点，封装了关于控件填充、渐变、边界、圆角、间隔的子节点信息，详细描述见表 3-10。

表 3-10　形状文件子节点描述表

序号	子　节　点	属　　性	说　　明
1	<solid>，填充	android:color	填充颜色值
2	<gradient>，渐变	android:startcolor	起始颜色
		android:endColor	结束颜色
		android:angle	渐变角度，必须为 45 的整数倍
		android:type	渐变类型，linear：线性渐变；radial：径向渐变
		android:gradientRadius	若径向渐变，则必须设置该值
3	<stroke>，边界	android:width	边界宽度
		android:color	边界颜色
		android:dashWidth	虚线及一个横线宽度
		android:dashGap	虚线时两个横线间的宽度
4	<corners>，圆角	android:radius	四个角的圆角弧度设定
5	<padding>，间隔	android:left，right，top，bottom	设置左、右、上、下间距

形状文件的根节点是 shape，表 3-10 中的 5 个标签都是 shape 的一个节点，其结构如下所示。

```
<shape>
    <solid 属性 1=值 1 …… 属性 n=值 n></solid>
    <gradient 属性 1=值 1 …… 属性 n=值 n></gradient>
    <stroke 属性 1=值 1 …… 属性 n=值 n></stroke>
    <corners 属性 1=值 1 …… 属性 n=值 n></corners>
    <padding 属性 1=值 1 …… 属性 n=值 n></padding>
</shape>
```

【例 3-24】　为 EditText 增加边框，且四个角是圆角。

① 主布局文件 main.xml。

```
<?xml version="1.0" encoding="utf-8"?>
<LinearLayout
    xmlns:android="http://schemas.android.com/apk/res/android"
    android:layout_width="match_parent"
    android:layout_height="match_parent">
```

```
    <EditText
        android:layout_width="100dp"
        android:layout_height="wrap_content"
        android:layout_marginLeft="20dp"
        android:layout_marginTop="20dp"
        android:textSize="30sp"
        android:background="@drawable/textshape"/>
</LinearLayout>
```

若调用形状文件，一定要设置 android:background 属性。示例中，该属性设置为“@drawable/textshape”，表明形状文件在 res\drawable\目录下，文件名为 textshape.xml。

② 形状文件 textshape.xml。

```
<?xml version="1.0" encoding="utf-8"?>
<shape xmlns:android="http://schemas.android.com/apk/res/android">
    <corners android:radius="10dp"></corners>
    <stroke android:width="1dp"></stroke>
</shape>
```

由以上代码可知：四个角的圆角半径都为 10dp，四条边的线宽都为 1dp。

③ MainActivity.java。

```
public class MainActivity extends AppCompatActivity {
    protected void onCreate(Bundle savedInstanceState) {
        super.onCreate(savedInstanceState);
        setContentView(R.layout.main);
    }
}
```

【例 3-25】 为 TableLayout 各单元格增加边框。

假设各单元格为 TextView，其实非常简单，只要设置各单元格的 background 属性为 shape 文件即可。布局文件如下所示(去掉例 3-24 中 textshape.xml 中的＜corners＞节点，仅保留＜stroke＞节点)。

```
<?xml version="1.0" encoding="utf-8"?>
<TableLayout xmlns:android="http://schemas.android.com/apk/res/android"
    android:layout_width="match_parent"
    android:layout_height="match_parent"
    android:stretchColumns="0,1,2">
    <TableRow>
```

```
        <TextView android:layout_width="wrap_content" android:layout_height=
"100dp"
            android:background="@drawable/textshape"/>
        <TextView android:layout_width="wrap_content" android:layout_height=
"100dp"
            android:background="@drawable/textshape"/>
        <TextView android:layout_width="wrap_content" android:layout_height=
"100dp"
            android:background="@drawable/textshape"/>
    </TableRow>
    <TableRow>
        <TextView android:layout_width="wrap_content" android:layout_height=
"100dp"
            android:background="@drawable/textshape"/>
        <TextView android:layout_width="wrap_content" android:layout_height=
"100dp"
            android:background="@drawable/textshape"/>
        <TextView android:layout_width="wrap_content" android:layout_height=
"100dp"
            android:background="@drawable/textshape"/>
    </TableRow>
</TableLayout>
```

本示例是以表格单元格为基本单位，四个框均有边界。若表格仅有上边界、下边界，又该如何设置？其实很简单，只要在 TableRow 节点上增加 background 属性，去掉各 TextView 节点的 background 属性即可。

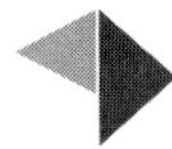

3.11　状态文件

从形状文件 shape 看，若控件背景属性设置成不同的形状文件，则控件就有多种展示形态。但是，布局文件中仅能设成一个背景值，没有进行条件判断，要实现随条件不同背景值设置也不同的功能，如何解决？状态文件出现了，它仍是 XML 文件结构，以配置文件的形式体现了条件不同，形状文件也不同的问题。

状态文件以<selector>为根节点，子节点仅由<item>节点组成，但<item>的属性众多。常用的状态属性见表 3-11。

表 3-11 常用的状态属性

序号	属性名称	说明
1	android:drawable	为控件指定资源
2	android:state_pressed	布尔值,true 指当用户单击或者触摸该控件的状态。默认为 false;一般用于按钮颜色/图片的设置
3	android:state_focused	布尔值,true 指当前控件获得焦点时的状态。默认为 false;一般用于 EditText
4	android:state_hovered	布尔值,true 表示光标移动到当前控件上的状态。默认为 false;光标是否悬停,通常与 state_focused 相同,一般用于 EditText
5	android:state_selected	布尔值,true 表示被选择的状态,例如,在一个下拉列表中从方向键下选择其中一个选项。 与 focus 的区别: selected 是 focus 不充分的情况。例如,一个 ListView 获得焦点(focus),而用方向键选择了其中一个 item (selected)
6	android:state_checkable	布尔值,true 表示可以被勾选的状态。仅当控件在被勾选和不被勾选的状态间转换时才起作用
7	android:state_checked	布尔值,true 表示当前控件处于被勾选(check)的状态
8	android:state_enabled	布尔值,true 表示当前控件处于可用的状态,比如可以被单击
9	android:state_activated	布尔值,true 表示当前控件被激活的状态

【例 3-26】 按钮状态文件示例,实现两个功能:按钮按下和松开后的背景不同;按钮按下和松开后的按钮文本显示的字体颜色不同。具体步骤如下所示。

① 创建两个背景图像 a.jpg、b.jpg,前者用于常态下的按钮背景,后者用于按下后的按钮背景。

② 创建两个选择器文件 myback.xml、mytext.xml。前者用于不同状态下的按钮背景选择,后者用于不同状态下的文本字体颜色选择。由于按钮背景与 background 属性有关,文本颜色与 textColor 属性有关,因此必须定义两个选择器文件。总之,若控件的某属性有多个选择,则必须让该属性值等于要选择的状态文件,一般不能把许多属性的状态值写到一个文件中。

```
//myback.xml:背景选择器
<?xml version="1.0" encoding="utf-8"?>
<selector xmlns:android="http://schemas.android.com/apk/res/android">
    <item android:drawable="@drawable/a"  android:state_pressed="true" />
    <item android:drawable="@drawable/b" />
</selector>
```

两个平齐的<item>子节点是互斥的，表达的含义是：当按钮按下（android:state_pressed=true）时背景选择 a.jpg；常态时背景选择 b.jpg。

```
//mytext.xml:字体颜色选择器
<?xml version="1.0" encoding="utf-8"?>
<selector xmlns:android="http://schemas.android.com/apk/res/android">
    <item android:color="#0000ff"  android:state_pressed="true" />
    <item android:color="#ffffff" />
</selector>
```

两个平齐的<item>子节点是互斥的，表达的含义是：当按钮按下（android:state_pressed=true）时文本颜色选择蓝色；常态时文本颜色选择白色。

③ 主布局文件 main.xml。

```
<?xml version="1.0" encoding="utf-8"?>
<LinearLayout xmlns:android="http://schemas.android.com/apk/res/android"
    android:layout_width="match_parent"
    android:layout_height="match_parent">
    <Button
        android:layout_width="wrap_content"
        android:layout_height="wrap_content"
        android:text="OK"
        android:textColor="@drawable/mytext"
        android:background="@drawable/myback"/>
</LinearLayout>
```

以上代码用了两个选择器：一个用于设置字体颜色，由 textColor 属性设置；一个用于设置背景颜色，由 background 属性设置。

④ 主应用程序 MainActivity.java。

```
public class MainActivity extends AppCompatActivity {
    protected void onCreate(Bundle savedInstanceState) {
        super.onCreate(savedInstanceState);
        setContentView(R.layout.main);
    }
}
```

【例 3-27】 不用状态文件，完全利用程序重新实现例 3-25 的功能。

有时，若在应用程序中写过多的状态文件，可能稍显凌乱，也不易维护。因此，在应用程序中实现状态文件功能有时是必须的，本示例的具体步骤如下所示。

① 主布局文件 main.xml。

```
<?xml version="1.0" encoding="utf-8"?>
<LinearLayout xmlns:android="http://schemas.android.com/apk/res/android"
    android:layout_width="match_parent"
    android:layout_height="match_parent">
    <Button
        android:id="@+id/mybtn"
        android:layout_width="wrap_content"
        android:layout_height="wrap_content"
        android:text="OK" />
</LinearLayout>
```

可以看出，本示例没有对 background、textColor 属性进行设置，其相应的功能完全由程序实现。

② MainActivity.java。

```
public class MainActivity extends AppCompatActivity {
    protected void onCreate(Bundle savedInstanceState) {
        super.onCreate(savedInstanceState);
        setContentView(R.layout.main);
        Button b = (Button)findViewById(R.id.mybtn);
        //设置背景状态列表功能
        StateListDrawable sd = new StateListDrawable();
        Drawable da = ResourcesCompat.getDrawable(getResources(),R.drawable.
a, null);
        Drawable db = ResourcesCompat.getDrawable(getResources(),R.drawable.
b, null);
        sd.addState(new int[]{android.R.attr.state_pressed}, da);
        sd.addState(new int[]{android.R.attr.state_enabled}, db);
        b.setBackground(sd);                              //为按钮设置背景
        //设置颜色状态列表功能
        ColorStateList cs = new ColorStateList(new int[][]{{android.R.attr.
state_pressed},
            {android.R.attr.state_enabled}}, new int[]{Color.RED,Color.BLUE});
        b.setTextColor(cs);                               //为按钮设置颜色
    }
}
```

对控件背景状态功能来说：首先，定义 StateListDrawable 对象(构造方法的参数为空),利用 ResourcesCompat 类中的静态方法 getDrawable()加载各 Drawable 对象;然后,利用 StateListDrawable 对象的 addState()方法将各 Drawable 对象与各种状态绑定;最后,设置控件的 background 属性为 StateListDrawable 对象即可。注意下面两个方法的原始定义。

- ResourcesCompat.getDrawable(Resources res,int id,Theme te),在本示例中,第三个参数 Theme 用 null 代替,读者只需记住即可。
- StateListDrawable 类中的方法: void addState(int[] state, Drawable d),其中 state 是系统控件状态标识数组,一般形如 android.R.attr_XXX,XXX 如表 3-11 中属性名称中的数据(去掉 android:)。

对颜色状态功能来说,设置起来相对简单,理解 ColorStateList 构造方法即可。

- ColorStateList(int [][]state,int[]color),其中 state 是二维数组,具体元素形如 android.R.attr_XXX,XXX 的含义前文已经描述。本构造方法的含义是:一行(可能多个)状态值对应一个颜色值。

习题 3

一、选择题

1. Android UI 视图的基类是(　　)。
   A. View　　B. ViewGroup　　C. JFrame　　D. JPanel
2. 下面包含两个正确状态开关组件的是(　　)。
   A. Button,Switch　　B. ImageButton,Switch
   C. Button,ImageButton　　D. ToggleButton、Switch
3. 下面哪一个是多选按钮?(　　)
   A. RadioButton　　B. CheckBox　　C. ImageButton　　D. Button
4. 关于 TextView 组件的描述正确的是(　　)。
   A. 仅能显示普通文本　　B. 不能显示图像
   C. 不支持 HTML 语句　　D. 支持 HTML 语句
5. 直接在 ListView 组件中设置数据源的属性是(　　)。
   A. android:entries　　B. android:src
   C. android:source　　D. android:path
6. 形状文件描述组件边界的节点是(　　)。
   A. <solid>　　B. <gradient>　　C. <stroke>　　D. <corners>

二、程序题

1. 单选多选按钮题目：定义一组单选按钮，数据项是语文、数学、英语；再定义一组多选按钮，数据项是音乐、体育、美术，当单击"确定"按钮时，利用 Toast 显示单选及多选内容。

2. TextView、EditText 题目：定义两个 EditText 组件，用于输入两个加数；定义一个 TextView 组件，当单击"确定"按钮时，将两个加数之和显示在 TextView 组件中。

3. ListView 题目：学生信息包括学号和姓名，仿真数据自己生成。定义两个 ListView 组件，用于显示学生信息；定义一个 ListView，用于显示一列（学号及姓名合成一个字符串）；定义一个 ListView，用于显示两列。

第 4 章

对话框与高级控件

本章介绍系统 AlertDialog 对话框的基本应用方法，以及日期时间控件、翻页类视图、增强型列表控件的具体实现和用法。

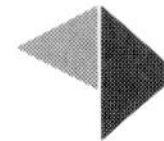

4.1 对话框

4.1.1 AlertDialog 简介

Android 系统对话框类是 AlertDialog。创建 AlertDialog 对象有很多方法，最常用的方法是应用 AlertDialog.Builder 类创建，该类常用的函数如下所示。

- void setTitle(CharSequence title)：设置对话框的标题。
- void setMessage(CharSequence msg)：设置对话框要传达的具体信息。
- void setIcon(int nID)：设置对话框的图标。
- void setCancelable(boolean mark)：单击对话框以外的区域是否让对话框消失，默认为 true。
- void setPositiveButton(CharSequence text，OnClicklistener listener)：设置正面按钮，表示“积极”“确认”的意思。text 为按钮上显示的文字，listener 是该按钮的 Clicklistener 事件回调对象。
- void setNegativeButton(CharSequence text，OnClicklistener listener)：设置反面按钮，表示“消极”“否认”“取消”的意思。
- AlertDialog show()：对话框显示函数，返回 AlertDialog 对象。
- void setItems(CharSequence items[]，View.OnClickListener listener)：设置列表对话框参数及响应事件。
- void setSingleChoiceItems(CharSequence items[]，int check， OnClickListener

listener)：设置单选列表对话框参数及响应事件。

- void setMultiChoiceItems (CharSequence items [ ], boolean states [ ], OnMultiChoiceClickListener listener)：设置多选列表对话框参数及响应事件。
- void setView(View view)：设置对话框内容为 view 对象内容。
- void setView(int layoutID)：设置对话框内容的资源 ID。

TextView、Button 等控件一般都固定在界面上，AlertDialog 与它们稍有不同，AlertDialog 在某个时机才会触发出来(例如，用户单击了某个按钮)。所以，AlertDialog 不需要在布局文件中创建，而是在代码中通过构造器(AlertDialog.Builder)构造标题、图标和按钮等内容，其应用步骤一般如下所示。

- 创建构造器 AlertDialog.Builder 的对象。
- 通过构造器的对象调用 setTitle()、setMessage()等方法构造对话框的标题、信息和图标等内容。
- 根据需要，设置正面按钮、负面按钮和中立按钮。
- 调用 show()方法，让对话框在界面上显示。

4.1.2 分类介绍

Android 常用的对话框有提示对话框，列表对话框、单选对话框、多选对话框、自定义对话框。下面通过示例加以说明。

1. 提示对话框

例如，对话框上显示“确认提交吗?”，以及“确认”“取消”两个按钮，实现该功能所需文件如下所示。

① 主界面 main.xml。

```
<?xml version="1.0" encoding="utf-8"?>
<LinearLayout xmlns:android="http://schemas.android.com/apk/res/android"
android:layout_width="match_parent"
android:layout_height="match_parent"
android:orientation="vertical">
    <Button
        android:id="@+id/mydlg"
        android:layout_width="wrap_content"
        android:layout_height="wrap_content"
        android:text="提示对话框"/>
</LinearLayout>
```

以上代码仅定义了一个按钮,id 为 mydlg,用于启动提示对话框。

② 主应用文件 MainActivity.java。

```
public class MainActivity extends AppCompatActivity{
    protected void onCreate(Bundle savedInstanceState) {
        super.onCreate(savedInstanceState);
        setContentView(R.layout.main);
        Button buttonNormal = (Button) findViewById(R.id.mydlg);
        buttonNormal.setOnClickListener(new View.OnClickListener() {
            public void onClick(View v) {  showNormalDialog();  }
        });
    }
    private void showNormalDialog(){
        final AlertDialog.Builder normalDialog =new AlertDialog.Builder
(MainActivity.this);
        normalDialog.setIcon(R.drawable.question);   //设置对话框图标
        normalDialog.setTitle("提示对话框");          //设置对话框标题
        normalDialog.setMessage("确认提交吗?");       //设置提示信息
        normalDialog.setPositiveButton("确定",        //设置确定按钮及消息响应
                new DialogInterface.OnClickListener() {
                    public void onClick(DialogInterface dialog, int which) { //   }
                });
        normalDialog.setNegativeButton("取消",        //设置取消按钮及消息响应
                new DialogInterface.OnClickListener() {
                    public void onClick(DialogInterface dialog, int which) { //   }
                });
        normalDialog.show();
    }
}
```

可以看出,对话框操作比较简单。showNormalDialog()函数中设置了对话框的图标、标题、提示信息串,并设置了确定、取消按钮及响应事件,最后调用 show()函数完成对话框的显示,如图 4-1 所示。

图 4-1　提示对话框

2. 列表、单选、多选对话框

这 3 类对话框的功能相似,分别与 4.1.1 节中的 setItems()、setSingleChoiceItems()、setMultiChoiceItems 相关。下面以多选对话框为例,关键代码如下所示。

```
private void showMultiDialog() {
    final String[] items = { "语文","数学","英语" };
    final boolean check[] = new boolean[3];
    AlertDialog.Builder multiDialog = new AlertDialog.Builder(MainActivity.
this);
    multiDialog.setTitle("多选对话框");
    multiDialog.setMultiChoiceItems(items, check, new DialogInterface
.OnMultiChoiceClickListener() {
        public void onClick(DialogInterface dialog, int which, boolean
isChecked) {
            check[which] = isChecked;
        }
    });
    multiDialog.setPositiveButton("确定", new DialogInterface
.OnClickListener() {
        public void onClick(DialogInterface dialog, int which) {
            String s = "";
            for(int i=0; i<check.length; i++){
                if(check[i]) s += items[i]+",";
            }
            Toast.makeText(MainActivity.this,s,Toast.LENGTH_LONG).show();
        }
    });
    multiDialog.show();
}
```

showMultiDialog()函数中定义了多选字符串数组 items 和多选初始状态 check 布尔数组,check 数组的大小必须与 items 数组的大小相同。本示例中,check 数组元素的初值均为 false,表明无一选中。

多选响应回调函数 onClick(DialogInterface dialog, int which, boolean isChecked)表明多选的第 which 项现在的状态是 isChecked 值。

在对话框确认按钮事件响应 onClick()函数中,遍历 check 数组,当元素值为 true,将对应的 items 数组元素挑选出来并进行累加。利用 Toast 将最终结果显示在界面上。

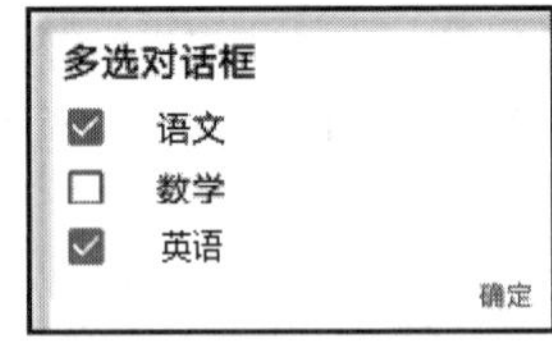

图 4-2　多选对话框

本示例对应的多选对话框如图 4-2 所示。

3. 自定义对话框

由 4.1.1 节中 Alert.Builder 类中有函数 setView(View view)可知,自定义对话框主要应用该函数,view 可自己创建,也可通过 XML 配置文件获得。那么,很容易想到以下两个关键问题:对话框中的功能按钮是自己创建还是用 AlertDialog.Builder 中的已有按钮?若是自己创建功能按钮,那么如何销毁对话框?带着这些思考,我们实现一个注册功能对话框,其涉及的文件如下所示。

① 自定义注册对话框界面资源 myregist.xml。

```
<?xml version="1.0" encoding="utf-8"?>
<LinearLayout xmlns:android="http://schemas.android.com/apk/res/android"
    android:orientation="vertical" android:layout_width="match_parent"
    android:layout_height="match_parent">
    <EditText
        android:id="@+id/myuser"
        android:layout_width="match_parent"
        android:layout_height="wrap_content"
        android:textSize="20sp"/>
    <EditText
        android:id="@+id/mypwd"
        android:layout_width="match_parent"
        android:layout_height="wrap_content"
        android:textSize="20sp"/>
    <Button
        android:id="@+id/myok"
        android:layout_width="wrap_content"
        android:layout_height="match_parent"
        android:text="注册"/>
</LinearLayout>
```

以上代码定义了两个 EditText 控件,id 分别为 myuser、mypwd,用于输入用户名和密码;一个 Button 功能按钮,id 为 myok,用于启动注册功能。很明显,在自定义对话框中一般不用 AlertDialog.Builder 中定义的各功能按钮。例如,当单击注册按钮时:若用户名或密码为空,则必须给出提示信息,对话框是存在的;当用户名或密码都不为空,则销毁对话框。也就是说,当单击功能按钮时,根据分支判断,对话框可能继续存在,也可能不存在,有多种情况,利用 AlertDialog.Builder 中的系统按钮很难实现。另外,其提供的系统功能按钮也可能与自定义对话框总体风格不一致。因此,在自定义对话框中一般不考虑系统功能按钮,完全自己创建或封装在配置文件中。

② 主界面文件 main.xml。

```
<?xml version="1.0" encoding="utf-8"?>
<LinearLayout xmlns:android="http://schemas.android.com/apk/res/android"
android:layout_width="match_parent"
android:layout_height="match_parent"
android:orientation="vertical">
    <Button
        android:id="@+id/mydlg"
        android:layout_width="wrap_content"
        android:layout_height="wrap_content"
        android:text="打开对话框"/>
</LinearLayout>
```

以上代码仅定义了一个按钮,id 为 mydlg,用于启动自定义注册对话框。

③ 主应用文件 MainActivity.java。

```
public class MainActivity extends AppCompatActivity{
    private void showDialog(){
        final AlertDialog.Builder build =new AlertDialog.Builder
(MainActivity.this);
        final View view = View.inflate(this, R.layout.myregist, null);
        build.setView(view);
        build.setPositiveButton("OK",null);
        final AlertDialog dialog = build.show();
        final Button positiveBtn = dialog.getButton(AlertDialog.BUTTON_
POSITIVE);
        positiveBtn.setVisibility(View.GONE);
        Button b = (Button)view.findViewById(R.id.myok);
        b.setOnClickListener(new View.OnClickListener() {
            public void onClick(View v) {
                EditText et = (EditText)view.findViewById(R.id.myuser);
                String user = et.getText().toString();
                EditText et2 = (EditText)view.findViewById(R.id.mypwd);
                String pwd = et.getText().toString();
                if(user.equals("")||pwd.equals(""))
                    Toast.makeText(MainActivity.this,"用户名或密码不能为空!",
                                          Toast.LENGTH_LONG).show();
```

```
                else{ positiveBtn.performClick(); }
            }
        });
    }
    protected void onCreate(Bundle savedInstanceState) {
        super.onCreate(savedInstanceState);
        setContentView(R.layout.main);
        Button buttonNormal = (Button) findViewById(R.id.mydlg);
        buttonNormal.setOnClickListener(new View.OnClickListener() {
            public void onClick(View v) {  showDialog();  }
        });
    }
}
```

showDialog()是实现自定义对话框显示和操作的核心函数。要对注册功能按钮添加事件响应及处理，本示例对用户名或密码为空的情况给出了提示信息。可以看出，关闭对话框是通过调用系统功能实现的：首先创建 Positive 按钮，名称为 OK，当调用 show()显示对话框时获得对话框 AlertDialog 对象 dialog；然后由 dialog 获得 OK 按钮对象 positiveBtn，并设置其为隐藏状态；最后，当单击自定义注册功能按钮，需要关闭对话框时，调用 positiveBtn.performClick()完成对话框的关闭。

总之，关闭对话框的核心思想是：要借助系统已有的 Positive 按钮，但不让其显示。

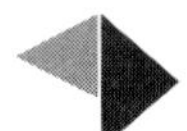

4.2 日期控件

日期和时间输入是 UI 的常用操作。Android 系统对应的日期控件是 DatePicker（用于输入年、月、日），时间控件是 TimePicker（用于输入时、分、秒）。由于这两个控件占用的空间都很大，因此一般不会将它们直接显示在界面上。为了解决这一情况，Android 又提供了两个对话框类 DatePickerDialog、TimePickerDialog，方便了日期、时间的输入。

本节简要说明 DatePickerDialog 的应用方法，功能是将选择的年、月、日信息显示在界面上。涉及的文件如下所示。

① 主界面文件 main.xml。

```
<?xml version="1.0" encoding="utf-8"?>
<LinearLayout xmlns:android="http://schemas.android.com/apk/res/android"
android:layout_width="match_parent"
android:layout_height="match_parent"
```

```
android:orientation="vertical">
    <Button
        android:id="@+id/mydlg"
        android:layout_width="wrap_content"
        android:layout_height="wrap_content"
        android:text="日期对话框"/>
    <TextView
        android:id="@+id/mydate"
        android:layout_width="match_parent"
        android:layout_height="wrap_content"
        android:textSize="20sp"/>
</LinearLayout>
```

以上代码定义了一个按钮，id 为 mydlg，用于启动日期对话框；一个 TextView 组件，id 为 mydate，用于显示选中的日期对话框的日期信息。

② 主应用文件 MainActivity.java。

```
public class MainActivity extends AppCompatActivity implements
DatePickerDialog.OnDateSetListener{
    private void showDialog(){
        Calendar c = Calendar.getInstance();        //获得当前时间
        int y=c.get(Calendar.YEAR);                 //获得年
        int m=c.get(Calendar.MONTH);                //获得月
        int d=c.get(Calendar.DAY_OF_MONTH);         //获得日
        DatePickerDialog dp = new DatePickerDialog(this,this,y,m,d);
                                                    //创建日期对话框对象
        dp.show();                                  //对话框显示
    }
    public void onDateSet(DatePicker view, int year, int month, int day) {
        String s = year+"-"+(month+1)+"-"+day;
                                       //month 的范围为[0,11],加 1 才代表真实月份
        TextView tv = (TextView)findViewById(R.id.mydate);
        tv.setText(s);
    }
    protected void onCreate(Bundle savedInstanceState) {
        super.onCreate(savedInstanceState);
        setContentView(R.layout.main);
        Button buttonNormal = (Button) findViewById(R.id.mydlg);
        buttonNormal.setOnClickListener(new View.OnClickListener() {
```

```
            public void onClick(View v) {  showDialog();  }
        });
    }
}
```

对于日期对话框类 DatePickerDialog，只要理解其构造方法即可，其形式如下所示。

DatePickerDialog(Context ctx, OnDataSetListener listener, int year, int month, int day)；ctx 是上下文对象；year、month、day 是初始化日期对话框的年、月、日信息；listener 是 OnDataSetListener 对象，当单击日期对话框中的"确定"按钮时，系统会回调 listener 的接口函数。

由代码 DatePickerDialog dp = new DatePickerDialog(this,this,y,m,d)可得：由于 DatePickerDiaog()构造方法的第 2 个参数实参为 this，表明 MainActivity 类必须实现 OnDataSetListener 接口，重写 onDataSet()函数。

日期对话框有确定、取消两个功能按钮，当单击"确定"按钮后，会回调 onDataSet()函数，在此函数中获得选中的年、月、日值，将其组合后显示在 id 为 mydate 的 TextView 组件中。

本示例的日期对话框如图 4-3 所示。

图 4-3　日期对话框

当然，时间选择对话框 TimePickerDialog 的用法与 DatePickerDialog 是相似的，请读者仿照 DatePickerDialog，自行完成 TimePickerDialog 的学习。

【例 4-1】　自行编制日期、时间选择对话框。

DatePickerDialog 仅能选择年、月、日，TimePickerDialog 仅能选择时、分、秒。若需要输入年、月、日、时、分、秒，该怎么办？很明显，先运行 DatePickerDialog，再运行

TimePickerDialog 不是一个好办法。一个好的办法是：采用自定义对话框形式，对话框内容由 DatePicker 日期组件＋自定义时间组件构成，当然也需要自定义接口回调函数。所涉及的功能文件如下所示。

① 自定义回调接口函数 OnMyDataSetListener()。

```
interface OnMyDataSetListener{
    void onMyDataSet(int y,int m,int d,int hh,int mm,int ss);
}
```

系统的 DatePickerDialog 中用到 OnDataSetListener，表明这是一个好的思路。那么，自定义日期、时间选择对话框也一定会用到类似的接口回调技术。至于如何应用，请读者继续往下看。

② 对话框界面模板文件 mydatetime.xml。

```
<?xml version="1.0" encoding="utf-8"?>
<LinearLayout xmlns:android="http://schemas.android.com/apk/res/android"
    android:orientation="vertical" android:layout_width="match_parent"
    android:layout_height="match_parent">
    <DatePicker
        android:id="@+id/mypicker"
        android:layout_width="wrap_content"
        android:layout_height="wrap_content">
    </DatePicker>
    <Spinner
        android:id="@+id/myhour"
        android:layout_width="wrap_content"
        android:layout_height="wrap_content"
        android:entries="@array/hour"></Spinner>
    <Spinner
        android:id="@+id/myminute"
        android:layout_width="wrap_content"
        android:layout_height="wrap_content"
        android:entries="@array/minute"></Spinner>
    <Spinner
        android:id="@+id/mysecond"
        android:layout_width="wrap_content"
        android:layout_height="wrap_content"
        android:entries="@array/second"></Spinner>
```

```
    <Button
        android:id="@+id/myok"
        android:layout_width="wrap_content"
        android:layout_height="wrap_content"
        android:text="确定"/>
</LinearLayout>
```

以上代码定义了一个 DatePicker 组件，id 为 mypicker，用以获得年、月、日的值；3 个 Spinner 下拉框组件，id 分别为 myhour、myminute、mysecond，用于选择时、分、秒的值。由于时、分、秒的下拉范围都是确定的，因此将它们都封装在资源文件 res/strings.xml 中，对应 3 个<string-array>节点，见表 4-1。

表 4-1　时、分、秒下拉框节点对应的配置文件节点表

小时值配置信息	分钟值配置信息	秒值配置信息
<string-array name="hour"> <item>1</item> <item>2</item> …… <item>24</item> </string-array>	<string-array name="minute"> <item>1</item> <item>2</item> …… <item>60</item> </string-array>	<string-array name="second"> <item>1</item> <item>2</item> …… <item>60</item> </string-array>

当然，本界面并不美观，读者在应用 DatePicker 组件的前提下，可以充分发挥想象力，将时、分、秒输入功能添加到对应的控件中。

③ 自定义对话框类文件 MyDateTimeDlg.java。

```
public class MyDateTimeDlg {
    Context ctx;
    OnMyDataSetListener listener;
    public MyDateTimeDlg(Context ctx,OnMyDataSetListener listener){
        this.ctx= ctx;
        this.listener = listener;
    }
    public void show() {
        AlertDialog.Builder build = new AlertDialog.Builder(ctx);
        final View view = View.inflate(ctx, R.layout.mydatetime, null);
        build.setView(view);
        build.setPositiveButton("OK", null);
        AlertDialog dialog = build.show();
```

```
        final Button positiveBtn = dialog.getButton(AlertDialog.BUTTON_
POSITIVE);
        positiveBtn.setVisibility(View.GONE);
        Button b = (Button) view.findViewById(R.id.myok);
        b.setOnClickListener(new View.OnClickListener() {
            public void onClick(View v) {
                DatePicker dp = (DatePicker)view.findViewById(R.id.mypicker);
                int y = dp.getYear();int m=dp.getMonth();int d=dp
.getDayOfMonth();
                Spinner sp = (Spinner)view.findViewById(R.id.myhour);
                int hh=Integer.parseInt(sp.getSelectedItem().toString());
                Spinner sp2 = (Spinner)view.findViewById(R.id.myminute);
                int mm=Integer.parseInt(sp2.getSelectedItem().toString());
                Spinner sp3 = (Spinner)view.findViewById(R.id.mysecond);
                int ss=Integer.parseInt(sp3.getSelectedItem().toString());
                positiveBtn.performClick();
                listener.onMyDataSet(y,m,d,hh,mm,ss);
            }
        });
    }
}
```

以上代码定义了两个成员变量：一个是上下文 Context 对象 ctx；一个是回调 OnMyDataSetListener 对象 listener。

show()函数完成自定义对话框的默认显示：主要通过加载 mydatetime.xml 配置文件实现。

show()函数完成对确定按钮功能的 OnClickListener 事件注册。在响应函数中，首先获得 DatePicker 对象 dp，进而获得确定的年、月、日值；然后分别获得 3 个 Spinner 对象 sp、sp2、sp3，进而获得时、分、秒值；最后调用 performClick()函数销毁对话框，调用回调函数 onMyDataSet()完成对年、月、日、时、分、秒值的传送。

④ 主界面文件 main.xml。

```
<?xml version="1.0" encoding="utf-8"?>
<LinearLayout xmlns:android="http://schemas.android.com/apk/res/android"
android:layout_width="match_parent"
android:layout_height="match_parent"
android:orientation="vertical">
    <Button
```

```
        android:id="@+id/mydlg"
        android:layout_width="wrap_content"
        android:layout_height="wrap_content"
        android:text="自定义日期、时间对话框"/>
    <TextView
        android:id="@+id/mydate"
        android:layout_width="match_parent"
        android:layout_height="wrap_content"
        android:textSize="20sp"/>
</LinearLayout>
```

以上代码定义了一个按钮，id 为 mydlg，用于启动自定义日期、时间对话框；一个 TextView 组件，id 为 mydate，用于显示选中的日期、时间对话框的日期、时间信息。

⑤ 主应用文件 MainActivity.java。

```
public class MainActivity extends AppCompatActivity implements
OnMyDataSetListener{
    private void showDialog(){
        MyDateTimeDlg dlg = new MyDateTimeDlg(this, this);
        dlg.show();
    }
    public void onMyDataSet(int y, int m, int d, int hh, int mm, int ss) {
        String t=y+"-"+m+"-"+d+ " "+hh+":"+mm+":"+ss;
        TextView tv = (TextView) findViewById(R.id.mydate);
        tv.setText(t);
    }
    protected void onCreate(Bundle savedInstanceState) {
        super.onCreate(savedInstanceState);
        setContentView(R.layout.main);
        Button buttonNormal = (Button) findViewById(R.id.mydlg);
        buttonNormal.setOnClickListener(new View.OnClickListener() {
            public void onClick(View v) {
                showDialog();
            }
        });
    }
}
```

showDialog()函数完成了对话框的显示与相关操作。onMyDataSet()回调函数返回了显示的日期及时间值，将它们组合成一个字符串，显示在 id 为 mydate 的 TextView 组件中。onCreate()函数主要完成对自定义日期、时间对话框功能按钮的注册与事件响应。

4.3 翻页控件

翻页控件对应 ViewPager 类。翻页的含义是：ViewPager 支持触摸左右滑动事件。当触摸左滑动时向左翻页，反之向右翻页。

实现翻页功能，主要是完成系统翻页适配器 PagerAdapter 子类的设计。PagerAdapter 是一个抽象类，在其子类中一般重写如下的 4 方法即可。

- int getCount()：返回需要翻页的总视图数目。
- void destroyItem(ViewGroup container, int position, Object object)：销毁 container 容器中指定位置的页面视图。
- Object instantiateItem(ViewGroup container, int position)：实例化指定位置的界面，将其添加到容器界面中。
- boolean isViewFromObject(View arg0, Object arg1)：判定当前页面视图 View 对象 arg0 与 Object 对象 arg1 是否相同。由于翻页视图经常涉及当前页面视图的销毁和添加，必然涉及当前 View 与视图集合对象元素的判定，因此该函数会经常用到。

【例 4-2】 实现如下功能的翻页视图：3 个不同背景颜色的视图，当用手在界面上左右滑动的时候，视图颜色的背景会发生变化，表明当前页视图对象发生了变化。实现该功能涉及的文件如下所示。

① 主界面文件 main.xml。

```
<?xml version="1.0" encoding="utf-8"?>
<LinearLayout xmlns:android="http://schemas.android.com/apk/res/android"
android:layout_width="match_parent"
android:layout_height="match_parent"
android:orientation="vertical">
    <android.support.v4.view.ViewPager
        android:id="@+id/mypage"
        android:layout_width="match_parent"
        android:layout_height="match_parent">
    </android.support.v4.view.ViewPager>
</LinearLayout>
```

以上代码定义了一个 ViewPager 翻页视图组件，id 为 mypage，用于在相同位置显示不同的视图。

② 翻页视图适配器文件 MyAdapter.java。

```
public class MyAdapter extends PagerAdapter {
    List<View> list;
    public MyAdapter(List<View> list){  this.list = list;  }
    public int getCount() {  return list.size();  }              //返回视图个数
    public boolean isViewFromObject(View view, Object object) {  return view==
object;  }
    public void destroyItem(ViewGroup container, int position, Object object) {
                                                                  //销毁视图
        container.removeView(list.get(position));
    }
    public Object instantiateItem(ViewGroup container, int position) {  //添加视图
        container.addView(list.get(position));
        return list.get(position);
    }
}
```

一般来说，翻页适配器要定义视图集合 List 成员变量 list，在构造函数中完成其初始化。重写的 4 个其他函数也都比较简单，在此不多论述。

③ 主应用程序 MainActivity.java。

```
public class MainActivity extends AppCompatActivity {
    protected void onCreate(Bundle savedInstanceState) {
        super.onCreate(savedInstanceState);
        setContentView(R.layout.main);
        List<View> list = new ArrayList();
        View v = new View(this);
        v.setBackgroundColor(Color.RED);
        View v2 = new View(this);
        v2.setBackgroundColor(Color.GREEN);
        View v3 = new View(this);
        v3.setBackgroundColor(Color.BLUE);
        list.add(v);list.add(v2);list.add(v3);
        ViewPager vp = (ViewPager) findViewById(R.id.mypage);
        vp.setAdapter(new MyAdapter(list));
```

```
        vp.setCurrentItem(0);
    }
}
```

以上代码首先产生了 3 个 View 对象，背景分别为红、绿、蓝，均添加到 list 集合对象中；然后创建 ViewPager 翻页对象 vp，并与 MyAdapter 适配器关联；最后利用 setCurrentItem(0)在翻页视图中显示第 1 页，即背景为红颜色的视图。

如果希望为翻页视图增加一个标题栏，也是可以的，需要应用<PagerTabStrip>标签，再为每个视图定义一个名称。需要改动的部分如下所示。

① 在主界面配置文件中增加<PagerTabStrip>节点。

```
<?xml version="1.0" encoding="utf-8"?>
<LinearLayout xmlns:android="http://schemas.android.com/apk/res/android"
android:layout_width="match_parent"
android:layout_height="match_parent"
android:orientation="vertical">
    <android.support.v4.view.ViewPager
        android:id="@+id/mypage"
        android:layout_width="match_parent"
        android:layout_height="match_parent">
        <android.support.v4.view.PagerTabStrip
            android:layout_width="wrap_content"
            android:layout_height="wrap_content">
        </android.support.v4.view.PagerTabStrip>
    </android.support.v4.view.ViewPager>
</LinearLayout>
```

PagerTabStrip 标签作为 ViewPager 标签的子节点加入即可。

② 翻页视图适配器文件 MyAdapter.java。

```
public class MyAdapter extends PagerAdapter {
    List<View> list;
    List<String> titleList;
    public MyAdapter(List<View> list,List<String>titleList){
        this.list = list;
        this.titleList = titleList;
    }
    public CharSequence getPageTitle(int position) {
        return titleList.get(position);
```

```
    }
    //其他所有代码同前文,略
}
```

与前文相比,这里增加了各视图名称的集合成员变量 titleList,并在构造函数中完成了初始化;重写了 getPageTitle()函数,返回了当前显示视图的名称。

③ 主应用文件 MainActivity.java。

```
public class MainActivity extends AppCompatActivity {
    protected void onCreate(Bundle savedInstanceState) {
        //此行之前的代码与前文完全相同,复制过来即可,略
        List<String> title = new ArrayList();
        title.add("红");title.add("绿");title.add("蓝");
        ViewPager vp = (ViewPager) findViewById(R.id.mypage);
        vp.setAdapter(new MyAdapter(list,title));
        vp.setCurrentItem(0);
    }
}
```

可以看出,onCreate()函数中仅增加了对各视图标题的处理,形成标题集合 title,并通过 MyAdapter 构造方法传入,其他代码与前文的 onCreate()函数完全一样。

当程序运行时,标题栏在视图的正上方,而视图名称在标题栏的水平中央。也就是说,视图名称的位置不是固定的,当它是焦点视图时,视图名称会移动到标题栏的正中央。

4.4 计时器控件

Chronometer 类是计时器控件类,由 TextView 派生,因此它具有 TextView 的所有属性,其常用函数如下所示。

- void setBase(long base):设置计时器基值。
- long getBase():获得计时器基值。
- void start():启动计时器。
- void stop():停止计时器。
- void setOnChronometerTickListener(OnChronometerTickListener listen):事件监听器,时间发生变化时可进行操作,一般 1s 调用该函数一次。

【例 4-3】 实现倒计时功能:从 10s 开始,逐秒递减显示,至 0s 结束。该功能涉及的文件如下所示。

① 主界面文件 main.xml。

```
<?xml version="1.0" encoding="utf-8"?>
<LinearLayout xmlns:android="http://schemas.android.com/apk/res/android"
android:layout_width="match_parent"
android:layout_height="match_parent"
android:orientation="vertical">
    <Chronometer
        android:id="@+id/mytime"
        android:layout_width="match_parent"
        android:layout_height="wrap_content" />
    <Button
        android:id="@+id/mystart"
        android:layout_width="wrap_content"
        android:layout_height="wrap_content"
        android:text="计时开始"/>
</LinearLayout>
```

以上代码定义了一个计时器 Chronometer 组件，id 为 mytime，用于显示倒计时所剩秒数；一个 Button 组件，id 为 mystart，用于启动倒计时功能。

② 主应用文件 MainActivity.java。

```
public class MainActivity extends AppCompatActivity implements Chronometer.
OnChronometerTickListener{
    int c = 10;
    Chronometer time;
    public void onChronometerTick(Chronometer meter) {
        time.setText(""+c);
        c -- ;
        if(c==-1) time.stop();
    }
    protected void onCreate(Bundle savedInstanceState) {
        super.onCreate(savedInstanceState);
        setContentView(R.layout.main);
        time = (Chronometer)findViewById(R.id.mytime);
        time.setText("");
        time.setOnChronometerTickListener(this);
        Button b = (Button)findViewById(R.id.mystart);
        b.setOnClickListener(new View.OnClickListener() {
```

```
                public void onClick(View v) {
                    time.start();
                }
            });
        }
    }
```

MaInActivity 类实现了 OnChronometerTickListener 接口，成员变量 c 是倒计时初始秒数值，time 是计时器对象。

onCreate()函数完成了开始计时按钮的事件响应。在响应函数中，利用 start()函数启动了计时器计时功能。于是，每间隔 1s 就调用 onChronometer()函数一次，在此函数中，利用 setText()函数将剩余描述显示在计时器组件中，同时计数值 c 减 1。当 c 等于 0 时，表明倒计时 10s 时间已到，利用 stop()函数停止计时功能。

4.5　增强型列表 RecyclerView 控件

4.5.1　简介

RecyclerView 是增强型列表控件，其功能远比 ListView、GridView 丰富。RecyclerView 的常用函数如下所示。

- void setAdapter(Adapeter ad)：设置列表项适配器，适配器是 RecyclerView.Adapter 的子类。
- void setLayoutManager(LayoutManager manager)：设置列表项布局管理器，包括线性布局管理器(LinearLayoutManager)、网格布局管理器(GridLayoutManager)、瀑布流网格布局管理器(StaggeredGridLayoutManager)等。
- void addItemDecoration(ItemDecoration decor)：设置列表项分割线。
- void removeItemDecoration()：移除列表项分割线。
- void addOnItemTouchListener(OnItemTouchListener listen)：添加列表项触摸监听器。
- void removeOnItemTouchListener()：删除列表项触摸监听器。

RecyclerView.Adapter 适配器类是一个抽象类，它在增强型列表 RecyclerView 中扮演重要的角色，负责数据视图与数据的绑定。实际应用中，一定要定义 RecyclerView.Adapter 的派生类，一般必须重写下列 3 个方法。

- void getItemCount()：返回列表项的总数目。

- ViewHolder onCreateViewHolder(ViewGroup parent, int viewType)：加载列表项布局。
- void onBindViewHolder(RecyclerView.ViewHolder holder, int position)：将列表项第 position 位置的数据与列表项显示视图绑定。

【例 4-4】 实现简单的 RecyclerView 视图：以列表形式显示 N 条数据。

应用 RecyclerView 与应用 ListView 的步骤相似：定义列表项布局；编制适配器类与主应用程序类。所需文件如下所示。

① 定义列表项布局文件 myitem.xml。

```
<?xml version="1.0" encoding="utf-8"?>
<LinearLayout xmlns:android="http://schemas.android.com/apk/res/android"
    android:layout_width="match_parent"
    android:layout_height="wrap_content"
    android:orientation="vertical">
    <TextView
        android:id="@+id/mytxt"
        android:layout_width="match_parent"
        android:layout_height="wrap_content"
        android:gravity="center"/>
</LinearLayout>
```

以上代码仅定义了一个 TextView 组件，用于显示每个列表项的数据。

② 适配器类文件 MyAdapter.java。

```
public class MyAdapter extends RecyclerView.Adapter<MyAdapter.MyHolder> {
    private List mList;            //列表项集合对象
    MyAdapter(List list) {         //通过构造函数完成列表项数据集合 mList 的初始化
        mList = list;
    }
    public MyHolder onCreateViewHolder(ViewGroup parent, int viewType) {
        View view = LayoutInflater.from(parent.getContext()).inflate
(R.layout.myitem, parent, false);
        MyHolder holder = new MyHolder(view);
        return holder;
    }
```

/*第 1 行代码的功能是加载已定义的列表项视图模板 myitem.xml，这决定了每一列表项的显示风格。第 2 行将 view 进一步封装成自定义类 MyHolder，定义 view 中与列表项相关的子组件控件。

由前文知，onCreateViewHolder（）的返回值是 RecyclerView. ViewHolder 类型对象，而该类是抽象类，必须用其子类，所以定义了 MyHolder 类。按照约束条件，也明白了类声明为什么写成如下形式的原因。

```
    public class MyAdapter extends RecyclerView.Adapter<MyAdapter.MyHolder>
    */
    public void onBindViewHolder(MyHolder holder, int position) {
        holder.textView.setText(mList.get(position).toString());
    }
    /*将列表项子视图与相应位置的列表项数据绑定*/
    public int getItemCount() {
        return mList.size();
    }
    class MyHolder extends RecyclerView.ViewHolder {
        TextView textView;
        public MyHolder(View itemView) {
            super(itemView);
            textView = (TextView) itemView.findViewById(R.id.mytxt);
        }
    }
}
```

MyHolder 类定义成内、外部类均可，其构造函数的参数 itemView 即 myitem.xml 对应的列表项子视图。将子视图中与数据项相关的组件在该类中都定义成相应的组件成员变量，并通过 findViewById()函数获得。

③ 主界面文件 main.xml。

```
<?xml version="1.0" encoding="utf-8"?>
<LinearLayout xmlns:android="http://schemas.android.com/apk/res/android"
    xmlns:app="http://schemas.android.com/apk/res-auto"
    android:layout_width="match_parent"
    android:layout_height="match_parent">
    <android.support.v7.widget.RecyclerView
        android:id="@+id/mylist"
        android:layout_width="match_parent"
        android:layout_height="match_parent" />
</LinearLayout>
```

以上代码仅定义了一个增强型列表 RecyclerView 组件，用于显示列表数据。

④ 主应用测试类文件。

```
public class MainActivity extends AppCompatActivity {
    private List mList;                         //数据项集合
    private MyAdapter ad;                       //适配器
    private RecyclerView rv;
    private LinearLayoutManager lm;             //布局管理器
    protected void onCreate(Bundle savedInstanceState) {
        super.onCreate(savedInstanceState);
        setContentView(R.layout.main);
        mList = new ArrayList();
        rv = (RecyclerView) findViewById(R.id.mylist);
        initData(mList);
        lm = new LinearLayoutManager(this, LinearLayoutManager.VERTICAL, false);
        ad = new MyAdapter(mList);
        rv.setLayoutManager(lm);
        rv.setAdapter(ad);
    }
    public void initData(List list) {
        for (int i = 1; i <= 6; i++) {
            list.add("item"+i);
        }
    }
}
```

onCreate()函数通过 initData()函数完成了数据初始化，均添加到 mlist 集合对象中。利用 mList，调用 MyAdapter 类构造方法，创建了适配器对象 ad。利用 findViewById()函数获得 RecyclerView 对象 rv。最后，利用 rv.setAdapter(ad)完成列表项数据的显示。

4.5.2 几个问题

运行例 4-4 后会发现，列表项之间没有分割线，也不能进行事件响应，如何解决该问题？下面一一论述。

1. 分割线问题

首先会想到利用 RecyclerView 提供的 addItemDecoration(ItemDecoration decor)函数，但系统并没有提供现成的 ItemDecoration 对象，而 ItemDecoration 是一个抽象类，需

要用其派生类重写 onDraw()、onDrawOver()、getItemOffsets()3 个函数，涉及图形绘制和一定的坐标计算，学完本书第 10 章图形和动画才可完成，因此就不采用此法了。

一个简单的方法是在前文 myitem.xml 中增加一个 TextView 组件，用设置背景的方法实现，如下所示。

```
<?xml version="1.0" encoding="utf-8"?>
<LinearLayout xmlns:android="http://schemas.android.com/apk/res/android"
    android:layout_width="match_parent"
    android:layout_height="wrap_content"
    android:orientation="vertical">
    <TextView
        android:id="@+id/mytxt"
        android:layout_width="match_parent"
        android:layout_height="30dp"
        android:gravity="center"
        android:text="数据" />
    <TextView
        android:layout_width="match_parent"
        android:layout_height="2dp"
        android:background="#d0d0d0"/>
</LinearLayout>
```

第 1 个 TextView 组件用于列表项数据项显示，第 2 个 TextView 组件用于显示边界，通过调整颜色值达到需要的状态。

2. 事件响应问题

大家知道，ListView 每一项都可以对 OnItemClick、OnItemLongClick 事件进行侦听和响应，但是 RecyclerView 却没有这一功能，可以利用接口回调技术实现相应事件的侦听和响应，下面仍以例 4-4 为例，为 RecyclerView 增加 OnItemClick 事件及响应，步骤如下。

① 定义自定义回调事件接口 OnRecyclerItemClickListener。

```
interface OnRecyclerItemClickListener{
    void onRecyclerItemClick(List list,int position);
}
```

接口函数 onRecyclerItemClick() 的参数 list 是数据项集合，position 是在 RecyclerView 中选择的项的索引(0 基)。该函数的功能是根据数据项集合及索引获得对应的具体数值。当然，由于是自定义接口，因此可根据需要灵活掌握接口参数的定义，不是一成不变的。

② 添加事件响应及回调机制的适配器文件 MyAdapter.java。

```
public class MyAdapter extends RecyclerView.Adapter<MyAdapter.MyHolder> {
    private List mList;                          //数据源
    OnRecyclerItemClickListener mListener;
    MyAdapter(List list,OnRecyclerItemClickListener listener) {
        mList = list;
        mListener = listener;
    }
    public MyHolder onCreateViewHolder(ViewGroup parent, int viewType) { //同例
4-4 一致,略}
    public void onBindViewHolder(MyHolder holder, int position) {
        holder.textView.setText(mList.get(position).toString());
        View view = holder.itemView;
        final int pos = position;
        view.setOnClickListener(new View.OnClickListener() {
            public void onClick(View v) {
                mListener.onRecyclerItemClick(mList,pos);
            }
        });
    }
    public int getItemCount() { //同例 4-4 一致,略}
    class MyHolder extends RecyclerView.ViewHolder {//同例 4-4 一致,略   }
}
```

可以看出，在例 4-4 的基础上增加了自定义回调接口 OnRecyclerItemClickListener 成员变量 mListener，并在构造方法中完成了初始化。

变化较大的是重写的 onBindViewHolder()函数。其实也容易理解，因为该函数的第 2 个参数 position 代表选中的数据项的索引位置，因此，若加消息响应，一定是在该函数中添加。大家知道，通过 MyHolder 对象的 itemView 属性可获得当前数据项所在的 View 对象 view。因此，对 view 可添加系统 OnClickListener 侦听，在其响应函数中利用回调函数 mListener.onRecycleItemClick(mList，pos)转向即可。

③ 主测试界面文件 main.xml：与例 4-4 一致，略。

④ 主应用文件 MainActivity.java。

```
public class MainActivity extends AppCompatActivity implements
```

```
OnRecyclerItemClickListener{
    //所有成员变量都与例 4-4 相同,略
    public void onRecyclerItemClick(List list, int position) {
        Toast.makeText(this, ""+position, Toast.LENGTH_LONG).show();
    }

    protected void onCreate(Bundle savedInstanceState) {
        super.onCreate(savedInstanceState);
        setContentView(R.layout.main);
        mList = new ArrayList();
        rv = (RecyclerView) findViewById(R.id.mylist);
        initData(mList);
        lm = new LinearLayoutManager(this, LinearLayoutManager.VERTICAL, false);
        ad = new MyAdapter(mList, this);
        rv.setLayoutManager(lm);
        rv.setAdapter(ad);
    }
    public void initData(List list) { //与例 4-4 相同,略   }
}
```

与例 4-4 相比，onCreate()函数中仅有一行代码不同，之前是 ad＝new MyAdapter(mList)，现在是 ad ＝ new Adapter(mList，this)。this 表明 RecyclerView 的 OnRecyclerItemClick 事件的响应函数在 MainActivity 中。因此，MainActivity 类必须实现 OnRecyclerItemClickListener 接口，重写 OnRecyclerItemClick()函数。

4.5.3　布局管理器

RecyclerView 之所以功能强大，主要在于它有强大的布局管理器设置功能。在适配器及列表项模板视图确定的情况下，可以随布局管理器设置的不同，界面显示的整体效果也不同。常用的 RecyclerView 布局管理器有线性布局管理器、网格布局管理器和瀑布流网格布局管理器。下面一一加以介绍。

1. 线性布局管理器

线性布局管理器，即 LinearLayoutManager，其构造方法如下所示。

- LinearLayoutManager(Context ctx，int orientation，boolean mark)：ctx 是上下文对象。orientation 表示列表方向，常用的值为 LinearLayout.HORIZONTAL，表示水平列表。LinearLayout.VERTICAL，表示垂直列表。mark 表示是否为相反

方向开始布局,默认为 false。若为 true,则垂直方向将从下向上布局,水平方向将从右向左布局。

2. 网格布局管理器

网格布局管理器,即 GridLayoutManager,其常用方法如下所示。

- GridLayoutManager(Context ctx, int cols):构造方法,设置网格为 cols 列。
- void setSpanSizeLookup(GridLayoutManager.SpanSizeLookup span):设置列表项占位规则,默认一项占一列。若有项占多列,则用到该函数。

3. 瀑布流网格布局管理器

瀑布流网格布局管理器,即 StaggeredGridLayoutManager,其构造方法如下所示。

- StaggereGridLayoutManager(int cols,int orientation):构造方法。cols 代表表格列数;orientation 代表方向,常用的值为 LinearLayout.HORIZONTAL,表示水平列表;LinearLayout.VERTICAL,表示垂直列表。该布局管理器的最大特点是每一项的高度或宽度可以是不同的,随实际大小变化而变化。

【例 4-5】 在列表项子视图及适配器不变的情况下,验证 RecyclerView 界面随设置成不同布局管理器的变化情况。

前提条件:将 6 幅大小不同的图片复制到工程的 res/drawable 目录下,假设名称为 a.jpg～f.jpg。本示例所需文件如下所示。

① 定义列表项布局文件 myitem.xml。

```
<?xml version="1.0" encoding="utf-8"?>
<LinearLayout xmlns:android="http://schemas.android.com/apk/res/android"
    android:layout_width="match_parent"
    android:layout_height="wrap_content"
    android:orientation="vertical">
    <ImageView
        android:id="@+id/myview"
        android:layout_width="wrap_content"
        android:layout_height="wrap_content" />
</LinearLayout>
```

很明显,列表项子视图是 ImageView 对象,要显示不同的图片。

② 适配器类文件 MyAdapter.java:仅列出与例 4-4 代码不同的部分。

```
public class MyAdapter extends RecyclerView.Adapter<MyAdapter.MyHolder> {
```

```
    //其他所有代码均与例4-4相同,略
    public void onBindViewHolder(MyHolder holder, int position) {
        Drawable draw = (Drawable)mList.get(position);
        holder.view.setImageDrawable(draw);
    }
    class MyHolder extends RecyclerView.ViewHolder {
        ImageView view;
        public MyHolder(View itemView) {
            super(itemView);
            view = (ImageView) itemView.findViewById(R.id.myview);
        }
    }
}
```

主要是onBindViewHolder()与MyHolder类与例4-4稍有区别,由对TextView对象操作变为对ImageView对象操作。

③ 主测试界面main.xml。

```
<?xml version="1.0" encoding="utf-8"?>
<LinearLayout xmlns:android="http://schemas.android.com/apk/res/android"
    xmlns:app="http://schemas.android.com/apk/res-auto"
    android:layout_width="match_parent"
    android:layout_height="match_parent"
    android:orientation="vertical">
    <EditText
        android:id="@+id/myin"
        android:layout_width="match_parent"
        android:layout_height="50dp" />
    <Button
        android:id="@+id/myok"
        android:layout_width="wrap_content"
        android:layout_height="wrap_content"
        android:text="OK"/>
    <android.support.v7.widget.RecyclerView
        android:id="@+id/mylist"
        android:layout_width="match_parent"
        android:layout_height="match_parent" />
</LinearLayout>
```

以上代码首先定义了一个 EditText 控件，用于输入整数特征值 mark。当 mark 为 1、2、3 时，分别显示线性、网格、瀑布流网格布局管理器下的 RecyclerView 显示情况；然后定义了一个 Button 组件，用以启动相应布局下的显示；最后定义了一个 RecyclerView 组件，用于显示列表数据。

④ 主测试文件 MainActivity.java。

```
public class MainActivity extends AppCompatActivity implements View
.OnClickListener{
    private List mList;                          //数据项集合
    private MyAdapter ad;                        //适配器
    private RecyclerView rv;
    private RecyclerView.LayoutManager lm;       //布局管理器
    public void onClick(View v) {
        EditText et = (EditText) findViewById(R.id.myin);
        int value = Integer.parseInt(et.getText().toString());
        if(value==1)
            lm = new LinearLayoutManager(this, LinearLayoutManager.VERTICAL,
false);
        if(value==2) {
            lm = new GridLayoutManager(this, 3);
            ((GridLayoutManager) lm).setSpanSizeLookup(new GridLayoutManager
.SpanSizeLookup() {
                public int getSpanSize(int position) {
                    if(position ==0) return 2;
                    return 1;
                }
            });
        }
        if(value==3)
            lm = new StaggeredGridLayoutManager(3,LinearLayout.VERTICAL);

        mList = new ArrayList();
        rv = (RecyclerView) findViewById(R.id.mylist);
        initData(mList);
        ad = new MyAdapter(mList);
        rv.setLayoutManager(lm);
        rv.setAdapter(ad);
    }
```

```
    protected void onCreate(Bundle savedInstanceState) {
        super.onCreate(savedInstanceState);
        setContentView(R.layout.main);
        Button b = (Button) findViewById(R.id.myok);
        b.setOnClickListener(this);

    }
    public void initData(List list) {
        list.add(this.getResources().getDrawable(R.drawable.a,null));
        list.add(this.getResources().getDrawable(R.drawable.b,null));
        list.add(this.getResources().getDrawable(R.drawable.c,null));
        list.add(this.getResources().getDrawable(R.drawable.d,null));
        list.add(this.getResources().getDrawable(R.drawable.e,null));
        list.add(this.getResources().getDrawable(R.drawable.f,null));
    }
}
```

initData()函数的功能是读取资源图像文件 a.jpg～f.jpg，生成相应的 Drawable 对象，并添加到成员变量 mList 集合对象中。

OK 按钮响应函数 onClick()中，根据输入的特征值 value 确定应用哪个布局管理器。当 value 为 1 时，则应用线性布局管理器；当 value 为 2 时，则应用网格布局管理器。本例应用了 setSpanSizeLookup()函数，重写了 SpanSizeLookup 接口函数 getSpanSize(int position)，代码表明数据项中的第 1 项(position＝0)数据占两列，其余项数据都占 1 列；当 value 为 3 时，则应用瀑布流网格布局管理器。

4.6 菜单控件

菜单是 Android 应用中常用的控件，主要分为选项菜单、上下文菜单、弹出菜单。下面一一加以说明。

4.6.1 选项菜单

选项菜单一般指 Activity 的主菜单。实现选项菜单有 3 个步骤：定义菜单资源；加载菜单；添加事件响应。下面通过一个具体实例，说明实现选项菜单的基本过程。

① 添加菜单资源文件 mymenu.xml。

在 res 目录下创建 menu 子目录(若已存在，则无须创建)，在该子目录下创建

mymenu.xml,其内容如下所示。

```
<?xml version="1.0" encoding="utf-8"?>
<menu xmlns:android="http://schemas.android.com/apk/res/android">
    <item android:id="@+id/mylogin" android:title="登录"/>
    <item android:id="@+id/myregist" android:title="注册"/>
    <item android:id="@+id/myexit" android:title="退出"/>
</menu>
```

可以看出,本示例中的菜单资源由根节点 menu 与多个菜单项 item 子节点组成。Item 的常用属性如下所示。

- android:id:定义资源 ID。
- android:title:定义菜单标题。
- android:icon:定义菜单项所要使用的图标。
- android:alphabeticShortcut:定义一个字符快捷键。
- android:numericShortcut:数字值,定义一个数字快捷键。
- android:checkable:布尔值,如果菜单项是可以复选的,那么就设置为 true。
- android:checked:布尔值,如果复选菜单项默认是被选择的,那么就设置为 true。
- android:visible:布尔值,如果菜单项默认是可见的,那么就设置为 true。
- android:enabled:布尔值,如果菜单项默认是可用的,那么就设置为 true。

② 主界面文件 main.xml。

```
<?xml version="1.0" encoding="utf-8"?>
<LinearLayout xmlns:android="http://schemas.android.com/apk/res/android"
    xmlns:app="http://schemas.android.com/apk/res-auto"
    android:layout_width="match_parent"
    android:layout_height="match_parent"
    android:orientation="vertical">
</LinearLayout>
```

③ 在主应用文件 MainActivity.java 中加载菜单及事件响应。

在重写 onCreateOptionsMenu(Menu menu)函数中完成 mymenu.xml 菜单资源文件的加载。在重写 onOptionsItemSelected(MenuItem item)函数中完成菜单项的消息响应。代码如下所示。

```
public class MainActivity extends AppCompatActivity{
    public boolean onCreateOptionsMenu(Menu menu) {
        getMenuInflater().inflate(R.menu.mymenu, menu);
```

```
        return true;
    }
    public boolean onOptionsItemSelected(MenuItem item) {
        int value  = item.getItemId();
        Toast.makeText(this, ""+item.getItemId(), Toast.LENGTH_LONG).show();
        return true;
    }
    protected void onCreate(Bundle savedInstanceState) {
        super.onCreate(savedInstanceState);
        setContentView(R.layout.main);
    }
}
```

在消息响应函数 onOptionsItemSelected()中仅获得菜单项的 ID 号并显示，根据该 ID 号可进一步写出不同的函数分支，完成相应的具体函数功能。本示例的运行结果如图 4-4 所示。

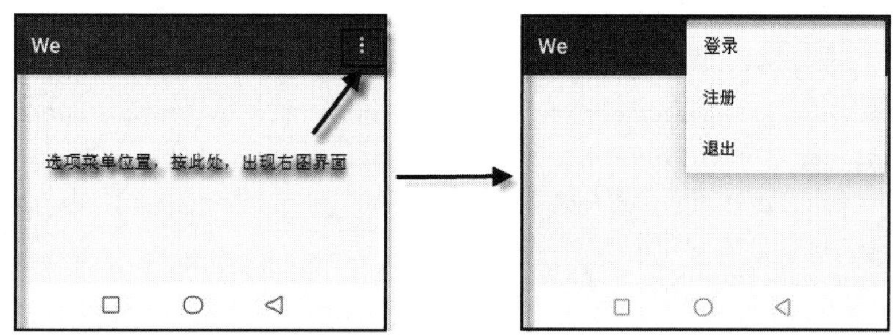

图 4-4　选项菜单界面图

本示例通过加载菜单资源 mymenu.xml 创建了菜单，也可用动态代码实现，相应的代码如下所示。

```
public boolean onCreateOptionsMenu(Menu menu) {
    menu.add(0,100,0,"登录");
    menu.add(0,101,0,"注册");
    menu.add(0,102,0,"退出");
    return true;
}
```

添加菜单子项 add()函数的原型如下所示。

void add(int group,int id,int order,CharSequence title);group 是组号，用以批量对

菜单子项进行处理;id 是菜单项 ID 号,是每个菜单子项的唯一标识;order 指菜单子项在选项菜单中的排列顺序;title 是菜单子项名称。

4.6.2 上下文菜单

一般针对 View 或其子类,当在该 View 中发生长按事件时出现的菜单叫作上下文菜单。其实现步骤如下所示。

- 给 View 注册上下文菜单 registerForContextMenu()。
- 重写 View 中的 onCreateContextMenu()函数,加载菜单。可以通过代码动态添加,也可以通过加载菜单资源 xml 文件添加。
- 重写 View 中的 onContextItemSelected();处理菜单项响应事件。

在 4.6.1 节实现选项菜单的基础上,增加上下文菜单功能。为实现该功能,所需文件如下所示。

① 菜单资源文件 mymenu.xml：与 4.6.1 节相同,略。

② 主界面文件 main.xml。

```
<?xml version="1.0" encoding="utf-8"?>
<LinearLayout xmlns:android="http://schemas.android.com/apk/res/android"
    xmlns:app="http://schemas.android.com/apk/res-auto"
    android:layout_width="match_parent"
    android:layout_height="match_parent"
    android:orientation="vertical">
    <View
        android:id="@+id/myview"
        android:layout_width="match_parent"
        android:layout_height="300dp"/>
</LinearLayout>
```

与 4.6.1 节的示例相比,这里增加了一个高为 300dp 的 View 节点。也就是说,要为 View 节点增加上下文菜单功能,当长按 View 节点区域时,会出现 mymenu.xml 代表的上下文菜单。

③ 主应用文件 MainActivity.java。

```
public class MainActivity extends AppCompatActivity{
    public boolean onCreateOptionsMenu(Menu menu) { //创建选项菜单,与 4.6.1 节同,
略 }
    public boolean onOptionsItemSelected(MenuItem item) { //选项菜单项响应函数,
```

```
与 4.6.1 节同,略}
    public void onCreateContextMenu(ContextMenu menu, View v,    //创建上下文菜单
                                    ContextMenu.ContextMenuInfo menuInfo) {
        super.onCreateContextMenu(menu, v, menuInfo);
        MenuInflater inflater = getMenuInflater();
        inflater.inflate(R.menu.mymenu, menu);
    }
    public boolean onContextItemSelected(MenuItem item) {//上下文菜单项事件响应
        int value = item.getItemId();
        Toast.makeText(this, "Hello"+value, Toast.LENGTH_LONG).show();
        return true;
    }
    protected void onCreate(Bundle savedInstanceState) {
        super.onCreate(savedInstanceState);
        setContentView(R.layout.main);
        View v = (View)findViewById(R.id.myview);
        registerForContextMenu(v);
    }
}
```

可以看出,本示例中既有选项菜单,又有上下文菜单,只不过它们显示的内容是一致的,都取自菜单资源文件 mymenu.xml。上下文菜单的显示位置随长按位置的不同而不同。上下文菜单的创建及响应与选项菜单的机理是一致的,只是函数名称不同。上下文菜单的创建及事件响应的函数分别是 onCreateContextMenu()、onContextItemSelected()。实现上下文菜单功能必须注册相应的 View 对象,选项菜单则无须注册。本示例对 View 对象进行上下文菜单的注册,见 onCreate()函数代码中的最后一行 registerForContextMenu(v),完整形式为 this.registerForContextMenu(v),含义是对 v 对象实现上下文菜单功能,其相应的重写函数在 this 中。this 是 MainActivity 对象,因此必须重写 MainActivity 类中的 onCreateContextMenu()和 onContextItemSelected()函数。

4.6.3 弹出菜单

从特点来说,选项菜单、上下文菜单的创建及事件响应都不是用户主动实现的。用户只能在重写的 onCreateXXXMenu()函数中完成菜单的创建,在重写的 onXXXItemSelected()函数中完成菜单项的事件响应。对弹出式菜单来说,用户均能主动地创建菜单及定义菜单项事件响应的目的函数。弹出菜单一般是响应 View 对象的某

事件而出现的,若空间足够大,它会显示在 View 的下方,否则显示在 View 的上方。

下面在 4.6.2 节的基础之上,为按钮增加弹出菜单及事件响应,所需文件如下所示。

① 菜单资源文件 mymenu.xml: 与 4.6.1 节相同,略。

② 主界面文件 main.xml。

```
<?xml version="1.0" encoding="utf-8"?>
<LinearLayout xmlns:android="http://schemas.android.com/apk/res/android"
xmlns:app="http://schemas.android.com/apk/res-auto"
android:layout_width="match_parent"
android:layout_height="match_parent"
android:orientation="vertical">
<View
    android:id="@+id/myview"  android:layout_width="match_parent"  android:
layout_height="300dp"/>
    <Button
        android:id="@+id/mypop"
        android:layout_width="wrap_content"
        android:layout_height="wrap_content"
        android:text="弹出菜单"/>
</LinearLayout>
```

与 4.6.2 节相比,这里仅增加了一个 Button 组件,用于启动弹出菜单。

③ 主应用文件 MainActivity.java。

```
public class MainActivity extends AppCompatActivity implements View.
OnClickListener,PopupMenu.OnMenuItemClickListener{
    public boolean onCreateOptionsMenu(Menu menu) { //选项菜单的创建,同 4.6.1
节,略 }
    public boolean onOptionsItemSelected(MenuItem item) { //选项菜单项响应,同
4.6.1 节,略}
    public void onCreateContextMenu(ContextMenu menu, View v,
          ContextMenu.ContextMenuInfo menuInfo) { //上下文菜单的创建,同 4.6.1
节,略 }
    public boolean onContextItemSelected(MenuItem item) {//上下文菜单项响应,
同 4.6.1 节,略 }
    public boolean onMenuItemClick(MenuItem item) {   //弹出菜单项事件响应
        int id = item.getItemId();
        Toast.makeText(this, "Hi"+id, Toast.LENGTH_LONG).show();
```

```
        return false;
    }
    public void onClick(View v) {                    //启动弹出菜单,并注册事件响应
        PopupMenu popupMenu = new PopupMenu(MainActivity.this,v);
        popupMenu.getMenuInflater().inflate(R.menu.mymenu,popupMenu
.getMenu());
        popupMenu.show();
        popupMenu.setOnMenuItemClickListener(this);
    }
    protected void onCreate(Bundle savedInstanceState) {
        super.onCreate(savedInstanceState);
        setContentView(R.layout.main);
        View v = (View)findViewById(R.id.myview);
        registerForContextMenu(v);                   //注册上下文菜单
        Button b = (Button)findViewById(R.id.mypop);
        b.setOnClickListener(this);                  //启动弹出菜单
    }
}
```

onClick()是“弹出菜单”按钮响应函数,包含弹出菜单的主要知识点:首先产生 PopupMenu 对象;然后加载 mymenu.xml 菜单资源,调用 show()函数显示弹出菜单;最后利用 setOnMenuItemClickListener()函数为弹出菜单添加消息响应事件。

习题 4

一、选择题

1. Android 系统最常用的对话框类是(　　)。
   A. AlertDialog　　B. Dialog
   C. JFrame　　D. AlertDialog.Builder
2. Android 常用的日期控件是(　　)。
   A. DatePicker　　B. DatePickerDialog
   C. TimePicker　　D. TimePickerDialog
3. 对于 RecyclerView 来说,哪一项是正确的?(　　)。
   A. 不能设置布局管理器　　B. 没有布局管理器
   C. 布局管理器很强大　　D. 无法确定

二、程序题

1. 对话框题目：在主界面定义一个 OK 按钮，当单击此按钮时弹出学生信息对话框，其中包含学号、姓名、出生日期、专业、学院信息。

2. 日期对话框题目：题目 1 中的出生日期应用 DatePickerDialog 对话框输入。

3. 菜单题目：为主界面增加一个选项菜单，其中包括学院、教师、专业 3 项；添加消息响应，并利用 Toast 验证消息是否响应。

第 5 章

Activity

Activity 是与用户交互的接口，提供了图形界面供用户操作。前文中每个应用程序都用到了 Activity：通过调用 setContentView()方法，为 Activity 生成一个图形界面。因此，Acrtivity 是 Android 系统最重要的功能之一。

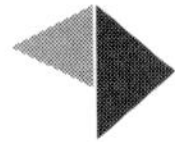

5.1 生命周期

在一个 Android 应用中可以有多个 Activity，这些 Activity 组成了 Activity 栈(Stack)，当前活动的 Activity 位于栈顶，之前的 Activity 被压入下面成为非活动 Activity，等待是否可能被恢复为活动状态。Activity 的 4 个重要状态见表 5-1。

表 5-1 Activity 的 4 个重要状态

序号	状　态	说　明
1	运行状态	当前 Activity，用户可见，可以获得焦点
2	暂停状态	失去焦点的 Activity，仍然可见
3	停止状态	该 Activity 被其他 Activity 覆盖，不可见，会保存所有状态和信息
4	销毁状态	该 Activity 结束

一个 Activity 从创建到消亡叫作一个生命周期，其中有很多有用的回调函数，方便截获添加有价值的处理功能。Activity 常用的回调函数见表 5-2。

表 5-2 Activity 常用的回调函数

序号	函数名称	说　明
1	onCreate()	创建 Activity 时被回调，执行更多的初始化功能，如添加消息响应

续表

序号	函数名称	说　　明
2	onStart()	启动 Activity 时被回调,也就是当 Activity 变为可见时被回调
3	onResume()	Activity 由暂停态变为运行态时被调用
4	onPause()	暂停 Activity 时被调用,通常用于持久保存数据
5	onRestart()	重新启动 Activity 时被调用,在 onStop()后运行
6	onStop()	停止 Activity 时被回调
7	onDestroy()	销毁 Activity 时被回调

对 Activity 来说,onCreate()与 onDestroy()函数都运行一次,其他回调函数可运行多次,主要有 4 个流程,如下所示。

流程 1:若焦点一直在 Activity 上,则流程为 onCreate()->onStart()->onResume()->正常运行界面的其他功能->onDestroy()正常销毁。

流程 2:焦点有切换,假设有两个 Activity,名称为 a、b。初始时焦点在 a 上,则流程为 onCreate()->onStart()->onResume()->正常运行界面的其他功能;当切换到名称为 b 的 Activity 时,对 a 来说,运行 onPause->onStop();当焦点再切回到名称为 a 的 Activity 时,对 a 来说,运行 onRestart()->onStart()->onResume->正常运行界面的其他功能。

流程 3:从 onPause()->onResume()函数,例如,当在某 Activity 中弹出某模态对话框后,Activity 运行 onPause()函数,当关闭模态对话框后,Activity 运行 onResume()函数。

流程 4:由 App 进程管理器直接销毁该 Activity 及对应的应用程序。

【例 5-1】 验证 Active 各回调函数的执行顺序。

本示例仅能验证上文所述的流程 1、流程 2 的情况。许多 Android 书都用日志监测 Activity 各回调函数的执行流程,在集成开发环境中看日志较方便,但将该应用安装在真实手机上看日志却并不方便。本示例利用图形用户界面显示 Activity 各回调函数响应的各种情况,在集成开发平台与在真实手机中均非常方便。思路是:将 Activity 响应函数名称显示在 TextView 控件中。各关键代码如下所示。

① 主布局文件:main.xml。

```
<?xml version="1.0" encoding="utf-8"?>
<LinearLayout xmlns:android="http://schemas.android.com/apk/res/android"
    android:layout_width="match_parent"
```

```
    android:layout_height="match_parent">
    <TextView
        android:id="@+id/mytext"
        android:layout_width="match_parent"
        android:layout_height="match_parent"
        android:textSize="30sp"/>
</LinearLayout>
```

以上代码采用线性布局，定义了 id 为 mytext 的 TextView 控件，用于显示 Activity 各回调函数的响应名称。

② MainActivity.java。

```
public class MainActivity extends AppCompatActivity {
    void show(String name) {
        TextView tv = (TextView) findViewById(R.id.mytext);
        String s = tv.getText().toString();
        tv.setText(s + "\n" +name);
    }
    protected void onCreate(Bundle savedInstanceState) {
        super.onCreate(savedInstanceState);
        setContentView(R.layout.main);
        show("onCreate");
    }
    protected void onStart() {super.onStart();show("onStart");}
    protected void onDestroy() {super.onDestroy();show("onDestroy");}
    protected void onStop() {super.onStop();show("onStop");}
    protected void onResume() {super.onResume();show("onResume");}
    protected void onPause() {super.onPause();show("onPause");}
    protected void onRestart() {super.onRestart();show("onRestart");}
}
```

关键理解 show(String name)函数，name 即回调函数字符串名称。该函数首先获得 TextView 组件上已有的内容，将其与当前 name 字符串相加，再重新设置回 TextView。

5.2　建立 Activity

5.2.1　入口 Activity 类

一个应用可以包含多个 Activity 活动页面，其入口 Activity 类对应应用的第一个界面，之前经常用的 MainActivity.java 就是入口 Activity 类，如何区分入口 Activity 类与其他普通 Activity 类？这涉及应用配置文件 Androidmanifest.xml(在 res 目录下)，打开该文件，可以看到与 Activity 类 MainActivity 对应的配置信息在<activity>标签内，如下所示。

```
<activity android:name=".MainActivity">
  <intent-filter>
    <action android:name="android.intent.action.MAIN" />
    <category android:name="android.intent.category.LAUNCHER" />
  </intent-filter>
</activity>
```

<activity>属性 name＝“.MainActivity”，标识了该 Active 类的名称为 MainActivity，在工程默认包下。

<activity>子标签<intent-filter>定义了过滤器信息，对初学者而言，只要知道<intent-filter>子标签<action>属性 name＝“android.intent.action.MAIN”，<category>属性 name＝“android.intent.category.LAUNCHER”，那么它对应的 Activity 类就位于优先级最高级，最先执行。

5.2.2 普通 Activity 类

从入口 Active 类类推，普通 Activity 类也应有如下内容：Activity Java 类、界面 XML 布局文件、Androidmanifest.xml 配置文件中增加的<activity>子标签内容。

【例 5-2】 最简单的 Activity 启动程序。一般来说，已有工程已经包含入口 Activity 类 MainActivity，现在要求再建立一个普通 Activity 类 SecondActivity。MainActivity 对应的主界面(main.xml)有一按钮 OK，当单击 OK 按钮时，启动 SecondActivity 对应的界面(main2.xml)；SecondActivity 对应的界面显示“This is second activity”，详细内容如下所示。

① Activity 类界面布局文件。

```
//main.xml
<?xml version="1.0" encoding="utf-8"?>
<LinearLayout xmlns:android="http://schemas.android.com/apk/res/android"
    android:layout_width="match_parent"
    android:layout_height="match_parent">
    <Button
        android:id="@+id/myok"
```

```
        android:layout_width="wrap_content"
        android:layout_height="wrap_content"
        android:text="ok"/>
</LinearLayout>
//main2.xml
<?xml version="1.0" encoding="utf-8"?>
<LinearLayout xmlns:android="http://schemas.android.com/apk/res/android"
    android:layout_width="match_parent"
    android:layout_height="match_parent">
    <TextView
        android:layout_width="match_parent"
        android:layout_height="wrap_content"
        android:text="This is second activity"/>
</LinearLayout>
```

② 配置文件 Androidmanifest.xml，以下仅列出与 Activity 有关的<activity>节点内容。

```
<activity android:name=".MainActivity">
  <intent-filter>
    <action android:name="android.intent.action.MAIN" />
    <category android:name="android.intent.category.LAUNCHER" />
  </intent-filter>
  </activity>
<activity android:name=".SecondActivity"></activity>
```

可以看出，每增加一个普通 Activity 类，在配置文件中就增加一个<activity>节点，而且内容较简单，设定其 name 属性等于 Activity 类名称即可。

③ 对应的两个 Activity 类。

```
//SecondActivity.java
public class SecondActivity extends AppCompatActivity {
    protected void onCreate(Bundle savedInstanceState) {
        super.onCreate(savedInstanceState);
        setContentView(R.layout.second);
    }
}
//MainActivity.java
public class MainActivity extends AppCompatActivity {
    protected void onCreate(Bundle savedInstanceState) {
```

```
            super.onCreate(savedInstanceState);
            setContentView(R.layout.main);
            Button b = (Button)findViewById(R.id.myok);
            b.setOnClickListener(new View.OnClickListener() {
                public void onClick(View v) {
                    Intent in = new Intent(MainActivity.this,SecondActivity.class);
                    startActivity(in);
                }
            });
        }
    }
```

可以看出,启动另一个 Activity 的关键函数是 startActivity(Intent intent),因此必须先产生 Intent 对象。Intent 一个常用的构造方法是 Intent(Context ctx, Class<?> cls),ctx 是上下文对象,cls 可以是待启动的 Activity 类的类名。

 5.3 Activity 通信

Activity 类间数据通信要应用 Intent 对象,Intent 保存数据的本质是 map 映射(键-值),键一般是字符串,值可以是多种类型,Intent 可读、可写,常用的函数如下所示。

- void putExtra(String key, XXX value),写单值函数,XXX 可以是 8 个基本数据类型:byte、short、int、long、float、double、char、boolean。
- void putExtra(String key, XXX value[]),写数组函数,XXX 可以是 8 个基本数据类型:byte、short、int、long、float、double、char、boolean。
- void putExtra(String key, CharSequence value),写字符串函数。
- void putExtra(String key, CharSequence value[]),写字符串数组函数。
- void putExtra(String key, Serializable value),写实现序列化对象函数。
- void putExtras(Bundle b),写捆绑对象。
- xxx getXxxExtra(String key, xxx default),读基本数据类型,3 个小写的"xxx"表示 8 种基本数据类型之一;首字符大写其余小写的"Xxx",表示 8 种基本数据类型名称首字符大写,其余小写。该函数的含义是:若读成功,则返回读的值,若读不成功,则返回 default 值。
- xxx[] getXxxArrayExtra(String key),读基本数据类型数组,若不成功,则返回 null。
- String getStringExtra(String key),读字符串数据。
- String[] getStringArrayExtra(String key),读字符串数组数据。

- Serializable getSerializableExtra(String key),读序列化对象数据。
- Bundle getExtras(String key),读捆绑对象。

假设有两个 Activity 类 A、B,A 启动 B,A、B 间传参有以下情况：A 向 B 传送基本数据类型;A 向 B 传送类对象类型;A 接收 B 销毁返回给 A 的数据。下面以关键代码的形式一一加以说明。

① A 向 B 传送基本数据类型,A 发送数据,B 解析数据,见表 5-3。

表 5-3 基本数据类型发送解析关键代码

	A 发送数据	B 解析数据

方法 1	Intent out =new Intent(A.this,B.class);
	int no=1000; String name = "zhang";
	out.putExtra("no",no);
	out.putExtra("name", name);
	startActivity(out);
		Intent in =B.this.getIntent();
		int no=in.getIntExtra("no",0);
		String name=in.getStringExtra("name");

方法 2	Bundle b =new Bundle();
	int no=1000;String name="zhang";
	b.putInt("no", no);
	b.putString("name", name);
	Intent out = new Intent(A.this,B.class);
	out.putExtras(b);
	startActivity(out);
		Intent in =B.this.getIntent();
		Bundle b = in.getExtras();
		int no = b.getInt("no");
		String name = b.getString("name");

② A 向 B 传送类对象类型,A 发送数据、B 解析数据。

假设自定义 Data 类如下所示。

```
class Data implements Serializable{
    int no;String name;
    Data(int no,String name){this.no=no; this.name=name;}
}
```

之所以实现 Serializable 接口,是因为要用到 putExtra(String key,Serializable value)函数,value 对应的类要求必须实现 Serializable 接口,其相应的发送、接收关键代码见表 5-4。

表 5-4 类对象数据发送、接收关键代码

A 发送数据	B 解析数据

Data d =new Data(1000,"zhang");
Intent out = new Intent(A.this,B.class);
out.putExtra("data",d);
startActivity(out);
	Intent in =this.getIntent();
	Data d = (Data) in.getSerializableExtra("data");

利用 Bundle 捆绑数据也能实现类对象数据在 Activity 间的发送和解析，本题中的自定义类也无须实现 Serializable 接口。请读者参照表 5-3 中的方法 2 实现相应的发送和接收数据。

③ A 接收 B 销毁返回给 A 的数据。

为了更好地理解此部分知识，还需要掌握系统 Activity 类的 3 个基本函数，如下所示。

- void startActivityForResult(Intent intent, int requestcode)，与 startActivity()函数不同，本函数是具有接收返回结果的 Activity 启动函数，requestcode 是请求识别码。
- void finish()，结束当前 Activity 对象，返回到前一个 Active 对象。
- void setResult(int resultcode, Intent data)，resultcode 是结果识别码，data 是准备返回前一级 Activity 对象的数据。

另外，还要知道：若某 Activity 对象接收下一级 Activity 对象的返回数据，则必须重写如下函数。

- protected void onActivityResult(int requestCode, int resultCode, Intent data);

很明显，data 是返回的数据，requestCode 是请求码，resultCode 是结果码。

实现 A 接收 B 销毁返回给 A 的数据的关键代码见表 5-5。

表 5-5 实现 A 接收 B 销毁返回给 A 的数据的关键代码表

A 中的关键代码	B 中的关键代码

//A 启动 B 代码
Intent out = new Intent (MainActivity. this, SecondActivity. class);
startActivityForResult(out,1);
//A 接收 B 的返回值，必须重载下面函数
protected voidonActivityResult(int requestCode,
    intresultCode, Intent data) {
super.onActivityResult(requestCode, resultCode, data);
if(requestCode==1 && resultCode==2){
int no = data.getIntExtra("no", 0);
    }
}
	//B 主动销毁时，直接调用以下函数
	onBackPressed()
	//重写 onBackPressed()函数
	public voidonBackPressed() {
	//super.onBackPressed();
	    Intent out = new Intent();
	    out.putExtra("no",1000);
	    setResult(2,out);
	    finish();
	}

B 销毁有两种情况：一是响应 B 中的界面子组件事件，如按钮事件等；二是响应手机固有的"返回"事件。这两种情况的响应函数本质上功能是一致的，因此本例重写了"返回"事件函数 onBackPressed()，完成了向上一级 Activity 对象设置返回值及销毁本级

Activity 对象功能。在 onBackPressed()函数内,注释了一行语句 super.onBackPressed(),若把注释打开,运行程序后会发现 A 中的 onActivityResult()函数不会接收 B 中传过来的数据。这是因为 B 中的 super.onBackPressed()函数已经完成了销毁 B 的功能,在该行代码下再设置向 A 发送的数据当然是无效了。

通过此例,再加深理解 startActivityForResult(intent,requestCode)及 setResult(resultCode,intent)中申请码 requestCode、结果码 resultCode 的作用,这两个值最终都体现在 onActivityResult(requestCode,resultCode,data)函数参数内:根据 requestCode 可知道是从哪个 Activity 对象返回的;根据 resultCode,可以知道该如何解析 data,因为不同的 resultCode 可能代表不同的、有用的数据类型。

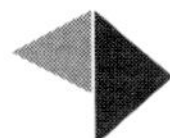

5.4　隐式启动 Activity

前文所述启动 Activity 都是显示方式,主要体现在建立 Intent 对象时明确体现了启动的 Activity 对象。例如,Intent in=new Intent(A.this, B.class),则 A 是源 Activity 类,B 是待启动的 Activity 类。隐式启动 Activity 的含义是:在程序中看不出启动的 Activity 类,它是通过匹配查找配置文件 Androidmanifest.xml 中与 Activity 有关的内容,找出应该启动哪个 Activity 类。

5.4.1　intent-filter

在 Androidmanidest.xml 中,与 Activity 有关的主要是<activity>标签的子标签<intent-filter>的内容,对隐式启动的 Activity 来说,必须配置<intent-filter>的内容。它主要包含<action>、<category>、<data>3 个子标签的内容设置,下面一一加以说明。

① <action>标签。

action 属性的值为一个字符串,它代表系统中已经定义了一系列常用的动作,形如<action android:name="XXX">,XXX 可以自定义,也可以应用系统定义的常量,如下所示。

- 自定义动作字符串。
- ACTION_MAIN:Android Application 的入口,每个 Android 应用必须且只能包含一个此类型的 Action 声明。
- ACTION_VIEW:系统根据不同的 Data 类型,通过已注册的对应 Application 显示数据。

- ACTION_EDIT：系统根据不同的 Data 类型，通过已注册的对应 Application 编辑数据。
- ACTION_DIAL：打开系统默认的拨号程序，如果 Data 中设置了电话号码，则自动在拨号程序中输入此号码。
- ACTION_CALL：直接呼叫 Data 中所带的号码。
- ACTION_ANSWER：接听来电。
- ACTION_SEND：由用户指定发送方式进行数据发送操作。
- ACTION_SENDTO：系统根据不同的 Data 类型，通过已注册的对应 Application 进行数据发送操作。

所有系统定义的常量都要加前缀"android.intent.action."，形如<action android:name="android.intent.action.ACTION_MAIN">。

② <category>标签

<category>标签用于指定当前 intent 被处理的环境，一个 intent-filter 可以包含多个 category 属性。

- 自定义种类字符串。
- CATEGORY_DEFAULT：Android 系统中默认的执行方式，按照普通 Activity 的执行方式执行。
- CATEGORY_HOME：设置该组件为 Home Activity。
- CATEGORY_PREFERENCE：设置该组件为 Preference。
- CATEGORY_LAUNCHER：设置该组件为在当前应用程序启动器中优先级最高的 Activity，通常与入口 ACTION_MAIN 配合使用。
- CATEGORY_BROWSABLE：设置该组件可以使用浏览器启动。
- CATEGORY_GADGET：设置该组件可以内嵌到另外的 Activity 中。

所有系统定义的常量都要加前缀"android.intent.category."，形如<category android:name="android.intent.category.CATEGORY_DEFAULT">。

③ <data>标签。

每个<data>标签都可以指定一个 URI 结构以及 data 的 MIME 类型。一个完整的 URI 由 scheme、host、port 和 path 组成，其结构如下所示。

```
<scheme>://<host>:<port>/<path>
```

其中，scheme 既可以是 Android 中常见的协议，也可以是自定义的协议。Android 中常见的协议包括 content 协议、http 协议、file 协议等，自定义协议可以使用自定义字符串。

<data>标签格式形如：<data android:scheme="XXX" android:host="XXX">。

5.4.2　自定义属性应用

自定义属性应用主要指<action>、<category>中的 android:name 属性值是自定义的，Android 系统要求在 Activity 隐式配置中必须有一个<action>标签，至少一个<category>标签。

考虑这样一个问题：在默认工程的基础上(已经包含 MainActivity.java，布局文件 main.xml)，再建立一个 Activity 类 Two，打开 Androidmanifest.xml，对 Two 类配置增加<action>标签内容，配置文件中关于 Activity 类的内容如下所示。

```
<activity android:name=".MainActivity">
  <intent-filter>
    <action android:name="android.intent.action.MAIN" />
    <category android:name="android.intent.category.LAUNCHER" />
  </intent-filter>
</activity>
<activity android:name=".Two" >
  <intent-filter>
    <action android:name="aaa"></action>
    <category android:name="android.intent.category.DEFAULT"></category>
  </intent-filter>
</activity>
```

由以上代码可知，MainActivity 是应用程序的入口 Activity 类，对 Activity 类 Two 增加了<action>标签，其属性值 android:name 是自定义的，本例中为字符串"aaa"。因此，程序中只有与<intent-filter>标签内容相匹配，对应的 Activity 类才会启动，假设由 MainActivity 类启动 Two 类，若有两段关键代码，其分析见表 5-6。

表 5-6　<action>自定义属性启动 Activity 类

关键代码	分　析

Intent in =new Intent();
in.setAction("aaa");
startActivity(in);
	可以看出，Intent 对象没有显示指定启动的 Activity 类，只是设定了一个自定义动作属性"aaa"，经查配置文件，明确是启动 Activity 类 Two

Intent in =new Intent();
in.setAction("bbb");
startActivity(in);
	可以看出，Intent 对象没有显示指定启动的 Activity 类，只是设定了一个自定义动作属性"bbb"，经查配置文件，没有发现有<action>标签动作属性值为"bbb"，因此程序出现异常

继续做实验，再增加一个 Activity 类 Three，action 标签属性值也设置为"aaa"，也就

是 Two 和 Three 类的配置相同。包含 Two、Three 类的内容如下所示。

```
<activity android:name=".Two" >
  <intent-filter>
    <action android:name="aaa"></action>
    <category android:name="android.intent.category.DEFAULT"></category>
  </intent-filter>
</activity>
<activity android:name=".Three">
  <intent-filter>
    <action android:name="aaa"></action>
    <category android:name="android.intent.category.DEFAULT"></category>
  </intent-filter>
</activity>
```

这时，再运行表 5-6 的第一段关键程序代码，会出现如图 5-1 所示的界面(单击 OK 按钮运行表 5-6 中的第 1 段关键代码)。

图 5-1　action 自定义属性运行界面

此类界面在手机使用时会经常遇到，这也是隐式启动 Activity 的特点之一，因为有许多 Activity 都满足启动条件，你只需要进一步选择哪个 Activity 应该启动。

对本示例而言，若不想出现如图 5-1 所示界面该如何呢？方法 1：在配置文件中，每个 Activity 类的 action 属性不同；方法 2：增加 category 种类属性。这种情况也很常见，例如都是输入功能，有的是文件输入，有的是小键盘输入。本示例采用方法 2，假设它们的动作属性相同，但种类不同，配置文件相关内容修改如下所示。

```
<activity android:name=".Two" >
  <intent-filter>
    <action android:name="aaa"></action>
  <category android:name="bbb"></category>
  <category android:name="android.intent.category.DEFAULT"></category>
    </intent-filter>
</activity>
<activity android:name=".Three">
```

```
    <intent-filter>
      <action android:name="aaa"></action>
      <category android:name="ccc    "></category>
      <category android:name="android.intent.category.DEFAULT"></category>
    </intent-filter>
</activity>
```

对 Two 类而言，增加了 category 标签，属性值设置为 bbb；对 Three 类而言，增加了 category 标签，属性值设置为 ccc。启动 Two、Three 类的关键代码表 5-7。

表 5-7 启动 Two、Three 类的关键代码

启动 Two 类的关键代码	启动 C 类的关键代码

Intent in =new Intent();
in.setAction("aaa");
in.addCategory("bbb");
startActivity(in);
	Intent in =new Intent();
	in.setAction("aaa");
	in.addCategory("ccc");
	startActivity(in);

对于 Activity 类来说，仅能设置一个动作，所以是 setAction()函数；但是可设置属于多个种类，所以是 addCategory()函数。

5.4.3 系统属性应用

对 Activity 属性来说，<action>、<category>、<data>都有许多固有的属性，见 5.4.1 节内容。这意味着可以在应用程序中直接调用系统已有资源，不必重新开发。

【例 5-3】 action 固有属性测试。

功能：将常用的 action 固有属性利用列表控件，当点击时启动相应的 Activity。详细步骤如下所示。

① 主布局文件 main.xml。

```
<?xml version="1.0" encoding="utf-8"?>
<LinearLayout xmlns:android="http://schemas.android.com/apk/res/android"
    android:layout_width="match_parent"
    android:layout_height="match_parent"
    android:orientation="vertical">
    <TextView
        android:id="@+id/myerror"
        android:layout_width="match_parent"
        android:layout_height="wrap_content" />
```

```
    <ListView
        android:id="@+id/mylist"
        android:layout_width="match_parent"
        android:layout_height="wrap_content"></ListView>
</LinearLayout>
```

id 为 mylist 的 ListView 控件用于显示 action 标签的固有动作属性。TextView 组件用于显示异常信息，因此将其 id 定义为 myerror。

② MainActivity.java。

```
public class MainActivity extends AppCompatActivity {
    String title[] = {"ACTION_VIEW","ACTION_EDIT","ACTION_DIAL","ACTION_CALL",
        "ACTION_ANSWER","ACTION_SEND","ACTION_SENDTO"};
    String action[] = {Intent.ACTION_VIEW,Intent.ACTION_EDIT,Intent.ACTION_
DIAL,Intent.ACTION_CALL,
        Intent.ACTION_ANSWER,Intent.ACTION_SEND,Intent.ACTION_SENDTO};
    protected void onCreate(Bundle savedInstanceState) {
        super.onCreate(savedInstanceState);
        setContentView(R.layout.main);
        ListView li = (ListView)findViewById(R.id.mylist);
        ArrayAdapter<String> ada = new ArrayAdapter<String>(this,android.R.
layout.simple_list_item_1,title);
        li.setAdapter(ada);
        li.setOnItemClickListener(new AdapterView.OnItemClickListener() {
            public void onItemClick(AdapterView<?> parent, View view, int
position, long id) {
                try {
                    Intent in = new Intent();
                    in.setAction(action[position]);
                    MainActivity.this.startActivity(in);
                }
                catch(Exception e){
                    TextView tv = (TextView)findViewById(R.id.myerror);
                    tv.setText(e.getMessage());
                }
            }
        });
    }
}
```

action 数组定义了 action 的系统固有动作，本示例仅添加了一部分，读者可进一步完善。title 数组定义了 ListView 中每行显示的内容。关键思路是：对 ListView 添加 OnItemClick 事件侦听，在响应函数中获取条目位置 position，进而获取待测的 action 动作属性为 action[position]，利用 Intent 设置该动作并启动 Activity。由于在加载 Activity 时可能出现异常，因此加了异常处理，将异常信息显示在 TextView 控件中。

【例 5-4】 data 属性测试。

data 是用一个 uri 对象表示的，uri 代表数据的地址，属于一种标识符。通常情况下，使用 action+data 属性的组合描述一个意图：做什么。使用隐式 Intent，不仅可以启动自己程序内的活动，还可以启动其他程序的活动，这使得 Android 多个应用程序之间的功能共享成为可能。例如，应用程序中需要展示一个网页，没有必要自己实现一个浏览器而是只利用系统的浏览器打开这个网页就行。关键代码如下所示。

```
Intent in = new Intent();
in.setAction(Intent.ACTION_VIEW);
Uri data = Uri.parse("http://www.sohu.com");
in.setData(data);
startActivity(in);
```

注意：data 的数据格式一定要精确，示例中的"http://www.sohu.com"是正确的，若写成"www.sohu.com"，运行时就会出现异常。

5.5 Fragment

5.5.1 引入 Fragment 的原因

众所周知，模块化是面向对象的重要思想，对 Activity 而言亦是如此。一个复杂的 Activity 界面复杂，功能复杂，若把全部界面写在一个文件，所有功能都写在一个类中，无疑增加了后续维护和二次开发的难度。对复杂 Activity 而言，一个好的架构如图 5-2 所示。

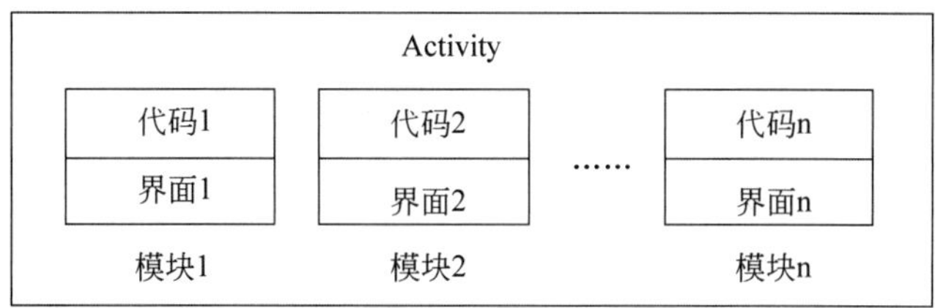

图 5-2　复杂 Activity 架构图

可以看出，一个复杂 Activity 是由多个分立的模块组成的，每个模块包含代码和与之对应的界面部分，也叫作 Fragment，即片段。也就是说，一个复杂 Activity 可以由若干 Fragment 片段组成。对界面主文件来说，可能就是几个空容器标签，具体的界面内容由相应的 Fragment 片段界面完成；对功能主 Activity 类而言，它只负责与各 Fragment 类传递参数，具体的功能由 Fragment 片段功能代码实现。

5.5.2 静态加载

Fragment 分为静态加载和动态加载，本节示例无具体含义，仅是讲解实现静态加载的基本原理，实现的功能如图 5-3 所示。

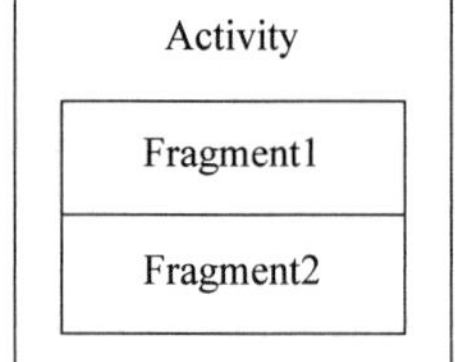

图 5-3 静态加载示例界面图

由上述可知，本 Activity 由两个片段组成，由于每个片段都由代码＋界面组成，所以包含的文件有主文件（MainActivity.java＋main.xml）、片段 1（MyFrag1.java＋frag1.xml）、片段 2（MyFrag2.java＋frag2.xml）。具体内容如下所示。

① 片段 1（MyFrag1.java＋frag1.xml）。

```
//frag1.xml
<?xml version="1.0" encoding="utf-8"?>
<LinearLayout xmlns:android="http://schemas.android.com/apk/res/android"
    android:orientation="vertical" android:layout_width="match_parent"
    android:layout_height="match_parent">
    <TextView
        android:layout_width="match_parent"
        android:layout_height="wrap_content"
        android:text="This is one"/>
</LinearLayout>
```

写 Fragment 片段布局文件与常规文件没有区别，不要人为加一些束缚，本布局文件很简单，仅加了一个 TextView 控件，显示内容为“This is one”。

```
//MyFrag1.java
public class MyFrag1 extends Fragment {
    public MyFrag1() { }
     public View onCreateView (LayoutInflater inflater, ViewGroup container,
Bundle savedInstanceState) {
        View v = inflater.inflate(R.layout.frag1,container,false);
                                                    //加载 frag1.xml 布局文件
    return v;
```

```
    }
}
```

每个 Fragment 类都需从系统 Fragment 类派生，Fragment 类在两个系统包里有(android.app.Fragment，android.support.v4.app.Fragment)，选择哪一个均可，但要保持一致。onCreateView 是重要的重载函数，主要负责生成片段页面。第 1 个参数 inflater 是已经可用的布局管理器对象，用于加载片段布局；第 2 个参数 container 是父容器，片段界面要添加到该容器中；第 3 个参数 savedInstanceState 一定要设置成 false。因为在 Fragment 内部实现中，会把该布局添加到 container 中，如果设为 true，就会重复做两次添加，否则会抛出如下异常：

Caused by：java. lang. IllegalStateException：The specified child already has a parent. You must call removeView() on the child's parent first.

② 片段 2(MyFrag2.java＋frag2.xml)：仿照前文片段 1 的论述即可写出，略。

③ 主文件(MainActivity.java＋main.xml)。

```
//main.xml
<?xml version="1.0" encoding="utf-8"?>
<LinearLayout xmlns:android="http://schemas.android.com/apk/res/android"
    android:layout_width="match_parent"
    android:layout_height="match_parent"
    android:orientation="vertical">
    <fragment
        android:name="com.example.dqjbd.we.MyFrag1"
        android:layout_width="match_parent"
        android:layout_height="wrap_content">
    </fragment>
    <fragment
        android:name="com.example.dqjbd.we.MyFrag2"
        android:layout_width="match_parent"
        android:layout_height="wrap_content">
    </fragment>
</LinearLayout>
```

<fragment>标签是用来静态加载的，其 name 属性等于待加载的 Fragment 类，之所以叫作静态加载，就是因为 name 属性设置了。本配置文件中仅定义了两个绑定 Fragment 类的空标签，非常简洁，几乎所有功能都是由绑定的 Fragment 类完成的(界面＋功能)。

```
//MainActivity.java
public class MainActivity extends AppCompatActivity {
    protected void onCreate(Bundle savedInstanceState) {
        super.onCreate(savedInstanceState);
        setContentView(R.layout.main);
    }
}
```

由于应用了 Fragment 片段技术，主类可以显得非常简洁。当运行 setContentView()函数时，加载 main.xml 布局文件内容。当遇到第 1 个<fragment>标签时，由于其 name 属性绑定了片段类 MyFrag1，因此加载该类，当运行到该类 onCreateView()函数时，加载片段 1 界面，将其添加到主页面中；当遇到第 2 个<fragment>标签时，由于其 name 属性绑定了片段类 MyFrag2，因此加载该类，当运行到该类 onCreateView()函数时，加载片段 2 界面，将其添加到主页面中。至此，界面全部生成完毕。

5.5.3 动态加载

通过上文可知，Fragment 静态加载是在<fragment>标签中利用 name 属性绑定片段类，name 值是静态的，那么不指定 name 值，在程序中设定 name 值，不就是动态加载吗？道理如此，但是动态加载步骤稍复杂，有一点要特别注意，动态加载 Fragment 用的标签一般是<FrameLayout>，不是<fragment>，只有静态加载时采用<fragment>标签。

下面是一个动态加载 Fragment 的例子，仿真功能如图 5-4 所示。

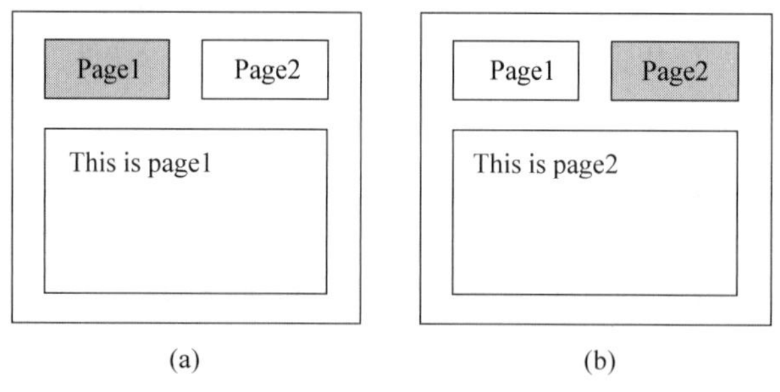

图 5-4　动态加载 Fragment 仿真功能图

功能：主界面有两个按钮 Page1、Page2 及下方的内容面板。当单击 Page1 按钮时，内容面板中显示对应的页面内容（本例中是 This is page1），如图 5-4(a)；当单击 Page2 按钮时，内容面板中显示对应的页面内容（本例中是 This is page2），如图 5-4(b)。

很明显，最终的文件有片段 1（Page1Fragment.java＋page1.xml）、片段 2（Page2Fragment.java＋page2.xml）、主文件（MainActivity.java＋main.xml）。具体内容如下所示。

① 片段 1(Page1Fragment.java＋page1.xml)。

```
//布局文件 page1.xml
<?xml version="1.0" encoding="utf-8"?>
<LinearLayout xmlns:android="http://schemas.android.com/apk/res/android"
    android:orientation="vertical" android:layout_width="match_parent"
    android:layout_height="match_parent">
    <TextView
        android:layout_width="match_parent"
        android:layout_height="wrap_content"
        android:text="This is page1"/>
</LinearLayout>
//MyPage1.java
public class Page1Fragment extends Fragment {
public Page1Fragment() {  }
    public View onCreateView(LayoutInflater inflater, ViewGroup container,
Bundle savedInstanceState) {
        return inflater.inflate(R.layout.page1, container, false);
    }
}
```

可以看出，动态加载和静态加载 Fragment 的代码和布局文件几乎是一致的。

② 片段 2(Page2Fragment.java＋page2.xml)：仿照前文片段 1 的论述即可写出，略。

③ 主文件(MainActivity.java＋main.xml)。

```
//main.xml
<?xml version="1.0" encoding="utf-8"?>
<LinearLayout xmlns:android="http://schemas.android.com/apk/res/android"
    android:layout_width="match_parent"
    android:layout_height="match_parent"
    android:orientation="vertical">
    <LinearLayout
        android:layout_width="match_parent"
        android:layout_height="wrap_content">
        <Button
            android:id="@+id/mypage1"
```

```
            android:layout_width="wrap_content"
            android:layout_height="wrap_content"
            android:text="Page1"/>
        <Button
            android:id="@+id/mypage2"
            android:layout_width="wrap_content"
            android:layout_height="wrap_content"
            android:text="Page2"/>
    </LinearLayout>
    <FrameLayout
        android:id="@+id/myfrm"
        android:layout_width="match_parent"
        android:layout_height="match_parent">
    </FrameLayout>
</LinearLayout>
```

动态添加 Fragment 界面用的标签一般是<FrameLayout>，事实证明用<LinearLayout>、<RelativeLayout>等代替都是可以的，但不能用<fragment>，它是静态加载 Fragment 的标志。

```
//MainActivity.java
public class MainActivity extends AppCompatActivity {
    void showPage(int id){
        Fragment f = null;
        if(id==R.id.mypage1)
                f = new Page1Fragment();
        if(id==R.id.mypage2)
                f = new Page2Fragment();
        FragmentManager manager = this.getFragmentManager();
        FragmentTransaction tran = manager.beginTransaction();
        tran.replace(R.id.myfrm, f);
        tran.commit();
    }
    protected void onCreate(Bundle savedInstanceState) {
        super.onCreate(savedInstanceState);
        setContentView(R.layout.main);
        Fragment fragment = new Page1Fragment();
        FragmentManager manager = this.getFragmentManager();
        FragmentTransaction tran = manager.beginTransaction();
```

```
        tran.add(R.id.myfrm, fragment);
        tran.commit();
        Button b = (Button)findViewById(R.id.mypage1);
        b.setOnClickListener(new View.OnClickListener() {
            public void onClick(View v) {
                showPage(R.id.mypage1);
            }
        });
        Button b2 = (Button)findViewById(R.id.mypage2);
        b2.setOnClickListener(new View.OnClickListener() {
            public void onClick(View v) {
                showPage(R.id.mypage2);
            }
        });
    }
}
```

可以看出，对动态加载 Fragment 来说，Activity 类变化是最大的，因为它必须体现出 Fragment 的增加、替换过程。一般来说，初始化动态加载默认 Fragment 的时候，完成的是增加操作，本例中 onCreate()函数中将 Page1Fragment 作为默认 Fragment，完成了它的添加工作；操作 Acticity 界面引起的 Fragment 变动是替换操作，本示例当单击 Page1 按钮时，将内容面板替换成 Page1Fragment，当单击 Page2 按钮时，将内容面板替换成 Page2Fragment。

Fragment 动态添加、替换操作是相似的，关键代码见表 5-8。

表 5-8　Fragment 动态添加、替换操作的关键代码

Fragment frag = new XXXFragment;	//利用多态产生片段对象
FragmentManager manager =this.getFragmentManager();	//产生片段管理器对象
FragmentTransaction tran = manager.beginTransaction();	//产生事务对象
tran.add(R.id.XXX, f);	//添加操作：将片段 f 插入 R.id.XXX 父容器内
//tran.replace(R.id.XXX,f);	//若替换，则运行注释行
tran.commit();	//开启事务运行

5.5.4　数据通信

数据通信主要涉及 3 方面：从 Activity 向 Fragment 传送数据；从 Fragment 向 Activity 传送数据；从 Fragement 向 Fragment 传送数据。下面分别加以说明。

① 从 Activity 向 Fragment 传送数据：详细说明如表 5-9 所示。

表 5-9 从 Activity 向 Fragment 传送数据关键代码

	代 码	说 明

Activity 发送	Fragment fragment =new XXXFragment();
	Bundle bb = new Bundle();
	bb.putString("name","zhang");
	fragment.setArguments(bb);
		建立 Fragment 对象时，利用 Bundle 捆绑传送的数据，再利用 setArguments()函数设置

Fragment 接收	Bundle b =this.getArguments(); String name = b.getString("name");	利用 getArguments()函数获得捆绑对象，再用 getXXX()函数解析即可

② 从 Fragment 向 Activity 传送数据：详细说明见表 5-10。

表 5-10 从 Fragment 向 Activity 传送数据关键代码

	类别	代 码	说 明
方法 1	Fragment 发送	MainActivity main = (MainActivity) getActivity(); main.send("Hello world");	利用 getActivity()获得 Activity 对象，再强制转换成 MainActivity 对象，调用其中的 send()函数，将参数传过去即可

	Activity 接收	void send(String s){
		……
		}
			在 Activity 类中实现 send()函数即可

方法 2	共享接口定义	interface IReceive{ void receive(String s) }	此接口在 Activity 和 Fragment 中均用到
	Fragment 发送	IReceive main = (IReceive) getActivity(); main.send("Hello world");	发送数据前，利用 getActivity()获得 Activity 对象，再强制转换成 IReceive 接口对象，调用 receive()函数发送数据即可

	Activity 接收	class MainActivity …… implements IReceive{
		……
		public void receive(String s){ ……
		}
		}
			Activity 类实现 IReceive 接口，重写 receive()函数即可

实现从 Fragment 向 Activity 通信利用方法 1 或方法 2 均可。方法 2 应用了接口回调技术，开发、维护更方便一些。

③ 从 Fragment 向 Fragment 传送数据。

有了①、②的基础，其实可以有多种实现 Fragment 间数据传送的方法。若有两个 Fragment 类 A、B：若实现 A 向 B 传送数据，可以先建立 B 对象，再建立 A 对象，将 B 对象作为 A 的成员变量即可；若实现 A、B 间双向通信，则 A 中有 B 的成员变量，B 中有 A 的成员变量，利用 setter()函数设置即可。

当然，也可采取"Fragment--->Activity--->Fragment"方式，利用 Activity 作为中介，实现 Fragment 间的数据通信。这样，每个 Fragment 类中无须定义待通信的 Fragment 类成员变量。前提条件可以有多种，常用的一种是知道各 Fragment 的 id 号。例如，若 Fragment 片段 A、B 对应的 id 资源号为 R.id.a，R.id.b，若实现从 A 向 B 的通信，则在 A 中的关键代码示例如下所示。

```
FragmentManager manager = this.getActivity().getFragmentManager();
                                          //获得片段管理器对象
Fragment frag = manager.findFragmentById(R.id.b);
                                          //获得 R.id.b 对应的 Fragment 对象
B obj = (B)frag;                          //强制转换成 B 对象
obj.send("Hello");                        //发送数据
```

关键思路是：在 FragmentManager 片段管理器查找到 R.id.b 对应的片段对象，将其强制转换成 B 对象，向 B 传送数据也就轻而易举了。由于在片段类 A 中写的通信代码，不能直接获得 FragmentManager 对象，因此必须由 getActivity().getFragmentManager()才能获得。

5.5.5　生命周期

Fragment 依附于 Activity，它也是有生命周期的，按流程顺序来说，如下所示。

- 创建阶段：onAttach()--->onCreate()--->onCreateView()-->onActivityCreated()，这些函数均只运行一次。
- 运行阶段：onStart()--->onResume()--->onPause()--->onStop

Fragment 起初时正常运行，由于其他原因引起暂停，然后又重启，因此是循环结构。

- 销毁阶段：onDestroyView()--->onDestroy()--->onDetach()，这些函数均运行一次。

可以看出，Fragment 与 Activity 相比，仅多了几个生命周期函数，下面一一介绍。

- onAttach：Fragment 生命周期第 1 个关键函数，用于初始化数据。
- onCreateView：创建 Fragment 片段 UI，但没有添加到 Activity 中。

- onActivityCreated：在 Activity 的 OnCreate()结束后，会调用此方法。所以，到这里的时候，Activity 已经创建完成，之后才可以使用 Activity 的所有资源。
- onDestroyView：如果 Fragment 即将被结束或保存，那么撤销方向上的下一个回调将是 onDestoryView()，它会将在 onCreateView()创建的视图与这个 Fragment 分离。
- onDetach：Fragment 生命周期中的最后一个回调是 onDetach()。调用它以后，Fragment 就不再与 Activity 相绑定，它也不再拥有视图层次结构，它的所有资源都将被释放。

深刻理解生命周期各函数的作用很关键。例如：考虑 Fragment 片段 A 消息响应添加问题。假设 A 中有一按钮，其 id 号为 R.id.btn，如何为其添加 click 事件？可以在 onCreateView()、onActivityCreated()中添加，但它们的代码是不一样的。这是因为运行 onCreateView()函数时，按钮还没有添加到 Activity 中，当运行到 onActivityCreated()函数时，按钮已经添加到 Activity 中，具体代码见表 5-11。

表 5-11 为 Fragment 添加消息响应比较表

<table>
<tr><td><pre>publicView onCreateView(LayoutInflater inflater,
ViewGroup container,Bundle b) {
    View v = inflater.inflate(......);
    Button b = v.findViewById(R.id.btn);
    b.setOnClickListener(new
        View.OnClickListener() {
            public voidonClick(View v) {}
        });
        return v;
}</pre></td><td><pre>public voidonActivityCreated(Bundle b) {
    super.onActivityCreated(b);
    Button btn = (Button)
    getActivity().findViewById(R.id.btn);
    btn.setOnClickListener(new
        View.OnClickListener() {
            public voidonClick(View v) {   }
        });
}</pre></td></tr>
<tr><td>按钮在 v 中，不在 Activity 中，只能从 v 获得 Button 对象</td><td>按钮已在 Activity 中，所以可从 Activity 中获得 Button 对象</td></tr>
</table>

习题 5

一、选择题

1. Activity 生命周期中由暂停态变为活动态时调用(　　)函数。

   A. onCreate()　　B. onPause()

   C. onStart()　　D. onResume()

2. Activity 类间通信最常用到的对象类是(　　)。

A. Map　　B. Set　　C. Intent　　D. Vector

3. 隐式启动 Activity 需要在配置文件中配置的最主要的节点是(　　)。

A. <action>　　B. <category>　　C. <data>　　D. <active>

4. Fragment 生命周期中的第 1 个关键函数是(　　)。

A. onCreate　　B. onCreateView()

C. onStart()　　D. onAttach()

二、程序题

1. 建立两个 Activity 类 One、Two。One 中可输入用户名、密码,当单击"确定"按钮时启动 Two,并将用户名、密码显示在 Two 中的 TextView 组件中。

2. 主页面中有两个按钮(Page1,Page2)及待添加的 Fragment 区域节点,当单击 Page1 按钮时,在 Fragment 区域显示学生信息(学号、姓名、专业、学院)录入界面;当单击 Page2 按钮时,在 Fragment 区域显示学生成绩(语文、数学、英语)录入界面。

第 6 章

网 络 通 信

众所周知，手机有强大的网络功能，本章主要讲解基于 http 协议的手机网络编程：介绍 URL、HttpURLConnection 类的基本应用方法；编制最简单的网络通信程序“Hello world”；指出多线程在网络编程中的重要性，并对编码、解码进行详细的讨论。

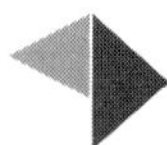

6.1 子线程刷新 UI 问题

网络通信一定会用到多线程技术，App 在子线程接收数据后，一般会刷新页面，读者可能很容易地写出下述代码。

```
public class MainActivity extends AppCompatActivity {
    protected void onCreate(Bundle savedInstanceState) {
        super.onCreate(savedInstanceState);
        setContentView(R.layout.main);
        new Thread(new Runnable() {
            public void run() {
                String s = "hello";
                TextView tv = (TextView)findViewById(R.id.myshow);
                tv.setText(s);
            }
        }).start();
    }
}
```

可以看出，在子线程中获得 TextView 组件对象，并对其进行了 setter 操作，这是最简单的在子线程操作 UI 的代码，但 Android 系统要求尽量不在子线程中操作 UI。上述代码可能在手机上都能正确运行，但随着线程中代码复杂性的增加，可能会出现运行异常，典型的异常信息是：Only the original thread that created a view hierarchy can touch its views，即只有创建视图层次结构的原始线程才能更新这个视图，也就是说，只有主线程才有权力更新 UI。因此，在子线程中刷新 UI 存在不确定性，这是 Android 系统与许多微机操作系统一个很大的不同。

那么，如何正确实现子线程的 UI 刷新呢？这要用到即将介绍的 Handler 类。

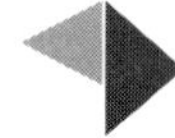

6.2　Handler 类

Handler 是 Android 的系统类，一般该类是与子线程 UI 刷新绑定的，归根结底也要转化为主线程更新 UI，原理如图 6-1 所示。

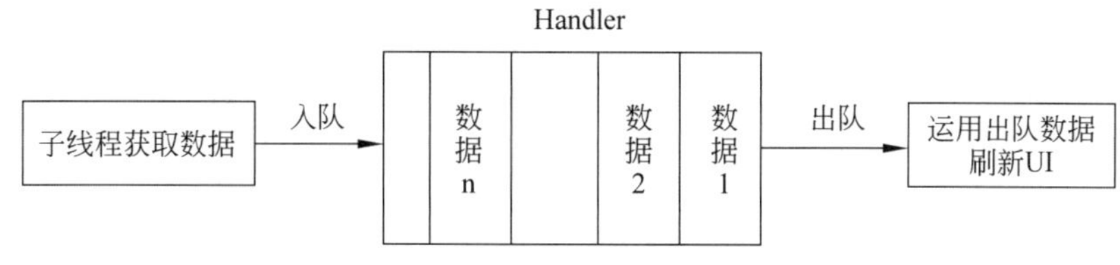

图 6-1　Handler 工作方式图

Handler 是一个数据队列，子线程将获得的数据不断地送到 Handler 队列中，当数据出队的时候，运用所需数据刷新 UI。所能操作的只是数据入队接口及刷新 UI 的响应函数，绝大多数功能都由 Android 系统完成了。

Handler 类常用的函数如下所示。

- public sendMessage(Message msg)：将消息 msg 放入队列中。
- public handleMessage(Message msg)：msg 出队后消息响应函数，需重写完成 UI 刷新。

Message 类用于封装各种数据，其关键的成员变量及方法如下所示。

- what 属性：int 类型，自定义属性值，主线程用来识别子线程发来的是什么消息。
- arg1、arg2 属性：int 类型，如果传递的消息类型为 int 型，可以将数字赋给 arg1、arg2。
- obj 属性：Object 类型，如果传递的消息是 String 或者其他，可以赋给 obj。
- Message obtain()：静态方法，常用此方法获得初始化的 Message 对象。

【例 6-1】　利用 Handler 实现子线程 UI 刷新功能。

程序运行后直接启动子线程，在子线程中设置字符串为 s＝“Hello”，在 TextView 控件中 x 显示字符串 s 的内容。具体步骤如下所示。

① 主布局文件 main.xml。

```
<?xml version="1.0" encoding="utf-8"?>
<LinearLayout xmlns:android="http://schemas.android.com/apk/res/android"
    android:layout_width="match_parent"
    android:layout_height="match_parent"
    android:orientation="vertical">
    <TextView
        android:id="@+id/myshow"
        android:layout_width="match_parent"
        android:layout_height="match_parent"/>
</LinearLayout>
```

② MainActivity.java。

```
public class MainActivity extends AppCompatActivity {
    Handler myHandler = new Handler(){
        public void handleMessage(Message msg) {
            String s = (String)msg.obj;
            TextView tv = (TextView)findViewById(R.id.myshow);
            tv.setText(s);
        }
    };
    protected void onCreate(Bundle savedInstanceState) {
        super.onCreate(savedInstanceState);
        setContentView(R.layout.main);
        new Thread(new Runnable() {
            public void run() {
                String s = "hello";
                Message msg = Message.obtain();
                msg.obj = s;
                myHandler.sendMessage(msg);
            }
        }).start();
    }
}
```

可以得出，利用 Handler 处理局部 UI 线程刷新的一般步骤如下所示。

- 在子线程中将数据封装成 Message 对象，传到 Handler 队列中。一般将数据封装到 Message.obj 属性中，注意产生初始 Message 对象一般用静态函数 obtain()，形如“Message msg = Message.obtain()”。

- 定义 Handler 的子类，重写 handleMessage(msg)函数，解析 msg 获取所需要的信息，刷新 UI 即可。

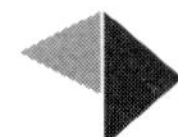

6.3　URL 类

URL 类是对统一资源定位符(Uniform Resource Locator)的抽象，使用 URL 创建对象的应用程序称为客户端程序，一个 URL 对象存放着一个具体资源的引用，表明客户端要访问这个 URL 中的资源。由于 URL 是对服务器资源的访问，因此应用 URL 的前提一般是服务器应用程序必须在运行中。与 socket、UDP 相比，URL 属于上层网络通信范畴。

一个 URL 对象通常包含 3 部分信息：协议、地址、资源，如 http、ftp 协议等。该类常用函数如下所示。

- public URL(String spec)throws MalformedURLException

根据 String 表示形式(包含：协议＋地址＋资源)创建 URL 对象，spec 形如“http://www.163.com”，如果该字符串指定的是未知协议，则创建对象时抛出 MalformedURLException 异常。

- public URL(String protocol, String host, int port, String file) throws MalformedURLException

根据协议 protocol、地址 host、端口号 port、资源名 file 创建 URL 对象，若 port 为-1，表明应用的是该协议的默认端口。

- public URL(String protocol, String host, String file)throws MalformedURLException

根据协议 protocol、地址 host、默认端口号、资源名 file 创建 URL 对象。

- public final InputStream openStream()throws IOException

打开到此 URL 的链接并返回一个用于从该链接读入的 InputStream。

【例 6-2】　返回 163 网站首页文本。

与例 6-1 相比，main.xml 无变化，MainActivity.java 大部分代码都是相似的，仅线程代码不同，涉及利用 URL 读取网站，另外，由于访问网络，因此要修改 Androidmanifest.xml 配置文件，增加网络访问允许信息，具体步骤如下所示。

① 在 Androidmanifest.xml 中增加网络允许设置。

```
<?xml version="1.0" encoding="utf-8"?>
<manifest xmlns:android="http://schemas.android.com/apk/res/android"
```

```
    package="com.example.dqjbd.we">
    <uses-permission android:name="android.permission.INTERNET" />
    <uses-permission android:name="android.permission.ACCESS_NETWORK_STATE " />
    <!—其他内容略-->
</manifest>
```

② 主布局文件 main.xml：同例 6-1。

③ MainActivity.java。

```
public class MainActivity extends AppCompatActivity {
    //其他代码与例 6-1 相同
    protected void onCreate(Bundle savedInstanceState) {
        super.onCreate(savedInstanceState);
        etContentView(R.layout.main);
        new Thread(new Runnable() {
            public void run() {
                try {
                    URL u = new URL("http://www.163.com");
                    BufferedReader in = new BufferedReader(new
InputStreamReader(u.openStream()));
                    String s = ""; String unit = "";
                    while ((unit = in.readLine()) != null) {
                        s += unit;
                    }
                    Message msg = Message.obtain();
                    msg.obj = s;  myHandler.sendMessage(msg);
                }
                catch(Exception e){
                    Message msg = Message.obtain();
                    msg.obj = e.getMessage(); myHandler.sendMessage(msg);
                }
            }
        }).start();
    }
}
```

网络通信调试代码时容易出现各种异常，因此本示例为代码增加了 try-catch 结构，正常获得的网络信息字符串或异常信息串，都利用了 Handler 技术，将信息显示在 TextView 组件中。

继续实验：去掉线程，将示例中 try-catch 的所有代码直接写在 onCreate()函数中的 setContentView(R.layout.main)代码下，也就是说，在主线程中读网络。运行程序后，发现不论等多长时间，屏幕都一片空白，这说明在 Android App 中读取网络必须是多线程。

进一步实验：保留原线程，将 try-catch 块中的 Handler 代码直接改为在子线程中局部刷新，见表 6-1。

表 6-1　Handler 代码改为局部刷新代码

位置	改前两行代码	改后代码

try	Message msg = Message.obtain();
	msg.obj = s;　myHandler.sendMessage(msg);
		TextView tv = (TextView)
		findViewById(R.id.myshow);
		tv.setText(s);

catch	Message msg = Message.obtain();
	msg.obj = e.getMessage();　myHandler.
	sendMessage(msg);
		TextView tv = (TextView)
		findViewById(R.id.myshow);
		tv.setText(e.getMessage());

程序运行后，会发现屏幕一片空白或者显示异常信息，这说明不应该在子线程中直接进行 UI 局部的刷新。

6.4　应用服务器

一个完整的手机网络程序一般包括 App+服务器部分，本书着重讲解 App 网络编程。因此，假设你已经会服务器编程，本书应用 Tomcat 服务器，服务器端程序都用简单的 JSP 实现。实现最简单的“Hello world”具体步骤如下所示。

① 服务器端部分。

- 利用 eclipse 建立动态 Web 工程 myapp，并建立 Test.jsp，代码如下所示。

```
<%
    out.print("Hello world");
%>
```

代码非常简单，向客户端直接输出“Hello world”字符串。

- 在 eclipse 中启动 tomcat 应用服务器即可。

② App 部分。

- 在 Androidmanifest.xml 中增加网络允许设置，同例 6-2。
- 主布局文件 main.xml：同例 6-1。
- MainActivity.java 如下所示。

```
public class MainActivity extends AppCompatActivity {
    //其他代码同例 6-2,略
    protected void onCreate(Bundle savedInstanceState) {
        super.onCreate(savedInstanceState);
        setContentView(R.layout.main);
            new Thread(new Runnable() {
                public void run() {
                    try {
                        URL u = new URL("http://192.168.1.101:8080/myapp/test
.jsp");
                        BufferedReader in = new BufferedReader(new
InputStreamReader(u.openStream()));
                        String s = ""; String unit = "";
                        while ((unit = in.readLine()) != null) {
                            s += unit;
                    }
                    Message msg = Message.obtain();
                    msg.obj = s;  myHandler.sendMessage(msg);
                }
                catch(Exception e){
                    Message msg = Message.obtain();
                    msg.obj = e.getMessage(); myHandler.sendMessage(msg);
                }
            }
        }).start();
    }
}
```

可以看出,代码中仅 URL u = new URL("http://192.168.1.101:8080/myapp/test.jsp")与之前的不同,原先指向 163 网站,现在指向 Tomcat 应用服务器的 test.jsp,其中"192.168.1.101"是应用服务器的 IP 地址,不同的情况下,IP 地址不同,URL 也就不同。

测试方法 1:将手机和应用服务器均与路由器相连,保证它们在同一个局域网,查出应用服务器的 IP 地址,写入上文的 URL 中即可。测试时要先运行 Tomcat 服务器,再运行手机 App 程序。

测试方法 2:若没有路由器,则可由两部手机(假设名为 A、B)及应用服务器完成测试。将手机 A 作为路由器,手机 B、应用服务器均通过 WiFi 与手机 A 相连。其余步骤同测试方法 1。

也许有读者问：一台手机＋应用服务器不也可以测试吗？只要应用服务器连接手机的 WiFi 不就解决了吗？但笔者做过实验，这样是不行的。

再看一个稍微复杂的例子，如下所示。

【例 6-3】 网络登录、注册程序。

App 包含的内容如下所示。

- 登录功能：界面部分包含账号、密码输入编辑框，登录按钮，注册按钮。输入账号、密码后，单击“登录”按钮，则请求服务器端进行校验，若通过，则转向主界面，否则给出提示信息。也可单击“注册”按钮，转向注册页面。
- 注册功能：界面部分包含账号、密码输入编辑框，注册按钮。输入账号、密码后，单击“注册”按钮，则请求服务器端进行校验，若通过，则转向登录界面，否则给出提示信息。

服务器完成相应的登录校验及注册功能。

综上，本示例涉及的代码文件如表 6-2 所示。

表 6-2　登录、注册功能各程序名称表

	名　称	说　明
App	Androidmanifest.xml	增加网络使能配置
	MainActivity.java main.xml	登录程序及界面文件
	RegistActivity.java regist.xml	注册程序及界面文件
	FramemActivity.java frame.xml	主界面及其程序
服务器	init.jsp	用户信息初始化
	regist.jsp	用户信息注册
	check.jsp	用户登录校验

App 端各代码的具体内容如下所示。

① Androidmanifest.xml，增加网络配置，同例 6-2。

② 登录程序及界面文件 MainActivity.java 及 main.xml。

```
//main.xml
<?xml version="1.0" encoding="utf-8"?>
<LinearLayout xmlns:android="http://schemas.android.com/apk/res/android"
android:layout_width="match_parent"
android:layout_height="match_parent"
android:orientation="vertical">
```

```
    <EditText
        android:id="@+id/user"
        android:layout_width="match_parent"
        android:layout_height="wrap_content" />
    <EditText
        android:id="@+id/pwd"
        android:layout_width="match_parent"
        android:layout_height="wrap_content" />
    <Button
        android:id="@+id/login"
        android:layout_width="match_parent"
        android:layout_height="wrap_content"
        android:text="登录"/>
    <Button
        android:id="@+id/regist"
        android:layout_width="match_parent"
        android:layout_height="wrap_content"
        android:text="注册"/>
    <TextView
        android:id="@+id/myerror"
        android:layout_width="match_parent"
        android:layout_height="wrap_content" />
</LinearLayout>
```

界面比较简单：第 1 个 EditText 控件用于输入用户名，第 2 个 EditText 控件用于输入密码，有登录、注册按钮，最后一个 TextView 控件用于显示登录失败信息。

```
//MainActivity.java
public class MainActivity extends AppCompatActivity {
    String user;
    String pwd;
    /*定义了用户名、密码两个成员变量，用于保存输入值*/
    Handler myHandler = new Handler(){
        public void handleMessage(Message msg) {
            String s = (String)msg.obj;
            TextView tv = (TextView)findViewById(R.id.myerror);
            tv.setText(s);
        }
    };
```

/＊利用 Handler 技术局部刷新页面,本示例用于将登录失败信息显示在界面上。信息保存在 msg.what 中,为字符串类型。＊/

```
    String response(String req){
        String s = ""; String unit="";
        try {
            URL u = new URL(req);
            BufferedReader in = new BufferedReader(new InputStreamReader
(u.openStream()));
            while ((unit = in.readLine()) != null) { s += unit; }
        }
        catch(Exception e){
            s = e.getMessage();
        }
        return s;
    }
```

/＊请求响应函数,req 是申请的服务器端 Http 页面,形如:http://IP:port/参数序列。该函数首先将 req 封装成 URL 对象,为了提高读取服务器响应结果的效率,将字节流封装成 BufferedReader 缓冲流,由于主要是读取文本字符串,因此采用循环行读 readLine()函数,直至读取完毕。＊/

```
    protected void onCreate(Bundle savedInstanceState) {
        super.onCreate(savedInstanceState);
        setContentView(R.layout.main);
        Button b = (Button)findViewById(R.id.login);
        b.setOnClickListener(new View.OnClickListener() {
            public void onClick(View v) {
                user = ((EditText)findViewById(R.id.user)).getText().toString();
                pwd = ((EditText)findViewById(R.id.pwd)).getText().toString();
                new Thread(new Runnable() {
                    public void run() {
                        String url = "http://192.168.1.103:8080/myapp/check.jsp?";
                        url += "user="+user+"&pwd="+pwd;
                        String result = response(url);
                        if(result.equals("YES")){
                            Intent in = new Intent(MainActivity.this,
FrameActivity.class);
                            startActivity(in);
                        }else{
                            Message msg = Message.obtain();
```

```
                        msg.obj = "登录失败"; myHandler.sendMessage(msg);
                    }
                }
            }).start();
        }
    });
    Button b2 = (Button)findViewById(R.id.regist);
    b2.setOnClickListener(new View.OnClickListener() {
        public void onClick(View v) {
            Intent in = new Intent(MainActivity.this,RegistActivity.class);
            startActivity(in);
        }
    });
  }
}
```

主程序主要是对登录、注册两个按钮增加事件响应。注册按钮对应的功能非常简单，直接启动注册界面。当单击“登录”按钮时：首先获得输入的账号 user 及密码值 pwd，形成带参数的 Http 请求字符串，请求的服务器页面是 check.jsp，且要把 user 及 pwd 动态添加到字符串 url 中；然后调用前文讲述过的 response(req)函数，获得服务器响应结果；若返回结果是 YES，则表明是合法用户，转向主应用界面；若返回结果是 NO，则表明是非法用户，将“登录失败”显示在界面中。

③ 注册程序及界面文件 RegistActivity.java 及 regist.xml。

```
//regist.xml
<?xml version="1.0" encoding="utf-8"?>
<LinearLayout xmlns:android="http://schemas.android.com/apk/res/android"
    android:layout_width="match_parent"
    android:layout_height="match_parent"
    android:orientation="vertical">
    <EditText
        android:id="@+id/user"
        android:layout_width="match_parent"
        android:layout_height="wrap_content" />
    <EditText
        android:id="@+id/pwd"
        android:layout_width="match_parent"
        android:layout_height="wrap_content" />
```

```
    <Button
        android:id="@+id/regist"
        android:layout_width="match_parent"
        android:layout_height="wrap_content"
        android:text="注册"/>
    <TextView
        android:id="@+id/myerror"
        android:layout_width="match_parent"
        android:layout_height="wrap_content" />
</LinearLayout>
```

第1个EditText控件用于输入注册用户名，第2个EditText控件用于输入注册密码，一个注册按钮，一个TextView控件用于显示注册失败信息。

```
//RegistActivity.java
public class RegistActivity extends AppCompatActivity {
    //其他所有代码同之前介绍的MainActivity.java一致，略
    protected void onCreate(Bundle savedInstanceState) {
        super.onCreate(savedInstanceState);
        setContentView(R.layout.regist);
        Button b = (Button)findViewById(R.id.regist);
        b.setOnClickListener(new View.OnClickListener() {
            public void onClick(View v) {
                user = ((EditText)findViewById(R.id.user)).getText().toString();
                pwd = ((EditText)findViewById(R.id.pwd)).getText().toString();
                new Thread(new Runnable() {
                    public void run() {
                        String url = "http://192.168.1.103:8080/myapp/regist.jsp?";
                        url += "user="+user+"&pwd="+pwd;
                        String result = response(url);
                        if(result.equals("YES"))
                            finish();
                        else{
                            Message msg = Message.obtain();
                            msg.obj = "注册失败"; myHandler.sendMessage(msg);
                        }
                    }
                }).start();
            }
```

```
        });
    }
}
```

主程序主要是对注册按钮增加事件响应。当按“注册”按钮时：首先获得注册的账号 user 和密码值 pwd，形成带参数的 Http 请求字符串，请求的服务器页面是 regist.jsp，且要把 user 及 pwd 动态添加到字符串 url 中；然后调用前文讲述过的 response(req)函数，获得服务器响应结果；若返回结果是 YES，则表明注册成功，返回登录界面；若返回结果是 NO，则表明注册失败，将“注册失败”串显示在界面中。

④ 功能主界面及程序 frame.xml 及 FrameActivity.java。

本示例中，主界面仅起到象征作用，即登录成功后能转到此界面，因此编制最简单的布局文件及源程序即可，在此略。

服务器端各代码的具体内容如下所示(利用 eclipse 建立动态 Web 工程 myapp)。

① init.jsp：用户信息初始化。

```
<%@ page import="java.util.*" %>
<%
    Map m = new HashMap();
    application.setAttribute("my", m);
%>
```

集合 Map 用于仿真用户信息表，键代表账号，值代表密码。该集合为所有用户共享，因此利用内置全局对象 application 进行设置，共享名称为 my。由于初始时没有用户注册，因此该集合元素为空。

② regist.jsp：注册页面。

```
<%@ page import="java.util.*" %>
<%
    String user = request.getParameter("user");
    String pwd = request.getParameter("pwd");
    Map m = (HashMap)application.getAttribute("my");
    if(m.get(user)==null){
        m.put(user, pwd);
        out.print("YES");
    }
    else
        out.print("NO");
%>
```

该页面首先解码从手机端传过来的 Http 请求，获得注册的用户名 user 及密码 pwd；然后从 application 域中获得用户信息 Map 集合 m，查询 m 中键 user 对应的值。若该值为空对象，则表明是一个新的用户 user，将 user 及 pwd 保存到 m 中，向手机客户端发送注册成功标志 YES；若该值非空，则表明 user 用户已经存在，向手机客户端发送注册失败标志 NO。

③ check.jsp：登录校验页面。

```
<%@ page import="java.util.* " %>
<%
    String user = request.getParameter("user");
    String pwd = request.getParameter("pwd");
    String value = (String)m.get(user);
    Map m = (HashMap)application.getAttribute("my");
    if(value==null || !value.equals(pwd))
        out.print("NO");
    else
        out.print("YES");
%>
```

该页面首先解码从手机端传过来的 Http 请求，获得登录的用户名 user 及密码 pwd；然后从 application 域中获得用户信息 Map 集合 m，查询 m 中键 user 对应的值 value。如果 value 为空对象，则表明 m 中没有该用户。如果 value 不为空，但是 value 与 pwd 不相同，则说明 m 中 user 值能匹配，但密码值不能匹配。这两种情况均表明登录失败，向客户端输出标识 NO。除以上情况外，user 及 pwd 均在集合 m 中，是合法用户，向客户端输出登录成功标志 YES。

该程序涉及手机 App 及服务器程序，因此必须使它们处于同一网段中。构建测试网络的方法详见例 6-3 之前的内容方法。测试时一定要先运行 Tomcat 服务器，之后运行 init.jsp，保证用户信息初始化(这一步非常关键，影响手机端的测试运行)，最后运行手机 App 即可。

当然，本示例还有许多待完善的地方，如用户名、密码不能为空，左右去空格；response()函数能否抽象成一个类，为多个 Activity 类所共享等，希望读者去完善。

其实，利用 URL 类已经能实现较复杂的网络通信，但仍有很多不足，如读、写 Http 请求头、响应头不方便，传输的 post 方式不易设置等，如何改进呢？这就是下面要讲的 HttpURLConnection 类。

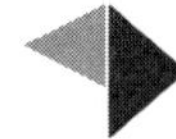

6.5 HttpURLConnection

6.5.1 简介

HttpURLConnection 是一种多用途、轻量级的 HTTP 客户端对象，使用它进行 HTTP 操作可以适用于大多数的网络应用程序。HttpURLConnection 无法直接实例化。一般通过调用 URL 对象的 openConnection()的返回值获得，例如 HttpURLConnection hurl=new URL(……).openConnection()。

HttpURLConnection 常用的函数包括：设置连接参数函数、设置请求头或响应头函数、发送请求函数、获取响应函数。下面一一加以介绍。

1. 设置连接参数函数

- void setAllowUserInteraction(boolean mark)，如果为 true，则在允许用户交互(例如弹出一个验证对话框)的上下文中对此 URL 进行检查。
- void setDoInput(boolean mark)，用于设置是否向连接中写入数据。如果参数值为 true，则写入数据，否则不写入数据。
- void setDoOutput(boolean mark)，用于设置是否从连接中写入数据。如果参数值为 true，则读取数据，否则不读取数据。
- void setUseCaches(boolean mark)，用于设置是否缓存数据，如果参数值为 true，则缓存数据，否则不缓存数据。

2. 设置请求头或响应头函数

- void setRequestProperty(String key, String value);
- void addRequestProperty(Stringkey,Stringvalue);

setRequestProperty()和 addRequestProperty()的区别是，setRequestProperty()会覆盖已经存在的 key 的所有 value，有清零重新赋值的作用，而 addRequestProperty 则是在原来 key 的基础上继续添加其他 value。

3. 发送请求函数

建立实际连接之后，就是发送请求，把请求参数传到服务器，这就需要使用 outputStream 把请求参数传给服务器。

- OutputStream getOutputStream()，获取输出流。

4. 获取响应函数

请求发送成功之后,即可获取响应的状态码。如果成功则可以读取响应中的数据。获取这些数据的方法主要包含以下函数。

- Object getContent(),获取内容对象。
- Object getContentLength(),获取内容总长度。
- String getHeaderField(String key),获取头信息。
- InputStream getInputStream(),获取输入流。

6.5.2　应用举例

【例 6-4】 GET、POST 方式的设置。我们知道,GET 传送方式的特点是参数附在 URL 的后面将请求传向服务器,这种方式有一定的局限性。对 Android 系统来说,若参数串长度超过 1024B,则发送失败。POST 方式的特点是:URL 仅包含服务器申请的页面,参数由数据区发送,因此可以发送较大数量的数据。

先利用 GET 传送方式编制如下功能:在 App 上输入一个整型数 m,单击 OK 按钮后,将 m 传到服务器端,产生 m 个字符'a',再传到 App 端显示这个字符串。根据此功能,涉及的文件有:配置文件 Androidmanifest.xml;主应用程序 MainActivity.java;主界面 main.xml;服务器端动态 Web 工程,名称为 myapp,页面为 calc.jsp,具体描述如下所示。

① Androidmanifest.xml,增加网络配置,同例 6-2。

② App 端主程序及界面文件 MainActivity.java 及 main.xml。

```
//main.xml
<?xml version="1.0" encoding="utf-8"?>
<LinearLayout xmlns:android="http://schemas.android.com/apk/res/android"
android:layout_width="match_parent"
android:layout_height="match_parent"
android:orientation="vertical">
    <EditText
        android:id="@+id/mynum"
        android:layout_width="match_parent"
        android:layout_height="wrap_content" />
    <Button
        android:id="@+id/myok"
        android:layout_width="match_parent"
```

```
        android:layout_height="wrap_content"
        android:text="OK"/>
    <TextView
        android:id="@+id/myinfo"
        android:layout_width="match_parent"
        android:layout_height="wrap_content" />
</LinearLayout>
```

EditText控件用于输入整型数m，表明要在服务器端产生m长度的字符串。OK按钮负责向服务器端发送Http请求，TextView控件用于显示从服务器端返回的字符串。

```
//MainActivity.java
public class MainActivity extends AppCompatActivity {
    Handler myHandler = new Handler(){
        public void handleMessage(Message msg) {
            String s = (String)msg.obj;
            TextView tv = (TextView)findViewById(R.id.myinfo);
            tv.setText(s);
        }
    };
    String response(String req,String para,String value){
        String s = ""; String unit="";
        try {
            req += "?"+para+"="+value;
            URL u = new URL(req);
            HttpURLConnection hurl = (HttpURLConnection) u.openConnection();
            BufferedReader in = new BufferedReader(new InputStreamReader(hurl.
getInputStream()));
            while ((unit = in.readLine()) != null) {
                s += unit;
            }
        }catch(Exception e){s=e.getMessage();}
        return s;
    }
    /* req是申请的服务器页面地址，形如http://IP:port/XXX.jsp。para是参数名，
value是参数值。按表意形式来说，首先将"para=value"附在req的后面，然后产生URL类对
象u，调用openConnection()获得URLConnection对象，将其强制转换成HttpURLConnection对
象hurl，最后调用getInputStream()获得响应输入流InputStream对象。为提高读入效率，再将
其封装成BufferedReader对象，按行读，直至结束。*/
```

```
    protected void onCreate(Bundle savedInstanceState) {
        super.onCreate(savedInstanceState);
        setContentView(R.layout.main);
        Button b = (Button)findViewById(R.id.myok);
        b.setOnClickListener(new View.OnClickListener() {
            public void onClick(View v) {
                new Thread(new Runnable() {
                    public void run() {
                        String num = ((EditText)findViewById(R.id.mynum))
.getText().toString();
                        String url = "http://192.168.1.114:8080/myapp/calc.jsp";
                        String result = response(url,"mynum",num);
                        Message msg = Message.obtain();
                        msg.obj = result;
                        myHandler.sendMessage(msg);
                    }
                }).start();
            }
        });
    }
}
```

当单击 OK 按钮时，直接启动多线程，从编辑框获得输入的整型数 num，url 为服务器端的 calc.jsp 页面，参数名为 mynum，参数值为 num，调用 response()请求响应函数，将结果存到 result 中，再利用 Handler 技术将 result 显示在界面上。

③ 服务器端(前述动态 Web 工程 myapp 即可)calc.jsp。

```
<%
    String value = request.getParameter("mynum");
    int n = Integer.parseInt(value);
    String s = "";
    for(int i=0; i<n; i++)
        s += "a";
    out.print(s);
%>
```

服务器端首先获得 mynum 参数对应的整型值 n，然后利用循环产生 n 个字符'a'的字符串 s，最后将 s 返回给手机客户端。

讨论 1：很明显，该示例中，response()是一个重要的与服务器通信的方法，若改为

POST 方式传送数据并接收响应，只修改该方法即可，如下所示。

```
String response(String req,String para,String value){
    String s = ""; String strPara=""; String unit="";
    try {
        URL u = new URL(req);
        strPara = para + "=" +value;
        HttpURLConnection hurl = (HttpURLConnection) u.openConnection();
        hurl.setRequestMethod("POST");
        OutputStream out = hurl.getOutputStream();
        out.write(strPara.getBytes());
        out.flush();
        BufferedReader in = new BufferedReader(new InputStreamReader(hurl
.getInputStream()));
        while ((unit = in.readLine()) != null) {
            s += unit;
        }
    }catch(Exception e){s=e.getMessage();}
    return s;
}
```

与 GET 方式相比，采用 POST 方式通信时，需注意以下几点：URL 申请串 req 仅包含地址，不包含参数；必须利用 setRequestMethod(type)设置传送方式，type 为“POST”或“GET”；必须获取手机端网络通信的输出流 OutputStream 对象，将参数作为数据发送到服务器端，这一点是与 GET 传送方式最大的不同；而后续的读取响应数据，两种传送方式是一致的。

讨论 2：上述读取网络响应数据采取的都是利用 BufferedReader 对象行读的办法，有没有其他办法呢？当然有，下述代码是方法之一。

```
String response(String req,String para,String value){
    String s = ""; String strPara=""; String unit="";
    try {
        URL u = new URL(req);
        strPara = para + "=" +value;
        HttpURLConnection hurl = (HttpURLConnection) u.openConnection();
        hurl.setRequestMethod("POST");
        OutputStream out = hurl.getOutputStream();
        out.write(strPara.getBytes());
```

```
        out.flush();
        int len = hurl.getContentLength();
        byte buf[] = new byte[len];
        int pos = 0; int size=0;
        InputStream in = hurl.getInputStream();
        while((size=in.read(buf,pos,len-pos))>0)
            pos += size;
        s = new String(buf, 0, pos);
    }catch(Exception e){s=e.getMessage();}
    return s;
}
```

关键思想是：利用 getContentLength()获得服务器返回的数据字节总长度 len，根据 len 值创建字节缓冲区数组 buf，将服务器返回的字节内容读至 buf 中，读取完毕后，将 buf 字节缓冲区转换成字符串即可。其中以下两点需要仔细分析。

- 应用 getContentLength()函数的位置。POST 方式下，一定是在参数作为数据发送之后才能应用，因此本示例在代码行“out.flush()”之后才能应用，加在之前的位置都会出现异常。GET 方式下，由于参数数据与 URL 串是绑定在一起的，所以当建立 URL 对象并强制将其转换成 HttpURLConnection 对象之后，即可应用此函数。
- 有读者认为既然字节缓冲区 buf 大小等于服务器响应的字节总长度，那么用 in.read(buf)读一次不就可以了吗？实验证明，当数据量小时是正确的，当数据量大时，in.read(buf)只是占用了 buf 的部分缓冲区，读者可以通过该函数的返回值验证。因此，必须用 while 循环读取多次，才能将 buf 缓冲区填满。

讨论 3：本示例仅是对一个参数的 URL 编制了 response()函数，能否更进一步，对多个参数进行设置，使该函数更通用？代码之一如下所示。

```
String response(String req,String para[],String value[]){
    String s = ""; String strPara=""; String unit="";
    try {
        URL u = new URL(req);
        strPara += para[0]+"="+value[0];
        for(int i=1; i<para.length; i++)
            strPara += "&" +para[i]+ "=" +value[i];
        HttpURLConnection hurl = (HttpURLConnection) u.openConnection();
        hurl.setRequestMethod("POST");
        OutputStream out = hurl.getOutputStream();
```

```
            out.write(strPara.getBytes()); out.flush();
            int len = hurl.getContentLength();
            byte buf[] = new byte[len];
            int pos = 0; int size=0;
            InputStream in = hurl.getInputStream();
            while((size=in.read(buf,pos,len-pos))>0) pos += size;
            s = new String(buf, 0, pos);
        }catch(Exception e){s=e.getMessage();}
        return s;
    }
```

其实非常简单，将参数名数组 para[]与值数组 value[]作为形参传入，动态形成参数最终串 strPara 即可，形如“para1=value1¶2=value2¶3=value3”。

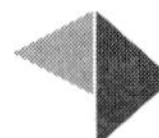 6.6 XML 解析

XML 是一种标准的网络传输通用格式，从服务器端返回 App 的信息一般定义为 XML 字符串，形如表 6-3。

表 6-3 XML 一般结构示例

内 容	说 明
<mydata> <stud> <name id="1000">zhang</name> <grade>80</grade> </stud> </mydata>	一个 XML 文件由多个层次的标签组成，每层标签起始于<标签名称>，结束于</标签名称>。越外围的标签，管理的数据越多

在 Android 中，解析 XML 串的工具有多种，本文介绍较常用的系统类 XmlPullParser 类，它是按事件流解析 XML 字符串的，其常用事件及相关函数如下所示。

- XmlPullParser.START_DOCUMENT，常量，解析开始事件标志。
- XmlPullParser.END_DOCUMENT，常量，解析结束事件标志。
- XmlPullParser.START_TAG，常量，解析标签事件开始标志，当遇到表 6-3 所示示例中的<mydata>、<stud>、<name>等时，会触发该事件。
- XmlPullParser.END_TAG，常量，解析标签事件结束标志，当遇到表 6-3 所示示例中的</mydata>、</stud>、</name>等时，会触发该事件。

- XmlPullParser newPullParser()，静态方法，返回 XmlPullParser 对象。
- void setInput(InputStream in, String encode)，对字节流 in 进行 xml 解析，其编码方式由 encode 指定。
- int getEventType()，获得当前的事件值。
- String getName()，获得标签的名称。
- String nextText()，获得标签的值。如表 6-3 所示示例中"<name id="1000">zhang</name>"中的 zhang 由该函数获得。
- String getAttributeValue(String namespace,String name)，获得标签属性值，并获得标签的值。如表 6-3 所示示例中"<name id="1000">zhang</name>"中的 1000 是由该函数获得的。

【例 6-5】 手机 App 端申请服务器端学生成绩信息页面，将返回的学生信息以列表的形式显示在界面中。

本示例先看服务器端(前述的动态 Web 工程 myapp)的 mydata.jsp 代码，如下所示。

```
<%
    String s = "<mydatas>"+
               "<stud><name>zhang</name><grade>80</grade></stud>"+
               "<stud><name>wang</name><grade>100</grade></stud>"+
        "</mydatas>";
    out.print(s);
%>
```

可以看出，以上代码将返回与表 6-3 相同结构的 xml 结构数据，具体包含两位学生的成绩信息。

手机 App 端涉及 Androidmanifest.xml 网络设置、主程序 MainActivity.java 及主界面 main.xml，如下所示。

① Androidmanifest.xml，增加网络配置，同例 6-2。

② 主界面 main.xml。

```
<?xml version="1.0" encoding="utf-8"?>
<LinearLayout xmlns:android="http://schemas.android.com/apk/res/android"
android:layout_width="match_parent"
android:layout_height="match_parent"
android:orientation="vertical">
    <ListView
        android:id="@+id/mylist"
        android:layout_width="match_parent"
```

```
        android:layout_height="wrap_content"></ListView>
</LinearLayout>
```

以上代码定义了 id 为 mylist 的 ListView 控件,用以显示从服务器端返回的学生成绩信息。

③ 主程序 MainActivity.java。

```
public class MainActivity extends AppCompatActivity {
    ArrayList<String> ary = new ArrayList();
    /* 定义了动态数组成员变量 ary,元素是 String 字符串,用于保存学生成绩信息,当然,将
学生相关信息都合并到一个字符串中,便于显示 ListView 控件 */
    Handler myHandler = new Handler(){
        public void handleMessage(Message msg) {
            String tag = "";
            String s = (String)msg.obj;
            ByteArrayInputStream in = new ByteArrayInputStream(s.getBytes());
            XmlPullParser parser = Xml.newPullParser();
            try{
                parser.setInput(in, "utf-8");
                int type = parser.getEventType();
                while(type != XmlPullParser.END_DOCUMENT){
                    switch(type){
                        case XmlPullParser.START_DOCUMENT:
                            break;
                        case XmlPullParser.START_TAG:
                            tag = parser.getName();
                            if(tag.equals("stud")) unit = "";
                            if(tag.equals("name")) unit += parser.nextText();
                            if(tag.equals("grade"))unit += "-" +parser.nextText();
                            break;
                        case XmlPullParser.END_TAG:
                            tag = parser.getName();
                            if(tag.equals("stud"))
                                ary.add(unit);
                            break;
                    }
                    type = parser.next();
                }
            }catch(Exception e){}
            ListView lv = (ListView)findViewById(R.id.mylist);
```

```
            ArrayAdapter<String> ada = new ArrayAdapter<String>(
                    MainActivity.this,android.R.layout.simple_list_item_1, ary);
            lv.setAdapter(ada);
        }
    };
```

/* handleMessage()函数中的字符串 s 即从服务器端返回的 XML 字符串,将该字符串的字节缓冲区进一步封装成 ByteArrayInput 对象 in。利用静态方法产生 XmlPullParser 对象 parser,利用其中的 setInput(in,"utf-8")")函数表明要对流对象 in 表示的字符串 s 进行 xml 解析,然后就是循环获得事件类型 type。

当 type 为 START_TAG 起始标签事件时,获得该标签的名称 tag,若 tag 为<stud>标签,表明是一个新学生信息的开始,由于本示例中将学生的姓名+成绩合并到 unit 字符串中,因此将 unit 初始化为空串;若 tag 为<name>标签,则通过 nextText()获得姓名值,添加到 unit 中;若 tag 为<grade>标签,则通过 nextText()获得成绩值,添加到 unit 中。

当 type 为 END_TAG 结束标签事件时,获得该标签的名称 tag。若 tag 为</stud>标签,则表明某一学生的姓名、成绩已保存到 unit 中。形如"zhang-80",则将该字符串保存到 ArrayList 成员变量 ary 中,为 ListView 控件准备好适配数据 */

```
    String response(String req){
        String s = "";
        try {
            URL u = new URL(req);
            HttpURLConnection hurl = (HttpURLConnection) u.openConnection();
            hurl.setRequestMethod("POST");
            OutputStream out = hurl.getOutputStream();
            int len = hurl.getContentLength();
            byte buf[] = new byte[len];
            int pos = 0; int size=0;
            InputStream in = hurl.getInputStream();
            while((size=in.read(buf,pos,len-pos))>0)
                pos += size;
            s = new String(buf, 0, pos);
        }catch(Exception e){s=e.getMessage();}
        return s;
    }
```

/* 请求响应函数,请求 req 没有参数,POST 传送方式,将服务器响应结果保存在字符串 s 中并返回 */

```
    protected void onCreate(Bundle savedInstanceState) {
        super.onCreate(savedInstanceState);
        setContentView(R.layout.main);
        new Thread(new Runnable() {
                public void run() {
```

```
                    String url = "http://192.168.1.126:8080/myapp/mydata.jsp";
                    String result = response(url);
                    Message msg = Message.obtain();
                    msg.obj = result;
                    myHandler.sendMessage(msg);
                }
            }).start();
        }
    }
```

主程序启动时直接运行多线程程序，申请的服务器页面是 mydata.jsp。

本示例中，Handler 匿名类中的 handleMessage()函数包含利用 XmlPullParser 解析 XML 字符串的全过程。可以看出，XmlPullParser 类中有一个重要函数 setInput(InputStream in, String encode)，第一个形参 in 是 InputStream 类型，这提供一个启示：响应服务器页面时(本示例的 response()函数中)，能否在接收数据的同时完成 XML 解析？无须先存成字符串，再完成 XML 解析？当然可以，程序改动不大，读者可试着完成。

6.7 JSON 解析

1. 简介

JSON，即 JavaScript Object Natation，一种轻量级的数据交换格式，与 XML 一样，它是一种被广泛采用的客户端和服务端交互的解决方案，具有良好的可读和便于快速编写的特性。与 XML 相比，描述相同的内容 JSON 的数据量更小，响应速度更快。

JSON 字符串的语法规则主要包含 3 部分：大括号、中括号、数据。大括号表示对象，中括号表示数组，大括号、中括号可以互相嵌套。数据由名称和值组成，名称和值间用冒号相隔，数据和数据之间用逗号相隔。下面列举几个典型的 JSON 串，将其与对应的 XML 格式比较，以加深对 JSON 格式的理解，见表 6-4。

表 6-4 几种 JSON 字符串

序号	JSON	等价 XML	说　明
1	{"stud": [1000,1001]}	<myapp> <stud>1000</stud> <stud>1001</stud> </myapp>	JSON 字符串可以省去最外层的 xml 标签<myapp>。 stud 是数组

续表

序号	JSON	等价 XML	说　　明

2	{
	"school"："lnnu"，
	"stud"：[1000,1001]
	}
		<myapp>
		<school>lnnu</school>
		<stud>1000</stud>
		<stud>1001</stud>
		</myapp>
			school 是单值，
			stud 是数组

3	{
	“school”：”lnnu”，
	“stud”：[
	{“id”：1000，“grade”：[60,70]}，
	{“id”：1001，“grade”：[80,90]}，
	]
	}
		<myapp>
		<school>lnnu</stud>
		<stud id=1000>
		<grade>60</grade>
		<grade>70</grade>
		</stud>
		<stud id=1001>
		<grade>80</grade>
		<grade>90</grade>
		</stud>
		</myapp>
			school 是单值，
			stud 是数组，每个数组元素包含两部分：第 1 部分是单值 id；第 2 部分是数组 grade

2. 常用的解析函数

Android 中，解析 JSON 字符串常用的类是 JSONObject 与 JSONArray。

JSONObject 类常用的函数如下所示。

- JSONObject(String s)，构造方法，当 JSON 字符串为大括号开始时，要利用该构造方法创建 JSONObject 对象。
- String getString(String name)，获得 name 对应的字符串值。
- JSONArray getJSONArray(String name)，获得 name 对应的 JSONArray 对象，当大括号内包含中括号(对象内包含数组类型的数据)时，常用此函数。

JSONArray 类常用的函数如下所示。

- JSONArray(String s)，构造方法，当 JSON 字符串为中括号开始时，要利用该构造方法创建 JSONArray 对象。
- JSONObject getJSONObject(int pos)；获得数组中 pos(0 基)位置的 JSONObject 对象。
- int length()，获得数据元素的长度。

【例 6-6】 写出解析表 6-4 中前两段 JSON 字符串的关键代码。

解析第 1 段 JSON 串的关键代码如下所示。

```
try {
    String s = "{'stud':[1000,1001]}";
    JSONObject obj = new JSONObject(s);
    JSONArray ary = obj.getJSONArray("stud");
    String unit="";
    for(int i=0; i<ary.length(); i++)
        unit = ary.getString(i);
}
catch(Exception e){}
```

Android 系统要求：解析 JSON 字符串一般需加异常处理代码。由于 s 是大括号开始，因此利用 JSONObject(s)获得 JSON 对象 obj；由于 stud 是数组，因此利用 getJSONArray()函数获得 JSONArray 对象 ary，遍历该数组，利用 getString()函数可获得具体值，最后将其保存在 unit 中。

解析第 2 段 JSON 串的关键代码如下所示。

```
try {
    String s = "{ 'school':'lnnu','stud':[1000,1001]}";
    JSONObject obj = new JSONObject(s);
    String school = obj.getString("school");
    JSONArray ary = obj.getJSONArray("stud");
    String unit="";
    for(int i=0; i<ary.length(); i++)
        unit = ary.getString(i);
    tv.setText(unit);
}
catch(Exception e){ }
```

由于 s 是大括号开始，因此利用 JSONObject(s)获得 JSON 对象 obj；由于 obj 中包含一个单值属性 school 及数组属性 stud，因此利用 getString()函数可获得学校名称并将其保存在 school 中，利用 getJSONArray()函数获得 JSONArray 对象 ary，遍历该数组，利用 getString()函数可获得具体值，并将其保存在 unit 中。

基于上述内容，可以修改例 6-5 中的 handleMessage()函数，使之解析 JSON 字符串，其余不必改动，完成一个真实的网络 JSON 解析实例。当然，服务器端 mydata.jsp 向手机端输出部分也要改为相应的 JSON 字符串。读者可尝试完成。

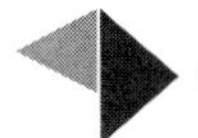

6.8 URL 编码

1. 为何需要编码

URL 包含了服务器端的请求地址及参数序列，是一个非常重要的值。URL 形如 http://.../a.jsp? para1=value1¶2=value2，其中“&”是一个很特殊的字符，它将相邻的两个参数/值割裂开来。但是，若 URL 串参数值中包含“&”，有时就会引起歧义。例如，想让 para1 的值等于 value1&，上述 URL 参数部分写为 para1=value1&¶2=value2，其中两个“&”连在了一起；再如，想让 para1 的值等于 para3=value3，上述 URL 参数部分写为 para1=para3=value3&¶2=value2，这不符合 URL 参数序列必须遵守“参数=值”的规范。

如何避免上述问题带来的歧义？解决之法就是对 URL 进行编码，URL 要求必须对保留字符、不安全字符进行编码。

URL 规定的保留字符包括“!、'、()、;、:、@、&、=、+、$、,、/、?、#、[]”。这些字符在某些情况下是有含义的。例如，冒号:用于分隔协议和主机组件，斜杠/用于分隔主机和路径，问号?用于分隔路径和查询参数，等等。

不安全字符包括空格。URL 在传输过程中，或者在用户排版过程中，或者在文本处理程序处理 URL 的过程中，都有可能引入无关紧要的空格，或者将那些有意义的空格去掉；引号＜＞和尖括号通常用于在普通文本中分隔 URL；#通常用于表示书签或者锚点；%本身用作对不安全的字符进行编码，是使用的特殊字符，因此本身需要编码。

那么，如何进行编码？利用下述介绍的系统类即可。

2. URLEncoder 和 URLDecoder

URLEncoder 是编码类，其中有一个静态方法 String encode(String src, String encode)，src 是待编码的字符串，encode 是编码特征字符串，返回值即编码后的字符串。

URLDecoder 是解码类，其中有一个静态方法 String decode(String src, String encode)，src 是待解码的字符串，encode 是编码特征字符串，返回值即解码后的字符串。

有编码就有解码，可以夸张地理解：编码相当于加密，解码相当于解密，两者缺一不可。字符串经 URL 编码后，源串中的保留字符及不安全字符被一定规则的安全字符替代了，其他字符是不变的。

【例 6-7】 分析保留字符“&=”的 URL 编码及解码的关键代码。

```
String s = "&=";
```

```
String t = URLEncoder.encode(s, "utf-8"); //t=%26%3d
```

由以上代码可知，编码后 t 为%26%3d，& 编码为%26，= 编码为%3d。也就是说，编码后，原先字符串中含有的保留字符或不安全字符都由一定规则的安全字符替代了，但最终还是要得到原始串。由 t 如何得到？利用 URLDecoder 即可，代码如下所示。

```
String t = "%26%3d";
String u= URLDecoder.decode(t, "utf-8");
```

在实际应用中一般包含两种情况：URL 发送，接收服务器响应数据，发送时用到编码，接收时用到解码。

URL 发送情况。主要是对值进行编码，若申请地址是变量 req，参数名是常量 para1，值是变量 value1，则：

- 若是 GET 方式传送，则传送最终的串是 String url = req + "? para1 = " + URLEncoder.encode(value1,"utf-8")。
- 若是 POST 方式传送，则不必对 req 进行编码，只对发送的参数进行编码即可。最终的参数串是 String para="para1=" + URLEncoder.encode(value1,"utf-8")。

接收服务器响应数据情况，也是对值进行解码，例如对形如下述的 JSON 响应字符串进行解码(s={'id': XXX,'name': XXX})，其中 XXX 表示值是动态变化的，代码如下所示。

```
JSONObject obj = new JSONObject(s);
String id = URLDecoder.decode(obj.getString("id"));         //获得解码后的 id 值
String name = URLDecoder.decode(obj.getString("name"));     //获得解码后的 name 值
```

3. Base64 编码

Base64 也是一种常见的 URL 编码和解码类，其作用与 URLEncoder、URLDecoder 类是相似的，只是编码、解码原理不同，其常用函数如下所示。

- String encodeToString(byte buf[], int flag)，静态编码方法，buf 是字符串某编码下的字节缓冲区，flag 是标识，一般取 Bas464.DEFAULT 即可，该函数的功能是返回已编码的字符串。
- byte[] decode(byte buf[],int flags)，静态解码方法，buf 是待解码的字符串字节缓冲区，flags 是标识，一般取 Bas464.DEFAULT 即可，该函数的功能是返回已解码的字符串字节缓冲区。

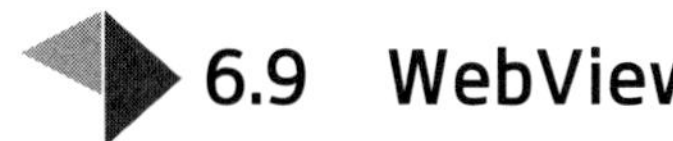

6.9　WebView

6.9.1　简介

WebView 是 Android 中一个非常重要的组件，其作用是展示一个 Web 页面，它使用的内核是 WebKit 引擎。Android 4.4 版本之后，直接使用 Chrome 作为内置网页浏览器。使用 WebView 之前，要在 Androidmanifest.xml 配置文件中声明网络访问权限，如下所示。

```
<uses-permission android:name="android.permission.INTERNET" />
```

常用的与 WebView 相关的类有 WebView、WebSettings、WebChromeClient，下面一一介绍。

WebView 类主要负责动态加载网页页面及与页面通信，常用的函数如下所示。

- void loadUrl(String url)，url 是具体申请的页面，会加载整个页面，类似于浏览器打开一样，渲染整个页面，包括排版布局。
- void loadData(String data, String mimeType, String encode)，data 为加载的字符串数据，mimeType 用于指定要显示内容的 MIME 类型，如果为 null，则默认为 text/html。
- void loadDataWithBaseURL(String baseURL, String data, String mimeType, String encode, String historyURL)，baseURL 用于指定当前页使用的基本 URL，如果为 null，则默认使用 about:blank，也就是空白页；data 为加载的字符串数据；mimeType 用于指定要显示内容的 MIME 类型，如果为 null，则默认为 text/html；encode 为字符编码类型；historyURL 用于指定当前页的历史 URL，也就是进入该页前显示页的 URL，如果为 null，则使用 about:blank。
- WebSettings getSettings()，获得网页属性设置对象。

WebSettings 类主要负责网页的应用属性设置，其常用的函数如下所示。

- void setDefaultTextEncodingName(String encode)，设置网页编码类型。
- void setJavascriptEnabled(Boolean mark)，mark 为 true，使能 JavaScript 代码；mark 为 false，禁止 JavaScript 代码。
- void addJavascriptInterface(Object obj, String name)，设置 JavaScript 代码可以调用 obj 类中的函数，name 是 obj 在 JavaScript 中的注册名称。

WebChromeClient 有很多用途，本书主要讲解它是如何解决 JavaScript 中 alert()、

confirm()、prompt()3 个弹出框正常运行问题的，这 3 个弹出框对应 WebChromeClient 类的 3 个重载函数：onJsAlert()、onJsConfirm()、onJsPrompt()。

6.9.2 应用举例

通过以下实例，主要从四方面深入理解 WebView：动态加载页面；Activity 调用 JavaScript 函数；JavaScript 调用 Activity 函数；使能 alert()、confirm()、prompt()弹出框。

1. 动态加载页面

【例 6-8】 利用 WebView 动态加载 my.html 页面。

正常来说，my.html 应放在服务器端，通过网络加载 my.html。但为了讲解方便，本示例将其放在 Android 工程的 assets 目录下。my.html 文件内容及说明见表 6-5。

表 6-5 my.html 文件内容及说明

内　　容	说　　明

<html>
<head>
<script type="text/javascript">
function jsToA(){
var obj = document.getElementById("myin");
obj.value = window.wv.jsToA();
alert("This is alert");
}
function aToJs(s){
var obj = document.getElementById("myin");
obj.value = s;
}
</script>
</head>
<body>
<div><input type='text' id='myin' /></div>
<div><input type='button' value='调用 Activity' onclick='jsToA()'/></div>
</body>
</html>
	jsToA()函数主要演示 JavaScript 调用 Activity 中的函数，将其返回值填充到 id 为 myin 的 input 标签中
	用于验证 WebChromeClient 类的作用
	aToJs()是 Activity 主动调用的 JavaScript 函数，它将从 Activity 直接传送的字符串 s 显示在 id 为 myin 的 input 标签中
	界面很简单：一个 id 为 myin 的 input 标签，一个按钮，onclick 事件响应函数是 jsToA()

动态加载该页面需要的代码如下所示。

① 主界面 main.xml。

```
<?xml version="1.0" encoding="utf-8"?>
<LinearLayout xmlns:android="http://schemas.android.com/apk/res/android"
android:layout_width="match_parent"
android:layout_height="match_parent"
android:orientation="vertical">
    <Button
        android:id="@+id/mybtn"
        android:layout_width="wrap_content"
        android:layout_height="wrap_content"
        android:text="调用 JavaScript"/>
    <WebView
        android:id="@+id/wv"
        android:layout_width="match_parent"
        android:layout_height="wrap_content">
    </WebView>
</LinearLayout>
```

以上代码定义了一个 id 为 mybtn 的按钮，为验证 Activity 直接调用 JavaScript 函数作准备；定义了一个 id 为 wv 的 WebView 组件，用以动态加载网页页面。

② 主应用 MainActivity.java。

```
public class MainActivity extends AppCompatActivity {
    protected void onCreate(Bundle savedInstanceState) {
        super.onCreate(savedInstanceState);
        setContentView(R.layout.main);
        WebView wv = (WebView) findViewById(R.id.wv);
        wv.getSettings().setDefaultTextEncodingName("utf-8");
        wv.loadUrl("file:///android_asset/my.html");
    }
}
```

由于 my.html 在 Android 工程目录 main/assets 下，因此它会保存在 App 中。代码中，wv.loadUrl()函数加载了该文件，但请注意 url 路径是“file:///android_asset/my.html”，不是“file:///assets/my.html”，读者记住即可。

当然，若 my.html 在服务器端，则可用 wv.loadUrl(“http://....../my.html”)加载页面。

读者会发现，手机 App 运行后，界面显示正常，但所有按钮都不响应，请继续往下实验。

2. Activity 调用 JavaScript 函数

Activity 调用 JavaScript 函数即实现界面中"调用 JS"按钮功能，调用的是 my.html（表 6-5）中的 aToJs(s) 函数，即将从 Activity 中传入的字符串 s 显示在 id 为 myin 的 <input>标签内。在主应用 onCreate() 函数中利用匿名类添加"调用 JavaScript"按钮 click 事件的代码如下所示。

```
Button btn = (Button) findViewById(R.id.mybtn);
btn.setOnClickListener(new View.OnClickListener() {
    public void onClick(View v) {
        final WebView wv = (WebView) findViewById(R.id.wv);
        wv.post(new Runnable() {
            public void run() {
                wv.loadUrl("javascript:aToJs('这是 Activity 调用 JavaScript')");
            }
        });
    }
});
```

可以看出，首先获得 WebView 组件对象 wv，然后应用 wv 中的 post(Runnable action)，重写 new Runnable() 匿名类的线程运行函数 run()，最重要的一行程序：wv.loadUrl("javascript: aToJs('这是 Activity 调用 JavaScript')")，表明调用了 JavaScript 函数 aToJs()，且传入了一个字符串参数。

当然，还要在 onCreate() 函数中加入一行使能 JavaScript 设置的代码，如下所示。

```
wv.getSettings().setJavaScriptEnabled(true);
```

这时，运行 App 后，"调用 JavaScript"按钮执行正确，但是"调用 Activity"按钮无反应，请继续实验。

3. JavaScript 调用 Activity 函数

在 MainActivity 类中增加函数 jsToA()，表明 JavaScript 函数要调用该函数，该函数的功能是将字符串返回给 JavaScript。注意，在函数上方写好注解，可清晰地看出调用方的类型。

```
@JavascriptInterface
public String jsToA(){
```

```
    return "JavaScript 调用 Activity";
}
```

当然,为了正确运行,还要使能 JavascriptInterface 设置,在 onCreate()中增加如下代码。

```
wv.addJavascriptInterface(this, "wv");
```

该行代码可以理解为:JavaScript 调用的代码在 this 中,this 是 MainActivity 对象,所以 JavaScript 可以调用 MainActivity 类中的函数,this(即 MainActivity 对象)在 JavaScript 中对应的对象名称是 wv。截取表 6-5 中的 jsToA()函数如下所示。

```
function jsToA(){
    var obj = document.getElementById("myin");
    obj.value = window.wv.jsToA();
    alert("This is alert");
}
```

重要的一行是:obj.value = window.wv.jsToA(),window.mv 表示 MainActivity 对象,mv 是上文中注册的名字。

运行 App,发现"调用 Activity"按钮运行正确,但是发现 jsToA()函数中最后一行代码"alert("This is alert")"没有运行,怎么办?请继续实验。

4. 使能 alert()、confirm()、prompt()弹出框

其实非常简单,只在 onCreate()函数中增加如下语句即可。

```
wv.setWebChromeClient(new WebChromeClient());
```

若想对弹出框进行修改,可产生 WebChromeClient 的子类,重写相应的 onJsAlert()、onJsConfirm()、onJsPrompt()函数即可。

习题 6

一、选择题

1. Android 中解析 JSON 字符串常用的两个类是(　　)。
   A. JSONObject,JSONArray　　B. JSONObject,XmlPullParser
   C. JSONArray,Properties　　D. XmlPullParser,Properties

2. 应用 WebView 组件，要在配置文件中声明哪个权限？（　　）

A. android.permission.INTERNET

B. android.permission.WRITE_EXTERNAL_STORAGE

C. android.permission.ACCESS_NETWORK_STATE

D. android.permission.WRITE_SETTINGS

二、程序题

1. 手机端实现学生信息（学号、姓名、专业、学院）添加界面，单击“确定”按钮后，将学生信息传到服务器端，在服务器端完成对数据库相应表信息的添加。

2. 拓展例 6-8，使之接收服务器端的 html 页面，同时使能 JavaScript 函数。

第 7 章

广播接收组件

广播接收组件是 Android 的重要组件之一。本章讲解了广播接收的基本原理，组件的静态注册与动态注册，普通广播与有序广播的不同，并对系统固有广播进行了讨论。

7.1 基本原理

简单地说，Android 广播机制的主要功能是实现一处发生事情，多处得到通知并加以处理。这种通知常常牵涉组件间、进程间通信，所以需要由 Android 系统集中管理。广播接收机制原理如图 7-1 所示。

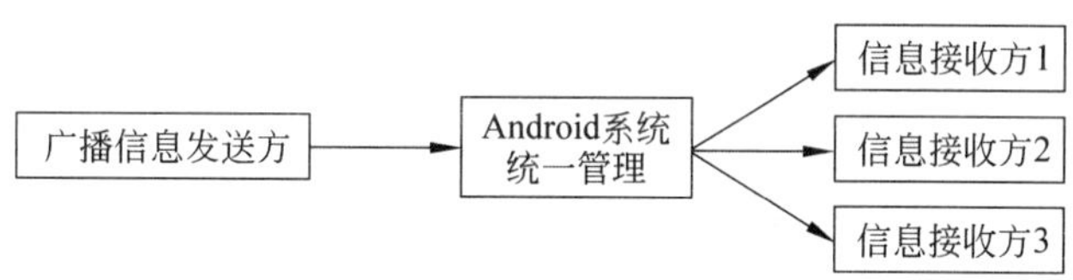

图 7-1 广播接收机制原理

广播信息发送方将信息发送给 Android 系统统一管理器，然后再由 Android 系统统一管理器发送给多个信息接收方接收处理。对开发者来说，仅需要编制广播信息发送方、信息接收方代码，以及定义它们的握手信号，最复杂的系统统一管理部分代码由 Android 系统完成。

类似图 7-1 的应用在现实中有很多。例如，手机电量低时，屏幕上会出现一条电量低的提示信息；再如，监测网络状态、开机、关机等。

常用的广播分为两种：普通广播和有序广播。普通广播是完全异步的，消息传递的效率比较高，但所有接收器的执行顺序不确定。缺点是：接收者不能将处理结果传递给下一个接收者，无法终止信息的传播，直到没有与之匹配的广播接收器为止。有序广播发送消息后，所有的广播接收器按优先级从高到低依次执行。当广播接收器接收到广播后，既可取得上一个广播接收器返回的结果，也可将结果传给下一个广播接收器。当然，也可让系统拦截该广播，并将其丢弃，使之不再继续传播。

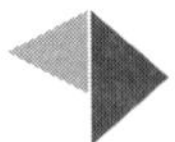

7.2 基本类

对广播发送来说，主要用 Context 类中的如下函数。

- void sendBroadcast(Intent in)：普通广播函数，信息封装在 Intent 对象中。
- void sendBroadcast(Intent in,String rcrPermission)：普通广播函数，信息封装在 Intent 对象中，rcrPermission 是对接收者 Receiver 做的权限设置，也就是 Receiver 必须拥有这个权限，才能处理它发出的广播。
- void sendOrderedBroadcast(Intent in,String rcrPermission)：有序广播函数，信息封装在 Intent 对象中，rcrPermission 是对接收者 Receiver 做的权限设置，也就是 Receiver 必须拥有这个权限，才能处理它发出的广播。
- void sendOrderedBroadcast(Intent intent,String receiverPermission,BroadcastReceiver resultReceiver, Handler scheduler, int initialCode, String initialData, Bundle initialExtras)：intent 是要发送的广播；receiverPermission 是发送的广播的权限，如果是 null，即认为没有，任何符合条件的接收器都能收到；resultReceiver 是最终处理数据的接收器，也就是自己定义的；scheduler 是自定义的一个 handler，用来处理 resultReceiver 的回调(其实就是设置运行这个接收器的线程)，如果为 null，默认在主线程；后面 3 个参数并不重要，通常情况下依次为－1，null，null。(Activity.RESULT_OK 即－1)
- void abortBroadcast()：停止广播函数。

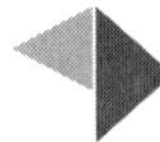

7.3 应用示例

主要掌握普通广播、有序广播编制方法及接收端静态注册、动态注册的特点。

7.3.1 普通广播＋静态注册

【例 7-1】 主界面有一 OK 按钮，当单击 OK 按钮时，广播一条消息“Hello”，有两个接收者，接到该消息，并利用 Toast 显示接收到的消息。

涉及的文件有：配置文件 Androidmanifest.xml；两个接收类 Receive1、Receive2，均

从 BroadcastReceiver 派生；主界面 main.xml；主应用 MainActivity 类，具体代码如下所示。

① 两个接收类 Receive1、Receive2。

```
//Receive1.java
public class Receive1 extends BroadcastReceiver {
    public void onReceive(Context context, Intent intent) {
        String msg = intent.getStringExtra("msg");
        Toast.makeText(context,"Receive1 接收:"+msg,Toast.LENGTH_LONG).show();
    }
}
```

接收端最主要的工作是重写 onReceive()函数，解析接收到的 intent，将获得的信息保存在变量 msg 中，并利用 Toast 将信息显示在屏幕上。

//Receive2.java：仿 Receive1 类，可写出 Receive2 类代码，略。

② 静态注册上述两个接收器类。

静态注册上述两个接收器类，即主要在配置文件 Androidmanifest.xml 中增加对接收类的配置，相关部分代码如下所示。

```
<receiver android:name=".Receive1">
    <intent-filter>
        <action android:name="myfirstbroad"></action>
        <category android:name="android.intent.category.DEFAULT"></category>
    </intent-filter>
</receiver>
<receiver android:name=".Receive2">
    <intent-filter>
        <action android:name="myfirstbroad"></action>
        <category android:name="android.intent.category.DEFAULT"></category>
    </intent-filter>
</receiver>
```

可以看出，静态注册是指对接收类的注册。一个广播接收类对应一个<receiver>标签，其配置内容与<activity>标签是一致的，主要是配置<intent-filer>过滤子标签的内容，特别是<action>标签的内容。例如，上述第 1 段<receiver>的含义是：动作属性为 myfirstbroad 的广播方与 Receive1 类是绑定的。上述第 2 段<receiver>的含义是：动作属性为 myfirstbroad 的广播方与 Receive2 类是绑定的。所以，动作属性为 myfirstbroad 的广播方与 Receive1、Receive2 相关联。

③ 主界面 main.xml。

```
<?xml version="1.0" encoding="utf-8"?>
<LinearLayout xmlns:android="http://schemas.android.com/apk/res/android"
android:layout_width="match_parent"
android:layout_height="match_parent"
android:orientation="vertical">
    <Button
        android:id="@+id/mybtn"
        android:layout_width="wrap_content"
        android:layout_height="wrap_content"
        android:text="OK"/>
</LinearLayout>
```

主界面很简单,仅定义了一个 id 为 mybtn 的 OK 按钮。

④ 主应用 MainActivity.java。

```
public class MainActivity extends AppCompatActivity {
    protected void onCreate(Bundle savedInstanceState) {
        super.onCreate(savedInstanceState);
        setContentView(R.layout.main);
        Button b = (Button)findViewById(R.id.mybtn);
        b.setOnClickListener(new View.OnClickListener() {
            public void onClick(View v) {
                Intent t = new Intent();
                t.putExtra("msg", "Hello");
                t.setAction("myfirstbroad");
                MainActivity.this.sendBroadcast(t);
            }
        });
    }
}
```

在按钮响应函数中,首先产生 Intent 对象 t,将广播信息 Hello 存入 t 中;然后利用 setAction()函数设置动作属性为 myfirstbroad,这是非常关键的一行,为将来能根据动作属性值查找配置文件得到与之匹配的接收方做好准备;最后将 Intent 对象广播出去,等待接收方响应。

讨论:读者看到这里可能会问:sendBroadcast(Intent t)单参数函数已经能完成广播功能,那么它与双参数 sendBroadcast(Intent t,String rcrPermission)函数有区别吗?第二个参数如何应用？这两个广播函数在功能上是一致的，后一个发送函数的第二个参数要求接收器权限更严格，一般来说，字符串 rcrPermission 是自定义权限字符串，若发送方函数中利用了该参数（假设为 test.permission.COMM），则在接收器所在的 App 配置文件 Androidmanifest.xml 中必须先声明该自定义许可条件，再应用该许可条件，即在配置文件中增加两条配置语句，如下所示。

```
<permission android:name="test.permission.COMM" android:protectionLevel="
normal"></permission>
<uses-permission android:name="test.permission.COMM" />
```

其他所有代码保持不变，即可正确运行。

7.3.2　普通广播+动态注册

动态注册即在程序中完成对接收方类的注册，无须在配置文件中注册，对例 7-1 而言，去掉配置文件 Androidmanifest.xml 中的<receiver>标签即可。在程序中实现注册功能的代码如下所示。

```
public class MainActivity extends AppCompatActivity {
    Receive1 rce1 = new Receive1();
    Receive2 rce2 = new Receive2();
    protected void onCreate(Bundle savedInstanceState) {
        super.onCreate(savedInstanceState);
        setContentView(R.layout.main);
        IntentFilter itf = new IntentFilter();
        itf.addAction("myfirstbroad"); itf.addCategory("android.intent
.category.DEFAULT");
        registerReceiver(rce1, itf);
        registerReceiver(rce2, itf);
        Button b = (Button)findViewById(R.id.mybtn);
        b.setOnClickListener(new View.OnClickListener() {
            //此处代码完全同例 7-1,略
        });
    }
    protected void onDestroy() {
        super.onDestroy();
        unregisterReceiver(rce1); unregisterReceiver(rce2);
    }
}
```

因为要在多个函数中操作两个接收器 Receive1、Receive2 类对象，因此直接将它们定义为成员变量 rce1、rce2。

接收器动态注册就是将<receiver>标签内容翻译成程序。产生 Intent-filter 对象 itf，利用 addAction()函数设置动作属性为 myfirstbroad，利用 addCategory()函数设置种类属性为默认属性，然后利用 registerReceiver()函数将接收器类对象 rce1、rce2 与 Intent-filter 过滤器对象绑定，完成注册功能。

当然，还要完成注册器对象的撤销功能。一般重载 onDestroy()函数，利用 unregisterReceiver()函数将其撤销。

7.3.3 有序广播+静态注册

静态注册指在 Androidmanifest.xml 中为接收器增加<receiver>标签配置的同时，为其指定优先级，范围为－1000～1000，值越大，优先级越高，越先响应。而广播时采用 sendOrderedBroadcast()函数，代表是有序广播。

【例 7-2】 主界面有一 OK 按钮，当单击 OK 按钮时，有序广播一条消息 Hello，启动两个接收器 A、B，A 的优先级高于 B。A 接到消息并显示该消息，然后再传送一条附加信息"I am Receive1"给 B，B 接收并显示广播及附加消息。

涉及的文件有：配置文件 Androidmanifest.xml；两个接收类 Receive1、Receive2，均从 BroadcastReceiver 派生，Receive1 的优先级小于 Receive2；主界面 main.xml；主应用 MainActivity 类，具体代码如下所示。

① 两个接收类 Receive1、Receive2。

```
//Receive1.java
public class Receive1 extends BroadcastReceiver {
    public void onReceive(Context context, Intent intent) {
        String msg = intent.getStringExtra("msg");
        Toast.makeText(context,"Receive1 接收:"+msg,Toast.LENGTH_LONG).show();
        Bundle b = new Bundle();
        b.putString("attach", "Receive1");
        this.setResultExtras(b);
    }
}
```

intent 对象是广播消息对象，解析它就可获得广播消息的具体内容。附加信息对象是由 setResultExtras(Bundle b)设置的，它要求形参必须是 Bundle 对象，附加信息绑定在 Bundle 对象中。所以，Receive1 接收器完成信息接收后，广播对象 intent 及附加信息对象 b 继续向优先级比 Receive1 低的接收器传播，对本示例来说，就是 Receive2 接收器。

```
//Receive2.Java
public class Receive2 extends BroadcastReceiver {
    public void onReceive(Context context, Intent intent) {
        String msg = intent.getStringExtra("msg");
        Bundle b = this.getResultExtras(true);
        String attach = b.getString("last");
        Toast toast = Toast.makeText(context,"Receive1 接收:"+msg+"\t 附加信息是:"+
                attach,Toast.LENGTH_LONG);
        toast.show();
    }
}
```

Receive2 解析包含两部分数据：一部分是解析 intent 对象获得广播信息；一部分是通过 getResultExtras()函数获得附加信息 Bundle 对象 b，再根据键值，获得附加信息字符串 attach，完成相应的显示功能。

若接收器 Receive2 不希望继续广播，则在 onReceive()中利用 abortBroadcast()停止广播。

② 静态注册上述两个接收器类。

即主要在配置文件 Androidmanifest.xml 中增加对接收类的配置，相关部分代码如下所示。

```
<receiver android:name=".Receive1">
    <intent-filter android:priority="500">
        <action android:name="myfirstbroad"></action>
    </intent-filter>
</receiver>
<receiver android:name=".Receive2">
    <intent-filter android:priority="100">
        <action android:name="myfirstbroad"></action>
    </intent-filter>
</receiver>
```

与例 7-1 相比，这里仅增加了<intent-filter>标签的 android:priority 优先级属性值设置。

Receive1 的优先级设置为 500，Receive2 的优先级设置为 100，Receive1 的优先级高于 Receive2。在同等响应条件的情况下，Receive1 接收器优先运行。

③ 主界面 main.xml：同例 7-1，略。

④ 主应用 MainActivity.java。

```
public class MainActivity extends AppCompatActivity {
    protected void onCreate(Bundle savedInstanceState) {
        super.onCreate(savedInstanceState);
        setContentView(R.layout.main);
        Button b = (Button) findViewById(R.id.mybtn);
        b.setOnClickListener(new View.OnClickListener() {
            public void onClick(View v) {
                Intent t = new Intent();
                t.putExtra("msg", "Hello");
                t.setAction("myfirstbroad");
                MainActivity.this.sendOrderedBroadcast(t,null);
            }
        });
    }
}
```

与例 7-1 的代码几乎相同，仅是利用 sendOrderedBroadcast ( ) 函数代替了 sendBroadcast()函数。

7.3.4 有序广播+动态注册

动态注册即在程序中完成对接收方类的注册，无须在配置文件中注册，对例 7-2 而言，去掉配置文件 Androidmanifest.xml 中的<receiver>标签即可。在程序中实现注册功能的代码如下所示。

```
public class MainActivity extends AppCompatActivity {
    Receive1 rce1 = new Receive1();
    Receive2 rce2 = new Receive2();
    protected void onCreate(Bundle savedInstanceState) {
        super.onCreate(savedInstanceState);
        setContentView(R.layout.main);
        IntentFilter itf = new IntentFilter();itf.setPriority(100);
         itf. addAction ( "myfirstbroad"); itf. addCategory ( " android. intent.
category.DEFAULT");
        registerReceiver(rce1, itf);
        IntentFilter itf2 = new IntentFilter(itf); itf2.setPriority(500);
```

```
        registerReceiver(rce2, itf2);
        Button b = (Button)findViewById(R.id.mybtn);
        b.setOnClickListener(new View.OnClickListener() {
            public void onClick(View v) {
                //代码同例 7-2 的有序广播+静态注册部分的代码
            }
        });
    }
    protected void onDestroy() {
        super.onDestroy();
        unregisterReceiver(rce1); unregisterReceiver(rce2);
    }
```

与“普通广播＋动态注册”相应部分的代码相比，这里仅多了对 IntentFilter 对象的优先级大小设置（利用 setPriority()函数），其余均是相近的。

7.3.5 其他广播

【例 7-3】 七参数 sendOrderedBroadcast()函数应用问题。

该函数的主要功能是：有序广播，且最后的接收器对象直接定义在该函数中。考虑这样一个应用：发送方注册了两个接收器 A、B，定义了一个最终的接收器 C。有序广播信息给 A、B，A、B 将自己对信息的表决结果（Yes 或 No）添加到广播信息，接收器 C 显示汇总的对广播信息表决的结果。

该功能涉及的文件有：两个有序广播接收器类 Receive1.java、Receive2.java；最终的接收器类 LastReceive.java；主界面 main.xml 及主应用 MainActivity.java。下面一一加以说明。

① 两个有序广播接收器类 Receive1.java、Receive2.java。

```
public class Receive1 extends BroadcastReceiver {
    public void onReceive(Context context, Intent intent) {
        Bundle b = new Bundle();
        b.putString("receive1","Yes");
        Toast.makeText(context, "Receive1:Yes", Toast.LENGTH_LONG).show();
        setResultExtras(b);
    }
}
public class Receive2 extends BroadcastReceiver {
    public void onReceive(Context context, Intent intent) {
```

```
            Bundle b = getResultExtras(true);
            b.putString("receive2","No");
            setResultExtras(b);
            Toast.makeText(context, "Receive2:No", Toast.LENGTH_LONG).show();
        }
    }
```

由于 Receive1 类的优先级高于 Receive2 类,所以在 Receive1 类中创建 Bundle 对象 b,将表决结果(Yes)放入其中,利用 setResultExtras()函数使广播继续传播;在 Receive2 类中获得 Bundle 对象 b,将表决结果(No)放入其中,利用 setResultExtras()函数使广播继续传播。

这两个接收器类必须注册。

② 最终接收器类 LastReceive。

该类不需要注册,但同样从 BroadcastReceiver 派生,如下所示。

```
public class LastReceive extends BroadcastReceiver {
    public void onReceive(Context context, Intent intent) {
        Bundle b = getResultExtras(true);
        String result1 = b.getString("receive1");
        String result2 = b.getString("receive2");
        String s = result1+","+result2;
        Toast.makeText(context, s, Toast.LENGTH_LONG).show();
    }
}
```

首先获得广播的 Bundle 对象 b,然后获得接收器 Receive1、Receive2 的表决结果,利用 Toast 将其显示在界面上。

③ 主界面 main.xml 及主应用 MainActivity.java。

```
//main.xml:该界面仅为象征意义,空白界面即可,略
//MainActivity.java
public class MainActivity extends AppCompatActivity {
    Receive1 rce1 = new Receive1();
    Receive2 rce2 = new Receive2();
    protected void onCreate(Bundle savedInstanceState) {
        super.onCreate(savedInstanceState);
        setContentView(R.layout.main);
        IntentFilter itf = new IntentFilter();                    //注册功能开始
        itf.addAction("myfirstbroad"); itf.setPriority(500);
```

```
        IntentFilter itf2 = new IntentFilter();
        itf2.addAction("myfirstbroad"); itf2.setPriority(100);
        registerReceiver(rce1, itf);registerReceiver(rce2, itf2);//注册功能结束
        Intent intent = new Intent();                              //发送功能开始
        intent.putExtra("msg", "hello");
        intent.setAction("myfirstbroad");
        sendOrderedBroadcast(intent,null,new LastReceive(),null,-1,null,null);
                                                                   //发送功能结束
    }
    protected void onDestroy() {
        super.onDestroy();
        unregisterReceiver(rce1);unregisterReceiver(rce2);
    }
}
```

由于要分别在两个函数中完成 Receive1、Receive2 类的注册和撤销，因此定义了 Receive1、Receive2 类成员变量 rce1、rce2。在 onCreate()函数中完成了对 rce1、rce2 的注册，在 onDestroy()函数中完成了对 rce1、rce2 对象的撤销功能。

LastReceive 也是接收器，但它并没有注册，而是写在了 sendOrderedBroadcast()函数的第 4 个参数(new LastReceive)中，表明 LastReceive 对象是最后的接收器，可以实现汇总结果、显示信息等功能。

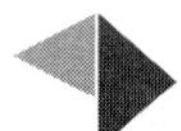

7.4　系统广播

可能读者有疑问：学完 7.1～7.3 节，只是知道普通广播、有序广播的抽象用途，具体的实际用途还是不清楚，本节通过系统广播的实例使读者加深对广播的理解。

【例 7-4】 手机电量的监测。

众所周知，手机电量很重要，当手机电量低到一定程度的时候，系统会给出一条提示信息，这条信息就是系统广播接收器的结果。那么，能否接收该广播信号，提取出电量信息，做一些自定义的具体应用？没有问题，要达到这一点，必须知道以下几点。

- 手机电量变化的系统信息是什么？

即广播的动作值 action 对应的系统固定字符串，它是广播接收器注册时用到的值。该值是 Intent 类中的静态变量 Intent.ACTION_BATTERY_CHANGED。

- 如何解析接收的 Intent 广播信息？

Intent 对象包含要解析的手机电量信息，但由于源信息是系统发送，不是用户主动发

送的,因此用户不能想当然地随意解析,必须查询材料获得正确的解析方式。对手机电量信息的获取,要按表 7-1 解析。

表 7-1　获取手机电量信息参考表

序号	BatteryManager 类字段名称	获取方法	说　明
1	EXTRA_SCALE	getIntExtra	电量刻度,一般为 100
2	EXTRA_LEVEL	getIntExtra	当前电量
3	EXTRA_TEMPERATURE	getIntExtra	当前温度
4	EXTRA_STATUS	getIntExtra	当前状态值
5	BATTERY_STATUS_UNKNOWN	X	未知
6	BATTERY_STATUS_CHARING	X	正在充电
7	BATTERY_STATUS_DISCHARING	X	正在断电
8	BATTERY_STATUS_NOT_CHARING	X	不在充电
9	BATTRY_STATUS_FULL	X	已经充满

表 7-1 仅列出了手机电池的部分信息。通过表 7-1 可知道,若获取电量刻度,则要用 int value=intent.getInntExtra(BatteryManager.EXTRA_SCALE);若获取电池充电状态,首先利用 int value=intent.getInntExtra(BatteryManager.EXTRA_STATUS),然后比较该值与表 7-1 中序号 5～8 的哪个具体值相等,从而确定出具体状态。

- 主配置文件 Androidmanifest.xml 是否需要添加允许信息?

这一条非常重要,对系统广播而言,一般要添加相应的允许信息;对手机电量监测而言,要添加如下信息。

```
<uses-permission android:name="android.permission.BATTERY_STATS"/>
```

另外要明确:手机电量广播信息是系统发出的,因此对系统广播而言,用户只做接收器部分即可。

综合上述,手机电量监测涉及的文件有:Androidmanifest.xml;主界面 main.xml;主应用 MainActivity.java;电量监测接收器类 Receive1.java,具体如下所示。

① Androidmanifest.xml:增加手机电量监测允许标识。

```
<uses-permission android:name="android.permission.BATTERY_STATS"/>
```

② 主界面 main.xml 及主应用 MainActivity.java。

```
//main.xml
```

```
<?xml version="1.0" encoding="utf-8"?>
<LinearLayout xmlns:android="http://schemas.android.com/apk/res/android"
android:layout_width="match_parent"
android:layout_height="match_parent">
    <TextView
        android:id="@+id/myinfo"
        android:layout_width="match_parent"
        android:layout_height="wrap_content" />
</LinearLayout>
```

以上代码定义了 id 为 myinfo 的 TextView 组件，用以将电量满刻度值、当前电量值、当前温度值显示在该组件中。

```
//MainActivity.java
public class MainActivity extends AppCompatActivity {
    Receive1 rce1 = new Receive1();
    protected void onCreate(Bundle savedInstanceState) {
        super.onCreate(savedInstanceState);
        setContentView(R.layout.main);
        IntentFilter itf = new IntentFilter();
        itf.addAction(Intent.ACTION_BATTERY_CHANGED);
        registerReceiver(rce1, itf);
    }
    protected void onDestroy() {
        super.onDestroy();
        unregisterReceiver(rce1);
    }
}
```

由于要分别在两个函数中完成 Receive1 类的注册和撤销，因此定义了 Receive1 类成员变量 rce1。在 onCreate()函数中完成了对 rce1 接收器的注册，注意 addAction()函数的参数是 Intent.ACTION_BATTERY_CHANGED，是电量监测的特征标识串，不能随意写；在 onDestroy()函数中完成了对 rce1 对象的撤销功能。

③ 电量监测接收器类 Receive1.Java。

```
public class Receive1 extends BroadcastReceiver {
    public void onReceive(Context context, Intent intent) {
        int scale = intent.getIntExtra(BatteryManager.EXTRA_SCALE,-1);
        int level = intent.getIntExtra(BatteryManager.EXTRA_LEVEL,-1);
```

```
            int temper = intent.getIntExtra(BatteryManager.EXTRA_TEMPERATURE,-1);
            Activity obj = (Activity)context;
            TextView tv = (TextView)obj.findViewById(R.id.myinfo);
            String s = "电量刻度:" + scale;
            s += "\n 当前刻度:" +level;
            s += "\n 当前温度:" +temper;
            tv.setText(s);
        }
    }
```

在接收器函数 onReceive()中,参考表 7-1,利用 getIntExtra()函数获得手机电量满刻度值 scale,当前手机电量 level,当前手机温度 temper。将此 3 个物理值显示在 TextView 组件中。

TextView 组件在主页面 main.xml 中定义,由于 onReceive()函数中的第 1 个形参是上下文对象 context,在本示例中它就是 Activity 对象(其实是 MainActivity 对象),因此将 context 强制转换成 Activity 对象,之后就可以获得用于显示的 TextView 组件了。

综合上述,当本示例 App 运行时,发现马上会在 TextView 组件上显示出电量满刻度、当前电量、温度物理值。这些值直到手机当前电量变化后,例如由 33 变到 34,才会接收到新的电量广播信息,才会更新 TextView 组件显示的内容。现在手机充电都比较快,在充电时,可以在 TextView 组件中看到电量、温度值的变化情况。

【例 7-5】 手机屏幕黑屏、亮屏消息的监测。

大家都有这样的体会,手机一段时间内没有操作,就会黑屏,这样的好处之一是省电。当操作手机时,手机屏幕又亮了起来。这两个消息也是系统广播消息,截获它们,人们可以做很多有益的工作。例如,手机是人们的必备工具,但经常过长时间看手机会对人眼及身体造成很大的伤害,那么能否有一个简易的手段监测人们看手机的时间?利用手机黑屏、亮屏事件可以做到一定程度的监测。原理是:一般来说,黑屏、亮屏信息是成对出现的,把每一次黑屏、亮屏的时间间隔累加起来,就是手机持有人的休息时间。时间越长,表明手机持有人休息的时间越长。

与本示例相关的文件有:主界面 main.xml;主应用 MainActivity.java;接收器类 Receive1.java。

① 主界面 main.xml,本示例中主界面仅起到象征作用,为空白即可,略。

② 主应用 MainActivity.java。

```
public class MainActivity extends AppCompatActivity {
```

```
    Receive1 rce1 = new Receive1();
    protected void onCreate(Bundle savedInstanceState) {
        super.onCreate(savedInstanceState);
        setContentView(R.layout.main);
        IntentFilter itf = new IntentFilter();
        itf.addAction(Intent.ACTION_SCREEN_ON);
        itf.addAction(Intent.ACTION_SCREEN_OFF);
        registerReceiver(rce1, itf);
    }
    protected void onDestroy() {
        super.onDestroy();
        unregisterReceiver(rce1);
    }
}
```

接收器类 Receive1 对象 rce1 要在多个函数中应用，因此将其定义为成员变量。在 onCreate()函数中完成了对 rce1 的注册，在 onDestroy()函数中完成了对 rce1 的撤销。

屏幕黑屏的特征动作字符串是 Intent.ACTION_SCREEN_OFF，屏幕亮屏的特征动作字符串是 Intent.ACTION_SCREEN_ON。注册过程中先产生 IntentFilter 过滤器对象 itf，然后利用 addAction()函数加入黑屏、亮屏动作特征字符串，最后利用 registerReceiver()函数将接收器对象 rce1 与过滤器对象 itf 绑定。

一个 IntentFilter 对象可以利用 addAction()函数添加多个特征动作字符串，这些动作是或的关系，不是并的关系。

③ 接收器类 Receive1.java。

```
public class Receive1 extends BroadcastReceiver {
    int total;
    long start,end;
    public void onReceive(Context context, Intent intent) {
        String action = intent.getAction();
        if(action.equals(Intent.ACTION_SCREEN_OFF)){
            Calendar c = Calendar.getInstance();
            start = c.getTimeInMillis();
        }
        else if(action.equals(Intent.ACTION_SCREEN_ON)){
            long  start,end;
            int hour,minute,second;
            Calendar c = Calendar.getInstance();
```

```
                end = c.getTimeInMillis();
                int value = (int)((end - start)/1000);
                total += value;
                hour = total/3600;minute = total%3600/60;second = total%60;
                String s=""+hour+":"+minute+":"+second;
                Toast.makeText(context,s,Toast.LENGTH_LONG).show();
            }
        }
    }
```

成员变量 total 是手机休眠的总秒数，start、end 是用于记录每次黑屏、亮屏的时间点数值。在 onReceive()函数中，首先获得该消息的特征动作字符串值 action，当 action 表示黑屏时，利用 Calendar 类获得黑屏时间点总毫秒数 start(距离 1970-01-01)；当 action 表示亮屏时，利用 Calendar 类获得亮屏时间点总毫秒数 end。此时(end-start)/1000 就是此次手机休眠的总秒数，与原有的 total 累加，即历次手机休眠的总秒数，将其转换为 hour、minute、second，利用 Toast 简单显示即可。

习题 7

一、选择题

1. 有序广播时采用的广播函数是(　　)。
   A. sendBroadcast()　　　　B. sendOrderedBroadcast()
   C. 两者皆可　　　　D. 两者皆错
2. 普通广播接收端要从哪个类派生，重写哪个函数？(　　)
   A. BroadcastReceiver，receive()
   B. BroadcastReceiver，onReceive()
   C. Broadcast，receive()
   D. Broadcast，onReceive()

二、程序题

定义 3 个接收器类 A、B、C。主界面上有普通广播、有序广播按钮。单击“普通广播”按钮，将“Hello”字符串广播给 3 个接收器；单击“有序广播”按钮，将“Hello”字符串有序广播给 3 个接收器。

第 8 章

Service

Service 是 Android 的重要组件之一。本章将讲解 Service 生命周期，启动 Service，绑定 Service 的特点及应用，并对跨进程调用 Service 进行深入的讨论。

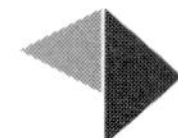

8.1 简介

Service 即服务，是能够在后台执行长时间运行操作并且不提供用户界面的应用程序组件。Service 通常分为两种类型，如下所示。

- 启动 Service：也可叫作未绑定 Service。当服务启动后，可无限期运行，只能单方向通信，即从启动方调用服务，不能将服务的结果返回启动方。而且，即使启动方组件已经销毁，服务仍能运行，因此必须考虑服务的自销毁问题。
- 绑定 Service：当服务启动后，启动方和服务可双向通信，可以将服务的结果返回给启动方。多个组件可同时绑定到服务，但当所有组件解绑服务后，服务也随即销毁，这是与启动 Service 很大的一个不同点。

每建立一个 Service，就要在 Androidmanifest.xml 配置文件中添加一个<service>节点，其格式示例如下所示。

```
<service android:enabled=["true" | "false"]
      android:exported=["true" | "false"]
      android:icon="drawable resource"
      android:label="string resource"
      android:name="string"
      android:permission="string"
      android:process="string">
   <intent-filter></intent-filter>
   <meta-data></meta_data>
```

```
</service>
```

其相关解释如下所示。

- android:enabled：表示该服务是否能够被实例化，如果设置为 true，则能够被实例化，否则不能被实例化，默认值是 true。如果有一个元素的 enabled 属性被设置为 false，则该服务就会被禁用，而不能被实例化。
- android:exported：表示该服务是否能够被其他应用程序组件调用或与它交互，如果设置为 true，则能够被调用或交互，否则不能。设置为 false 时，只有同一个应用程序的组件或带有相同用户 ID 的应用程序才能启动或绑定该服务。
- android:icon：定义代表服务的图标，它必须引用一个包含图片定义的可绘制资源。如果这个属性没有设置，则会使用<application>元素的 icon 属性所设定的图标代替。
- android:label：设定一个要显示给用户的服务的名称。如果没有设置这个属性，则会使用<application>元素的 label 属性值代替。
- android:name：指定实现该服务的 Service 子类的类名。它应该是完整的 Java 类名(如 com.example.dqjbd.we.MyService)。也可以使用简写(如.MyService)，系统会把<manifest>元素中 package 属性所设定的值添加到简写名称的前面。
- android:permission：定义了要启动或绑定服务的实体必须要有的权限。如果调用者的 startService()、bindService()和 stopService()方法没有被授予这个权限，那么这些方法就不会工作，并且 Intent 对象也不会发送给该服务。
- android:process：设定服务所运行的进程名称。通常，应用程序的所有组件都运行在应用程序创建的进程中，进程名与应用程序的包名相同。<application>元素的 process 属性能够给应用程序的所有组件设置一个不同的默认名称。但是，每个组件自己的 process 属性都能覆盖这个默认值，这样允许把应用程序分离到不同的多个进程中。

8.2 启动 Service

8.2.1 生命周期

对有意义的启动 Service 来说，要从系统类 Service 派生，重写 Service 类定义的相关函数，因此，深刻理解各重写函数在生命周期的位置就非常重要。一般来说，对启动 Service 来说，经常重写的 3 个函数是 onCreate()、onStartCommand()和 onDestroy()，具体描述如下所示。

- onCreate()：当创建服务时调用，若该服务已存在，则不再调用。也就是说，该函数在 Services 生命周期中仅调用一次。
- onStartCommand()：每次调用该服务时，都会运行该函数，因此，该函数包含服务的所有实际功能，是最重要的重载函数。
- onDestroy()：顾名思义，当执行该函数时，服务即将销毁，该函数主要具有资源释放等功能。

【例 8-1】 编程验证启动 Service 生命周期的调用顺序。

主界面有两个按钮，功能是创建 Service 和销毁 Service。在自定义 Service 中，重写 onCreate()、onStartCommand()、onDestroy()函数，分别用 Toast 显示函数名称即可。运行时，比较第一次及后续多次单击"创建"按钮各 Toast 的显示顺序，当单击"销毁"按钮时，看 Toast 的显示情况。

涉及的文件有：主界面 main.xml；主应用 MainActivity.java；自定义 Service 类 MyService.java；配置文件 Androidmanifest.xml 新增的 MyService 类信息。下面一一说明。

① 主界面 main.xml。

```
<?xml version="1.0" encoding="utf-8"?>
<LinearLayout xmlns:android="http://schemas.android.com/apk/res/android"
android:layout_width="match_parent"
android:layout_height="match_parent">
    <Button
        android:id="@+id/mystart"
        android:layout_width="wrap_content"
        android:layout_height="wrap_content"
        android:text="create"/>
    <Button
        android:id="@+id/mystop"
        android:layout_width="wrap_content"
        android:layout_height="wrap_content"
        android:text="destroy"/>
</LinearLayout>
```

文本为 create 的按钮，功能是创建 Service，id 为 mystart；文本为 destroy 的按钮，功能是销毁 Service，id 为 mystop。

② 主应用 MainActivity.java。

```
public class MainActivity extends AppCompatActivity {
```

```
    protected void onCreate(Bundle savedInstanceState) {
        super.onCreate(savedInstanceState);
        setContentView(R.layout.main);
        Button b = (Button)findViewById(R.id.mystart);
        Button b2= (Button)findViewById(R.id.mystop);
        b.setOnClickListener(new View.OnClickListener() {
            public void onClick(View v) {
                Intent t = new Intent(MainActivity.this, MyService.class);
                startService(t);
            }
        });
        b2.setOnClickListener(new View.OnClickListener() {
            public void onClick(View v) {
                Intent t = new Intent(MainActivity.this, MyService.class);
                stopService(t);
            }
        });
    }
}
```

启动 Service 需要两步：首先产生 Intent(Context ctx,Class des)对象 t,ctx 是上下文对象,des 是带启动的自定义 Service 类,本例中是 MyService.class;然后利用 startService(t)启动 Service。

销毁 Service 需要两步：首先产生 Intent(Context ctx,Class des)对象 t,ctx 是上下文对象,des 是带启动的自定义 Service 类,本例中是 MyService.class;然后利用 stopService(t)销毁 Service。

③ 自定义 Service 类 MyService.java。

```
public class MyService extends Service {
    public MyService() {}
    public IBinder onBind(Intent intent) {
        throw new UnsupportedOperationException("Not yet implemented");
    }
    public void onCreate() {
        super.onCreate();
        Toast.makeText(getApplicationContext(),"Create",Toast.LENGTH_LONG).
show();
    }
```

```
    public int onStartCommand(Intent intent, int flags, int startId) {
        Toast.makeText(getApplicationContext(),"StartCommand",Toast.LENGTH_
SHORT).show();
        return super.onStartCommand(intent, flags, startId);
    }
    public void onDestroy() {
        Toast.makeText(getApplicationContext(),"Service Destroy",Toast
.LENGTH_LONG).show();
        super.onDestroy();
    }
}
```

这里重写了 onCreate()、onStartCommand()、onDestroy()函数,仅是用 Toast 显示了函数名称,方便看 Service 生命周期。

在 Service 类中获得上下文对象要用到 getApplicationContext()函数。

注意:onBind()不是启动 Service 需要的函数,其实它是绑定 Service 需要的函数。但必须重写该函数,否则编译通不过,这和 Android 系统有关。

④ 配置文件 Androidmanifest.xml 新增的 MyService 类信息。

```
<service
    android:name=".MyService"
    android:enabled="true"
    android:exported="true">
</service>
```

<service>标签有众多属性,但最基本的是上面这 3 个属性。name 指明了要调用的 Service 类是 MyService;enabled 为 true,表明 MyService 类可实例化;exported 为 true,表明该服务可被其他应用程序或组件调用。

本示例可按如下步骤测试:运行时第 1 次单击 create 按钮,发现 Toast 显示的顺序是 onCreate(),onStartCommand();之后不论单击多少次 create 按钮,Toast 仅显示调用了 onStartCommand();当单击 destroy 按钮时,Toast 显示调用了 onDestroy()函数。这时若再单击 create 按钮,发现执行顺序是 onCreate(),onStartCommand(),Service 又重新创建并运行了。

8.2.2 几个知识点

1. Service 隐式启动

例 8-1 采用显示启动方式启动 Service。其实,有了前面的 Activity、BroadcastReceiver 隐

式启动的例子,也就不难写出 Service 隐式启动的关键代码。仍以例 8-1 为例,Service 隐式启动关键代码见表 8-1。

表 8-1　Service 隐式启动关键代码示例表

配置文件的 MyService 部分	隐式启动关键代码

<service
android:name=".MyService"
android:enabled="true"
android:exported="true">
<intent-filter>
<action android:name="myfirstservice">
</action>
<category android:name=
"android.intent.category.DEFAULT">
</category>
</intent-filter>
</service>
	Intent t =new Intent();
	t.setPackage("com.example.dqjbd.we");
	t.setAction("myfirstservice");
	startService(t);

若实现 Service 隐式启动:一是在<service>标签中配置<intent-filter>相关信息,主要是 action 动作信息,这是隐式启动的关键;二是在关键代码中建立空 Intent 对象 t,利用 setPackage()函数设置工程包名,这样才能根据包名找到相应响应的 Service 类。利用 setAction()函数设置动作名称字符串。最后利用 startService()函数隐式启动 Service。

2. 与 Activity 是同一线程

由前文可知,Service 是能够在后台执行长时间运行操作并且不提供用户界面的应用程序组件。那么,以例 8-1 为例,重写 onStartCommand()代码如下所示。

```
public int onStartCommand(Intent intent, int flags, int startId) {
    boolean mark = true;
    while(mark){  }
    return super.onStartCommand(intent, flags, startId);
}
```

上述代码稍微有些夸张,可以在 while 循环中完成相应的后台工作,而且保持长时间运行,符合 Service 的特点。但当实际运行时,第 1 次单击 create 按钮创建了 MyService,onStartCommand()函数运行了。当再单击 create 按钮时,发现按钮背景无变化过程,根本不响应,等一会屏幕出现“XXX 无响应。是否将其关闭?”对话框,这说明 Service 与

Activity 在同一个主线程内，若 Service 内有一处是 while 死循环，则界面其他部分无法响应。因此，若在 Service 内能长期工作，那么它一定是启动了一个线程，增加多线程代码如下所示。

```
public int onStartCommand(Intent intent, int flags, int startId) {
    new Thread(new Runnable() {
        public void run() {
            boolean mark = true;
            while(mark){   }
        }
    }).start();
    return super.onStartCommand(intent, flags, startId);
}
```

运行 App 后会发现，create、destroy 按钮均可多次操作，流程运行正确。因此，在 Service 内运行多线程代码，才能保证服务长期在后台运行。

那么有读者会问：既然 Service 也是启动多线程完成相应任务，那还要 Service 干什么，直接利用 Activity 启动多线程不就可以了吗？这样是不行的，因为当 Activity 创建线程，假设线程可长时间工作，这时若销毁 Activity 组件(一直按手机返回键)，则工作线程也随之销毁，而 Service 却不会因为 Activity 组件销毁而销毁，它会长期存在，这就是下面要论述的 Service 的单独存在性。

3. 单独存在性

Service 可以由 Activity 启动，但不会因为 Activity 销毁而销毁，这保证了 Service 可为其他组件和应用程序所共享，这就是 Service 独立性的特点。如何验证？仍以例 8-1 为例，做到以下两点即可。

- 在 MainActivity 类中，重写 onDestroy()函数，如下所示。

```
protected void onDestroy() {
    Toast.makeText(this, "Activity  destroyed",Toast.LENGTH_SHORT).show();
    super.onDestroy();
}
```

- 在 MyService 类中，重写 onStartCommand()函数，如下所示。

```
public int onStartCommand(Intent intent, int flags, int startId) {
    new Thread(new Runnable() {
        public void run() {
```

```
                try {
                    Thread.sleep(20000);
                    MyService.this.stopSelf();
                }catch(Exception e){}
            }
        }).start();
        return super.onStartCommand(intent, flags, startId);
    }
```

在线程中利用休眠函数休眠 20s，之后调用 stopSelf()函数销毁该服务，得出：若在自定义 Service 类实现销毁服务功能，则利用 Service 类中的 stopSelf()函数；若在上下文 Context 类中销毁 Service，则利用 stopService()函数。

实验时，单击 create 按钮，保证 MyService 运行，它将在 onStartCommand()函数中休眠 20s。这时不停地按手机返回键，当 MainActivity 销毁时，运行其生命周期的最后一个函数 onDestroy()，由于已经重写，所以在屏幕上看到利用 Toast 显示的“Activity destroyed”字符串，表明 MainActivity 已经销毁完毕。MyService 休眠 20s 后主动调用 MyService.this.stopSelf()函数，引起 MyService 类中的 onDestroy()函数运行，直到此时才在屏幕上显示“Service destroy”字符串。这表明 MainActivity 销毁时，MyService 仍然存在，证明了它的独立性特点。

8.2.3 应用示例

【例 8-2】 简易的 MP3 播放器。

Android 提供了常见的音频、视频的编码、解码机制。借助多媒体类 MediaPlayer 的支持，开发人员可以很方便地在应用中播放音频、视频。MediaPlayer 类常用函数分类及介绍如下所示。

① 设置数据源函数。

- void setDataSource(String path)：通过一个媒体资源的地址指定 MediaPlayer 的数据源，这里的 path 可以是一个本地路径，也可以是网络路径。
- void setDataSource(Context context，Uri uri)：通过一个 Uri 指定 MediaPlayer 的数据源，这里的 Uri 可以是网络路径或本地路径。
- void setDataSource(FileDescriptor fd)：通过一个 FileDescriptor 指定一个 MediaPlayer 的数据源。

② 播放函数。

- void start()：开始播放或恢复播放。

- void stop()：停止播放。
- void pause()：暂停播放。

③ 资源释放函数。

- void release ()：回收流媒体资源，当不再播放时，要调用该函数，释放占有的内存资源。

④ 数据源属性或播放设置函数等。

- int getDuration()：获取流媒体的总播放时长，单位是毫秒。
- int getCurrentPosition()：获取当前流媒体的播放的位置，单位是毫秒。
- void seekTo(int msec)：设置当前 MediaPlayer 的播放位置，单位是毫秒。
- void setLooping(boolean looping)：设置是否循环播放。
- boolean isLooping()：判断是否循环播放。
- boolean isPlaying()：判断是否正在播放。
- void prepare()：同步的方式装载流媒体文件。
- void prepareAsync()：异步的方式装载流媒体文件。
- void release ()：回收流媒体资源。
- void setAudioStreamType(int streamtype)：设置播放流媒体的类型。
- void setWakeMode(Context context, int mode)：设置 CPU 唤醒的状态。
- void setNextMediaPlayer(MediaPlayer next)：设置当前流媒体播放完毕，下一个播放的 MediaPlayer 对象。

本示例实现的功能是：MP3 数据源固定为 a.mp3，且在服务器端，手机 App 端通过网络播放 a.mp3，且有暂停和停止功能。

本示例涉及服务器端和 App 编程：服务器端即动态 Web 工程 myapp，将 a.mp3 保存在工程根目录下即可。App 端涉及：配置文件 Androidmanifest.xml；主界面文件 main.xml；主应用文件 MainActivity.java，自定义 Service 类文件 MyService.java。App 端各具体代码详细描述如下所示。

① 配置文件 Androidmanifest.xml：由于涉及网络，因此必须设置网络访问允许权限。由于 MP3 数据流可能涉及手机外部存储的读写，因此最好加上允许外部读写权限，如下所示。

```
<uses-permission android:name="android.permission.INTERNET" />
<uses-permission android:name="android.permission.ACCESS_NETWORK_STATE" />
<uses-permission android:name="android.permission.READ_EXTERNAL_STORAGE" />
<uses-permission android:name="android.permission.WRITE_EXTERNAL_STORAGE" />
```

② 主界面文件 main.xml。

```
<?xml version="1.0" encoding="utf-8"?>
<LinearLayout xmlns:android="http://schemas.android.com/apk/res/android"
android:layout_width="match_parent"
android:layout_height="match_parent">
    <Button
        android:id="@+id/play"
        android:layout_width="wrap_content"
        android:layout_height="wrap_content"
        android:text="play"/>
    <Button
        android:id="@+id/pause"
        android:layout_width="wrap_content"
        android:layout_height="wrap_content"
        android:text="pause"/>
    <Button
        android:id="@+id/stop"
        android:layout_width="wrap_content"
        android:layout_height="wrap_content"
        android:text="stop"/>
</LinearLayout>
```

id 为 play 的按钮对应播放按钮,id 为 pause 的按钮对应暂停按钮,id 为 stop 的按钮对应停止按钮。

③ 主应用文件 MainActivity.java。

```
public class MainActivity extends AppCompatActivity {
    protected void onCreate(Bundle savedInstanceState) {
        super.onCreate(savedInstanceState);
        setContentView(R.layout.main);
        Button b = (Button)findViewById(R.id.play);
        Button b2= (Button)findViewById(R.id.pause);
        Button b3= (Button)findViewById(R.id.stop);
        b.setOnClickListener(new View.OnClickListener() {
            public void onClick(View v) {
               Intent t = new Intent(MainActivity.this, MyService.class);
                t.putExtra("cmd",1);
                startService(t);
```

```
            }
        });
        b2.setOnClickListener(new View.OnClickListener() {
            public void onClick(View v) {
                Intent t = new Intent(MainActivity.this, MyService.class);
                t.putExtra("cmd",0);
                startService(t);
            }
        });
        b3.setOnClickListener(new View.OnClickListener() {
            public void onClick(View v) {
                Intent t = new Intent(MainActivity.this, MyService.class);
                stopService(t);
            }
        });
    }
}
```

Button 对象 b、b2 对应的播放、暂停歌曲功能，都是通过 startService()函数显示启动服务的。由于它们在 MyService 中走的流程大致相同，为了区分是播放还是暂停功能，通过设置 Intent 对象的值加以区分，播放功能的特征值是 1，暂停功能的特征值是 0。Button 对象 b3 对应停止按钮，用于停止播放及结束服务。

④ 自定义 Service 类文件 MyService.java。

```
public class MyService extends Service {
    MediaPlayer player;
    public MyService(){  }
    public IBinder onBind(Intent intent) {
        throw new UnsupportedOperationException("Not yet implemented");
    }
    public void onCreate() {
        super.onCreate();
        String url = "http://192.168.1.100:8080/myapp/a.mp3";
        try {
            player = new MediaPlayer();
            player.setDataSource(this, Uri.parse(url));
            player.prepare();
        }catch(Exception e){ }
```

```
    }
    /* 当产生 MyServcice 实例时会调用 onCreate()函数，直接建立 MediaPlayer 对象
player，利用 setDataSource()函数设置服务器的播放数据源为 a.mp3，利用 prepare()函数
做好播放准备。
    总之，onCreate()是 Service 生命周期中仅运行一次的函数，在此一般完成所需功能的初
始化工作 */
    public int onStartCommand(Intent intent, int flags, int startId) {
        int cmd = intent.getIntExtra("cmd", -1);
        if(cmd==1) player.start();
        if(cmd==0) player.pause();
        return super.onStartCommand(intent, flags, startId);
    }
    /* 每执行一次 startService()函数，onStartCommand()函数就运行一次，因此 Service
服务功能一般都写在该函数中。本例中，首先通过解析 Intent 对象 intent 获得传入的命令值
cmd。当 cmd 为 1 时，执行歌曲播放命令；当 cmd 为 0 时，执行歌曲播放暂停命令 */
    public void onDestroy() {
        super.onDestroy();
        player.stop();player.release();
    }
}
```

当 Service 服务销毁时，首先要通过 stop()函数停止播放音乐，再用 release()函数释放媒体播放器所占资源。onDestroy()函数是 Service 生命周期最后执行的函数，请读者记住在此处一定要释放相应的资源。

实验时，要保证手机、服务器在同一局域网，先运行服务器，再运行手机 App。

【例 8-3】 跨进程访问例 8-2 中的简易播放器 Service 类 MyService。

本示例实现的功能如图 8-1 所示。

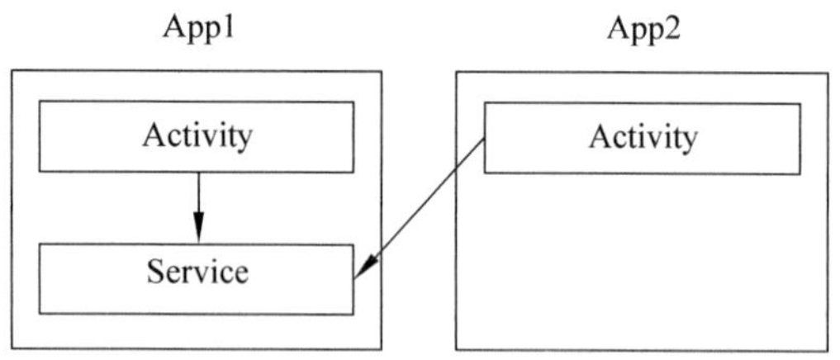

图 8-1 跨进程访问 Service 图

两个 Android 工程 App1、App2。App1 中包含 Activity 及 Service，通过 Activity 控制 Service；App2 工程中希望通过 Activity 跨进程访问 App1 中的 Service。App1 就是例

8-2 中的工程，App2 是新建的一个工程。

实现例 8-3 仅需要主界面 main.xml 和主应用程序 MainActivity.java。下面一一说明。

① 主界面 main.xml。

```
<?xml version="1.0" encoding="utf-8"?>
<LinearLayout xmlns:android="http://schemas.android.com/apk/res/android"
android:layout_width="match_parent"
android:layout_height="match_parent">
    <Button
        android:id="@+id/play"
        android:layout_width="wrap_content"
        android:layout_height="wrap_content"
        android:text="play"/>
</LinearLayout>
```

以上代码仅定义了 id 为 play 的播放按钮，主要验证当歌曲暂停后，单击该按钮时，看歌曲能否继续歌唱。当然，也可增加暂停、停止播放按钮，读者可继续完成。

② 主应用程序 MainActivity.java。

```
public class MainActivity extends AppCompatActivity {
    protected void onCreate(Bundle savedInstanceState) {
        super.onCreate(savedInstanceState);
        setContentView(R.layout.activity_main);
        Button b = (Button)findViewById(R.id.myok);
        b.setOnClickListener(new View.OnClickListener() {
            public void onClick(View v) {
                Intent t = new Intent();
                t.putExtra("cmd",1);
                t.setPackage("com.example.dqjbd.we");
                t.setAction("myfirstservice");
                startService(t);
            }
        });
    }
}
```

在本示例中，因为是跨进程访问 MyService，因此一定是隐式启动，要看一下 MyService 在例 8-2 工程中配置文件 Androidmanifest.xml 中的相关配置，配置内容如之前的表 8-1 所示。根据配置内容写出相应的隐式启动代码，一般来说，要利用 setPackage()函数设置 MyService 对应原工程中的包名，利用 setAction()函数设置动作属性特征字符串，最后利用 startService()函数隐式启动 Service。

当然，在设置 Intent 对象 t 中，要注意传参必须与例 8-2 中的传参规则一致，如代码中的 t.put("cmd",1)，否则即使启动了 MyService，执行的功能也是错误的。

实验前：将例 8-2、例 8-3 工程安装到手机上。实验时：先启动服务器，运行例 8-2 工程 App，按 play 播放键，歌曲播放。按 pause 键，歌曲停止播放；然后切换到例 8-3 工程 App（保证例 8-2 工程 App 仍运行），按 play 播放键，发现歌曲继续播放。这说明例 8-3 工程 App 可访问例 8-2 工程中的 MyService，跨进程访问成功。

8.3 绑定 Service

8.3.1 生命周期

对有意义的绑定 Service 来说，都要从系统类 Service 派生。重写 Service 类定义的相关函数，因此深刻理解各重写函数在生命周期的位置就非常重要。一般来说，对绑定 Service 来说，经常重写的 4 个函数是 onCreate()、onBind()、onUnbind()和 onDestroy()，这些函数在绑定 Services 生命周期中均只执行一次。具体描述如下所示。

- onCreate()：当创建服务时调用，若该服务已存在，则不再调用。
- onBind()：绑定函数，通过 IBinder 接口将调用端与 Service 捆绑在一起，使得调用端获得 Service 的实例，可以访问 Service 类中的函数。
- onUnbind()：解绑函数，切断调用端与 Service 的关联。
- onDestroy()：顾名思义，当执行该函数时，服务即将销毁，该函数主要具有资源释放等功能。

注意：绑定 Service 与 onStartCommand()函数无关，若在绑定 Service 中重写了该函数，可能引起 App 运行崩溃。

总之，绑定 Service 具有如下特点。

- 绑定 Service 调用者和服务之间是典型的 Client-Server 的接口，即调用者是客户端，Service 是服务端，Service 就一个，但是连接绑定到 Service 上面的客户端 Client 可以多个。
- 客户端可以通过 IBinder 接口获取 Service 的实例，从而可以实现在 Client 端直接调用 Service 中的方法，以实现灵活的交互，并且可借助 IBinder 实现跨进程的 Client-Server 的交互，这在纯 startService()启动的 Service 中是无法实现的。

• 不同于 startService()启动的服务默认无限期执行，绑定 Service 服务的生命周期与其绑定的 client 息息相关。当 client 销毁的时候，client 会自动与 Service 解除绑定。当然，client 也可以通过明确调用 Context 的 unbindService()方法与 Service 解除绑定。当没有任何 client 与 Service 绑定的时候，Service 会自行销毁。

【例 8-4】 验证绑定 Service 生命周期。

由于绑定 Service 要实现调用端——Service 间通信，因此其生命周期代码要比启动 Service 复杂。涉及的文件包括：配置文件 Androidmanifest.xml；绑定 Service 类文件 WorkService.java；主界面 main.xml；主应用 MainActivity.java。下面一一说明。

① 配置文件 Androidmanifest.xml 中与 Service 相关节点内容。

```
<service
    android:name=".WorkService"
    android:enabled="true"
    android:exported="true"></service>
```

② 绑定 Service 类文件 WorkService.java。

```
public class WorkService extends Service {
    public class WorkBinder extends Binder{    }
    public WorkService() {    }
    public void onCreate() {
        super.onCreate();
        Toast.makeText(getApplicationContext(),"create",Toast.LENGTH_SHORT).
show();
    }
    public IBinder onBind(Intent intent) {
        Toast.makeText(getApplicationContext(),"bind",Toast.LENGTH_SHORT)
.show();
        return new WorkBinder();
    }
    public boolean onUnbind(Intent intent) {
        Toast.makeText(getApplicationContext(),"unbind",Toast.LENGTH_SHORT).
show();
        return super.onUnbind(intent);
    }
    public void onDestroy() {
        super.onDestroy();
        Toast.makeText(getApplicationContext(),"destroy",Toast.LENGTH_
```

```
SHORT).show();
    }
}
```

以上代码重写了绑定 Service 生命周期的 4 个重要函数：onCreate()、onBind()、onUnbind()、onDestroy()。下面着重分析 onBind()函数，该函数的返回值是 IBinder 对象，该对象要返回到调用端，因此，调用端和 Service 端都有 IBinder 对象，可方便实现调用端和 Service 端的通信。由于 IBinder 是系统接口，因此在 WorkService 类中利用内部类定义了 Binder(IBinder 接口的实现类)的派生类 WorkBinder()，代码最后直接返回 new WorkBinder()。

当然，本示例仅是为了验证生命周期流程，MyBinder 类内容为空，还有待完善。

③ 主界面 main.xml。

```
<?xml version="1.0" encoding="utf-8"?>
<LinearLayout xmlns:android="http://schemas.android.com/apk/res/android"
    android:layout_width="match_parent"
    android:layout_height="match_parent">
    <Button
        android:id="@+id/mybind"
        android:layout_width="wrap_content"
        android:layout_height="wrap_content"
        android:text="bind"/>
    <Button
        android:id="@+id/myunbind"
        android:layout_width="wrap_content"
        android:layout_height="wrap_content"
        android:text="unbind"/>
</LinearLayout>
```

以上代码定义了两个按钮：一个 id 为 mybind 的按钮，用于绑定 WorkService；另一个 id 为 myunbind 的按钮，用于解绑 WorkService。

④ 主应用 MainActivity.java。

```
public class MainActivity extends AppCompatActivity {
    public ServiceConnection sc = new ServiceConnection() {
        public void onServiceConnected(ComponentName name, IBinder service) {
            Toast.makeText(MainActivity.this,"Connect",Toast.LENGTH_SHORT)
.show();
        }
```

```
        public void onServiceDisconnected(ComponentName name) {
            Toast.makeText(MainActivity.this,"disconnect",Toast.LENGTH_
SHORT).show();
        }
    };
    protected void onCreate(Bundle savedInstanceState) {
        super.onCreate(savedInstanceState);
        setContentView(R.layout.main);
        Button b = (Button)findViewById(R.id.mybind);
        b.setOnClickListener(new View.OnClickListener() {
            public void onClick(View v) {
                Intent t = new Intent(MainActivity.this, WorkService.class);
                bindService(t, sc, Service.BIND_AUTO_CREATE);
            }
        });
        Button b2 = (Button)findViewById(R.id.myunbind);
        b2.setOnClickListener(new View.OnClickListener() {
            public void onClick(View v) {
                Intent t = new Intent(MainActivity.this, WorkService.class);
                unbindService(sc);
            }
        });
    }
}
```

MainActivity 与 WorkService 通信要借助 ServiceConnection 对象,它是 Android 的一个系统类,该类的主要作用是截获 Service 的绑定对象。ServiceConnection 类主要有如下两个重载函数。

- onServiceConnected():系统调用该函数,传送在 Service 的 onBind()中返回的 IBinder 对象,若该函数运行,表明绑定成功。
- onServiceDisconnected():Service 在正常运行和销毁时并不调用该函数,仅在某些意外情况时调用,例如当 Service 崩溃或被系统杀死的情况下。

绑定 Service 的具体函数是 bindService(Intent t,ServiceConnection sc, int flags)。t 是 Intent 对象,sc 是 ServiceConnection 对象,flags 是标志位,表明绑定中的操作,它一般应是 BIND_AUTO_CREATE,这样就会在 Service 不存在时创建一个,其他可选的值是 BIND_DEBUG_UNBIND 和 BIND_NOT_FOREGROUND,不想指定时设为 0 即可。

解绑 Service 的具体函数是 unbindService(ServiceConnection sc)。

App 运行后,实验过程如下所示。

单击 bind(绑定)按钮,根据 Toast 显示,发现其走的流程是:首先是 Service 端的 onCreate()->onBind(),然后是 Activity 端的 onServiceConnected()。再次单击 bind 按钮,发现无任何反应,表明上述执行流程仅走一次。

单击完 bind 按钮,再单击 unbind(解绑)按钮,根据 Toast 显示,发现其走的流程是:Service 端的 onUnbind()->onDestroy()。

单击 bind 按钮后,再按手机返回键,根据 Toast 显示,发现其走的流程是:Service 端的 onUnbind()->onDestroy()。其流程与单击 unbind 按钮的相应流程相同。

【例 8-5】 在例 8-4 的基础上,编制最简单的绑定 Service 类 WorkService。其功能是:在 WorkService 中定义函数 func(),在调用端直接调用,将其返回值利用 Toast 显示。其具体代码如下所示。

① 主界面 main.xml。

```
<?xml version="1.0" encoding="utf-8"?>
<LinearLayout xmlns:android="http://schemas.android.com/apk/res/android"
    android:layout_width="match_parent"
    android:layout_height="match_parent">
    <Button
        android:id="@+id/demo"
        android:layout_width="wrap_content"
        android:layout_height="wrap_content"
        android:text="demo"/>
</LinearLayout>
```

以上代码定义了 id 为 demo 的按钮,单击此按钮,调用 WorkService 中的 func()函数。

② Service 文件 WorkService.java。

```
public class WorkService extends Service {
    public WorkService() { }
    public class WorkBinder extends Binder{
        WorkService getService(){
            return WorkService.this;
        }
    }
    public IBinder onBind(Intent intent) {
        return new WorkBinder();
```

```
    }
    public String func(){ return "Hello";}
}
```

该类略去了无关的生命周期函数，仅重写了 onBind()，用以向调用端返回 WorkBinder 对象。WorkBinder 类中编写了自定义函数(非重载)getService()，直接返回 WorkService 对象，该函数是调用端调用的函数。

③ 主应用文件 MainActivity.java。

```
public class MainActivity extends AppCompatActivity {
    WorkService ws;
    ServiceConnection serviceConnection = new ServiceConnection() {
        public void onServiceConnected(ComponentName name, IBinder binder) {
            ws = ((WorkService.WorkBinder)binder).getService();
        }
        public void onServiceDisconnected(ComponentName name) {}
    };
    protected void onCreate(Bundle savedInstanceState) {
        super.onCreate(savedInstanceState);
        setContentView(R.layout.main);
        Intent service = new Intent(MainActivity.this, WorkService.class);
        bindService(service, serviceConnection, Context.BIND_AUTO_CREATE);
        Button b = (Button)findViewById(R.id.demo);
        b.setOnClickListener(new View.OnClickListener() {
            public void onClick(View v) {
                Toast.makeText(MainActivity.this, ws.func(), Toast.LENGTH_
LONG).show();
                unbindService(sc);
            }
        });
    }
}
```

以上代码定义了 WorkService 成员变量 ws，表明要通过服务的方式获得 WorkService 实例，是在 onServiceConnected()函数中获得的：在该函数内必须先将形参 binder 强制转换为 WorkBinder 对象，然后再通过 getService()函数获得 WorkService 实例对象 ws，这样才能调用 WorkService 类中的函数。

执行流程是：程序一启动，就通过 bindService()将 MainActivity 与 WorkService 建

立绑定关系。当单击 demo 按钮时,调用服务端的 func()函数,将其字符串返回值利用 Toast 显示,然后利用 unbindService()函数直接完成解绑工作。

该示例是非常普通的绑定 Service 示例,关键思想是在调用方获得 Service 实例,因此一定要定义其为成员变量。试想,若跨 App 访问,是不可能将 XXX Service 定义为成员变量的,因此本示例有较大的局限性,如何解决? Messenger 技术是较好的方案之一,见下文。

8.3.2 Messenger 技术

在 6.2 节学习了 Handler 技术,Messenger 类可以对 Handler 进一步封装,用法和 Handler 基本一致,请看例 8-6。

【例 8-6】 最简单的 Messenger 发送、接收消息示例,涉及的文件如下:主界面 main.xml;主应用 MainActivity.java。下面一一介绍。

① 主界面:main.xml,仅是象征意义,空白界面即可。

② 主应用:MainActivity.java。

```
public class MainActivity extends AppCompatActivity {
    protected void onCreate(Bundle savedInstanceState) {
        super.onCreate(savedInstanceState);
        setContentView(R.layout.main);
        Messenger msgs = new Messenger(new Handler(){
            public void handleMessage(Message msg) {
                super.handleMessage(msg);
                Toast.makeText(MainActivity.this,(String)msg.obj,Toast
.LENGTH_SHORT).show();
            }
        });
        try {
            Message msg = Message.obtain();
            msg.obj = "Hello";
            msgs.send(msg);
        }catch(Exception e){}
    }
}
```

Messenger 常用的构造方法是 Messenger(Handler handler),本例用匿名类建立了 Handler 对象,并重写了 handlerMessage()函数。

程序执行时：首先建立 Message 对象 msg，信息设置为 Hello，然后利用 msgs.send(msg)将消息发送出去，接收函数在获得的字符串 handlerMessage()函数中直接将获得的字符串利用 Toast 显示在界面上。

本例与 Service 无关，Messenger 如何与 Service 关联呢？Messenger 还有另外形式的构造方法 Messenger(IBinder binder)及一个重要的函数 IBinder getBinder()，IBinder 与 Service 是相关的，因此在 Service 中一定能用到 Messenger。但为了便于理解，继续做实验，仍不与 Service 关联。替换实例中 onCreate()函数中的 try-catch 代码，见表 8-2。

表 8-2　发送消息代码替换对比表

try-catch 块源代码	try-catch 块替换后的代码
Message msg= Message.obtain(); msg.obj = "Hello"; msgs.send(msg);	Messenger msgs2 =new Messenger(msgs.getBinder()); Message msg = Message.obtain(); msg.obj = "Hello"; msgs2.send(msg);

可以发现，代码替换前后的运行结果是一致的，由于 msgs、msgs2 都是 Messenger 对象，msgs2 对象建立的过程是：首先利用 msgs 的 getBinder()函数获得 IBinder 对象，再以此构建 Messenger 对象 msgs2。也就是说，msgs、msgs2 对象的 getBinder()是同一个 IBinder 对象，因此用 msgs 或者 msgs2 发送消息，它们的消息接收函数一定是一样的，因此结果相同。

其实，明白了上述道理，也就能编制利用 Messenger 实现的 Service 功能的类了。

【例 8-7】 利用 Messenger 重新实现例 8-5 的功能。

例 8-5 的功能是调用端调用服务器端的 func()函数，将字符串值返回给服务器端并显示。为了解说方便，将此功能分为两个阶段：调用端传送消息到服务器端；服务器端传送消息到调用端。

首先，实现调用端传送消息到服务器端，涉及的文件有服务类 WorkService.java；主界面 main.xml；主应用 MainActivity.java。下面一一说明。

① 服务类 WorkService.java。

```
public class WorkService extends Service {
    Messenger msgServer = new Messenger(new Handler(){
        public void handleMessage(Message msg) {
            super.handleMessage(msg);
            String s = (String)msg.obj;
            Toast.makeText(getApplicationContext(),s,Toast.LENGTH_LONG).show();
        }
```

```
    });
    public IBinder onBind(Intent intent) {
        return msgServer.getBinder();
    }
    String func(){ return "Hello";}
}
```

服务器端定义了 Messenger 成员变量 msgServer，用于接收调用方传过来的消息。onBind()函数中将 msgServer 绑定的 IBinder 对象返回给调用端。

② 主界面 main.xml：与例 8-5 相同，略。

③ 主应用 MainActivity.java。

```
public class MainActivity extends AppCompatActivity {
    Messenger msgServer;
    ServiceConnection sc = new ServiceConnection() {
        public void onServiceConnected(ComponentName name, IBinder binder) {
            msgServer = new Messenger(binder);
        }
        public void onServiceDisconnected(ComponentName name) {  }
    };
    protected void onCreate(Bundle savedInstanceState) {
        super.onCreate(savedInstanceState);
        setContentView(R.layout.main);
        Intent t = new Intent(this, WorkService.class);
        bindService(t,sc,Service.BIND_AUTO_CREATE);
        Button b = (Button)findViewById(R.id.demo);
        b.setOnClickListener(new View.OnClickListener() {
            public void onClick(View v) {
                try {
                    Message msg = Message.obtain();
                    msg.obj = "func";
                    msgServer.send(msg);
                }catch(Exception e){}
            }
        });
    }
}
```

以上代码定义了 Messenger 成员变量 msgServer。onServiceConnected()函数中的

关键语句"msgServer = new Messenger(binder)"表明 MainActivity、WorkService 中两个同名 msgServer 对象有相同的 binder 对象。因此,当单击 demo 按钮,利用 msgServer 传送字符串"func"时,在服务器 WorkService 端的 handleMessage()函数就会接收到该字符串,并利用 Toast 显示在界面上。

下面在完成第一阶段(调用端传送消息到服务器端)功能的基础上,再实现第二阶段的功能:服务器端传送消息到客户端。

很明显,调用端也要有单独的 Messenger 对象,用于接收服务器端传过来的消息。下面仅列出与上述 MainActivity.java、WorkService.java 的差异代码。

① MainActivity.java。

```
public class MainActivity extends AppCompatActivity {
    Messenger msgServer;
    Messenger msgClient = new Messenger(new Handler(){
        public void handleMessage(Message msg) {
            super.handleMessage(msg);
            Toast.makeText(MainActivity.this,(String)msg.obj,Toast.LENGTH_
LONG).show();
        }
    });
    ServiceConnection sc = new ServiceConnection() {//与前文完全相同,略}
    protected void onCreate(Bundle savedInstanceState) {
        super.onCreate(savedInstanceState);
        setContentView(R.layout.main);
        Intent t = new Intent(this, WorkService.class);
        bindService(t,sc,Service.BIND_AUTO_CREATE);
        Button b = (Button)findViewById(R.id.demo);
        b.setOnClickListener(new View.OnClickListener() {
            public void onClick(View v) {
                try {
                    Message msg = Message.obtain();
                    msg.replyTo = msgClient;
                    msg.obj = "func";
                    msgServer.send(msg);
                }catch(Exception e){}
            }
        });
    }
}
```

以上代码定义了 Messenger 成员变量 msgClient，重写了 handleMessage()函数，用于接收从服务器端传送的字符串数据，并利用 Toast 将其显示在界面上。

当单击 demo 按钮设置消息时，多了一行语句"msg.replyTo＝msgClient"，表明该消息发送到服务器端后，服务器端回发消息时，回发消息的接收器 replyTo 是 msgClient 重写的 handleMessage()函数。

② WorkService.java。

```
public class WorkService extends Service {
    Messenger msgServer = new Messenger(new Handler(){
        public void handleMessage(Message msg) {
            super.handleMessage(msg);
            String s = (String)msg.obj;
            if(s.equals("func")){
                try {
                    Message msg2 = Message.obtain();
                    msg2.obj = func();
                    msg.replyTo.send(msg2);
                }catch(Exception e){}
            }
        }
    });
    //其他代码同之前的代码,略
}
```

可以看出，在服务端接收消息时，由于传过来的是字符串，因此要进行判断，根据该特征值执行相应的函数。本例中传过来的是 func，执行的是 func()函数。当向调用端回传结果信息时，用到关键的一行语句"msg.replyTo.send(msg2)"，其中 msg.replyTo 是在调用端设置的接收响应 Messenger 对象 msgClient，它附着在消息中传到了服务端，服务端直接应用即可。

总之，利用 Messenger 对象实现了调用端-服务端的通信，优点是：省去了自定义 Binder 派生类的编制，方便跨进程的访问通信，通用性更好。不足是：在调用方-服务器端有过多的字符串解析与判别，程序显得不易理解，维护不方便。

8.3.3 AIDL 技术

利用 Messenger 可以实现进程通信，但涉及太多的字符串操作。是否有更简洁的替代技术？AIDL 技术是重要的手段，AIDL 是 Android interface definition language(安卓接口定义语言)的简称,它可以简化进程间数据格式转换和数据交换的代码,允许在不同进程的调用者和 Service 之间相互传递数据,属于轻量级的进程通信机制。

AIDL 的核心思想是公用接口技术,其基本构架如图 8-2 所示。

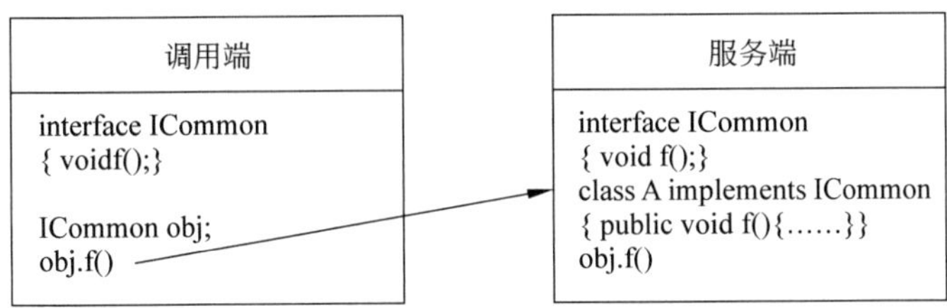

图 8-2　AIDL 跨进程访问原理简图

可以看出,跨进程的调用端和服务端有共同的接口 ICommon,接口的实现类 A 在服务端。调用端通过某种机制可以产生 ICommon 对象 obj,当 obj 调用接口函数 f()时,它调用的其实是服务端的 f()函数。

应用 AIDL 创建远程服务一般分为 3 步:使用 AIDL 定义远程服务接口;通过继承 Service 类实现远程服务;绑定和使用远程服务。

【例 8-8】 利用 AIDL 在单进程中实现例 8-5 的功能。具体步骤如下所示。

① 定义 AIDL 接口文件 IWorkService.aidl。

```
package com.example.dqjbd.we2;
interface IWorkService {
    String func();
}
```

可以看出,AIDL 文件的扩展名必须是 aidl,其定义与普通接口基本一致,注意一定要写上全包名。本示例所有程序文件都在 com.example.dqjbd.we2 包下,但唯独该文件写上了包名,主要是强调该文件移植到其他工程中包名随工程不同而不同。

当产生该文件后,会发现生成了与之匹配的 Java 文件 IWorkServicew.java,其主要内容如下所示。

```
public interface IWorkService extends android.os.IInterface
{
    public static abstract class Stub extends android.os.Binder implements
          com.example.dqjbd.we2.IWorkService
    { //代码略 }
    public java.lang.String func() throws android.os.RemoteException;
}
```

该文件内部有一个抽象类——Stub 类（而且是一个 Binder 类）及定义的 func()函数。

② 服务类 WorkService.java。

```
public class WorkService extends Service {
    private final IWorkService.Stub binder = new IWorkService.Stub() {
        public String func() throws RemoteException {
            return "Hello";
        }
    };
    public IBinder onBind(Intent intent) {
        return binder;
    }
}
```

以上代码主要实现了 IWorkService.Stud 抽象类定义的各个方法，由此可得，若想调用服务端的 N 个方法，则将这些抽象方法均定义在 IWorkService.aidl 文件中，在服务端的 IWorkService.Stub 匿名类或子类中一一实现即可。

③ 主界面 main.xml：与例 8-5 相同，略。

④ 主应用 MainActivity.java。

```
public class MainActivity extends AppCompatActivity {
    IWorkService ws ;
    ServiceConnection sc = new ServiceConnection() {
        public void onServiceConnected(ComponentName name, IBinder binder) {
            ws = IWorkService.Stub.asInterface(binder);
        }
        public void onServiceDisconnected(ComponentName name) {     }
    };
    protected void onCreate(Bundle savedInstanceState) {
        super.onCreate(savedInstanceState);
        setContentView(R.layout.main);
        Intent t = new Intent(this, WorkService.class);
        bindService(t,sc,Service.BIND_AUTO_CREATE);
        Button b = (Button)findViewById(R.id.demo);
        b.setOnClickListener(new View.OnClickListener() {
            public void onClick(View v) {
                try {
                    String s = ws.func();
                    Toast.makeText(MainActivity.this,s,Toast.LENGTH_LONG)
```

```
.show();
                }catch(Exception e){ }
            }
        });
    }
}
```

以上代码定义了 AIDL 接口 IWorkService 成员变量 ws，ws 的具体值是在 onServiceConnected()函数中通过代码 ws = IWorkService.Stub.asInterface(binder)获得的。在按钮响应事件中调用服务器端函数，形式为 ws.func()，非常简洁，与平时的普通类方法调用无异。

【例 8-9】 跨进程访问例 8-8 中的 WorkService。

由于跨进程访问，因此一定是隐式访问。例 8-8 的配置文件 Androidmanifest.xml 应该增加<intent-filter>设置，如下所示。

```
<service
    android:name=".WorkService"
    android:enabled="true"
    android:exported="true">
    <intent-filter>
        <action android:name="myfirstaidl"></action>
    </intent-filter>
</service>
```

以上代码增加了 action 动作设置，将其 name 属性设置为 myfirstaidl，其他进程将根据该属性访问 WorkService。

上述工作完成后，开始例 8-9 的代码编制，其实它的代码几乎与例 8-8 中的代码相同，只是没有 WorkService 类而已，下面一一论述所需的代码文件。

① IWorkService.aidl 代码文件。

```
package com.example.dqjbd.we2;
interface IWorkService {
    String func();
}
```

接口定义与例 8-8 中的完全相同，只是包名随工程不同而不同。

② 主界面与例 8-8 中的 main.xml 完全一致，略。

③ 主应用 MainActivity.java。

```
public class MainActivity extends AppCompatActivity {
        //其他所有代码同,略
    protected void onCreate(Bundle savedInstanceState) {
        super.onCreate(savedInstanceState);
        setContentView(R.layout.main);
        Intent t = new Intent();
        t.setAction("myfirstaidl");
        t.setPackage("com.example.dqjbd.we2");
        bindService(t,sc,Service.BIND_AUTO_CREATE);
        Button b = (Button)findViewById(R.id.demo);
        b.setOnClickListener(new View.OnClickListener() {
            public void onClick(View v) {
                try {
                    String s = ws.func();
                    Toast.makeText(MainActivity.this,s,Toast.LENGTH_LONG)
.show();
                }catch(Exception e){ }
            }
        });
    }
}
```

onCreate()函数几乎与例 8-8 中的完全相同,只是 Service 由显示启动变成了隐式启动,setAction(para)函数设置动作属性时,para 要与例 8-8 中系统配置文件中 WorkService 对应的动作属性一致,setPackage(package)函数设置包名时,package 指的是例 8-8 中 WorkService 所在的包路径,与例 8-9 工程的包路径无关。

实验时:先运行例 8-8 的工程,这时 WorkService 服务已经建立;然后切换界面,运行例 8-9 的工程,单击 demo 按钮,可看到利用 Toast 显示的调用 WorkService 中的 func()函数返回的字符串"Hello",表明跨进程访问服务成功。

【例 8-10】 利用 AIDL 实现求两个数的和。

该功能可划分为 3 部分,如图 8-3 所示。

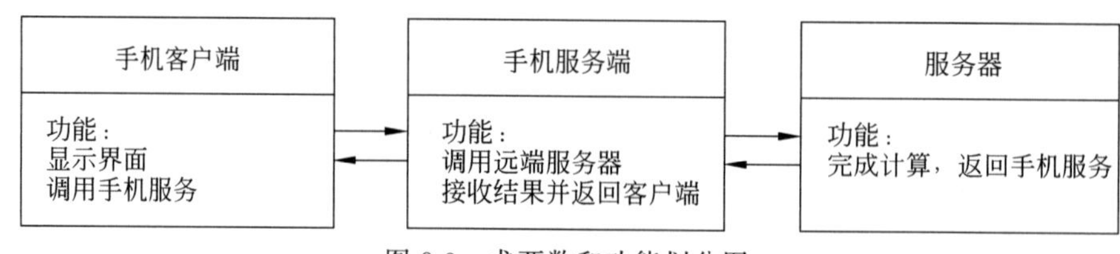

图 8-3 求两数和功能划分图

求两数和功能共划分为手机客户端、手机服务端、服务器 3 部分。手机客户端负责显示界面及调用手机服务(计算功能)。手机服务端接收客户端发出的计算请求,并继续向服务器端发送计算请求,服务器端接收计算请求,完成计算,将结果返回给手机服务端,手机服务端将结果返回给手机客户端。手机客户端完成最终的结果显示。

本示例涉及的文件有:配置文件 Androidmanifest.xml;AIDL 接口文件 IWorkService.aidl;手机客户端界面 main.xml 及应用程序 MainActivity.java;手机服务端程序 WorkService.java;服务器端的 add.jsp 文件。下面一一说明。

① 配置文件 Androidmanifest.xml:增加网络访问允许设置,如下所示。

```
<uses-permission android:name="android.permission.INTERNET"></uses-
permission>
```

② AIDL 接口文件 IWorkService.aidl。

```
package com.example.dqjbd.we2;
interface IWorkService {
    int calc(int num1,int num2);
}
```

以上代码定义了 calc()函数,其含义是计算 num1 与 num2 的和。

③ 手机客户端界面 main.xml 及应用程序 MainActivity.java。

```
//main.xml
<?xml version="1.0" encoding="utf-8"?>
<LinearLayout xmlns:android="http://schemas.android.com/apk/res/android"
    android:layout_width="match_parent"
    android:layout_height="match_parent"
    android:orientation="vertical">
    <EditText
        android:id="@+id/mynum1"
        android:layout_width="match_parent"
        android:layout_height="wrap_content" />
    <EditText
        android:id="@+id/mynum2"
        android:layout_width="match_parent"
        android:layout_height="wrap_content" />
    <TextView
        android:id="@+id/myresult"
        android:layout_width="match_parent"
```

```
        android:layout_height="wrap_content" />
    <Button
        android:id="@+id/mybtn"
        android:layout_width="wrap_content"
        android:layout_height="wrap_content"
        android:text="Add"/>
</LinearLayout>
```

两个 EditText 控件用于输入两个加数,id 分别为 mynum1、mynum2;一个 TextView 控件,用于显示两个加数的和,id 为 myresult;一个 Button 控件,id 为 mybtn,用于调用手机服务程序。

```
//应用程序 MainActivity.java
public class MainActivity extends AppCompatActivity {
    IWorkService ws ;
    Handler handle = new Handler(){
        public void handleMessage(Message msg) {
            super.handleMessage(msg);
            int result = msg.arg1;
            TextView tv = (TextView)findViewById(R.id.myresult);
            tv.setText(""+result);
        }
    };
    ServiceConnection sc = new ServiceConnection() {
            public void onServiceConnected(ComponentName name, IBinder binder) {
            ws = IWorkService.Stub.asInterface(binder);
        }
            public void onServiceDisconnected(ComponentName name) {      }
    };
    protected void onCreate(Bundle savedInstanceState) {
        super.onCreate(savedInstanceState);
        setContentView(R.layout.main);
        Intent t = new Intent(this, WorkService.class);
        bindService(t,sc,Service.BIND_AUTO_CREATE);
        Button b = (Button)findViewById(R.id.mybtn);
        b.setOnClickListener(new View.OnClickListener() {
            public void onClick(View v) {
                new Thread(new Runnable() {
                    public void run() {
```

```
                    EditText et = (EditText)findViewById(R.id.mynum1);
                    int num1 = Integer.parseInt(et.getText().toString());
                    EditText et2 = (EditText)findViewById(R.id.mynum2);
                    int num2 = Integer.parseInt(et2.getText().toString());
                    try {
                        int n = ws.calc(num1, num2);
                        Message msg = Message.obtain();
                        msg.arg1 = n;
                        handle.sendMessage(msg);
                    }catch(Exception e){ }
                }
            }).start();
        }
    });
  }
}
```

依据 AIDL 文件，定义了 IWorkService 成员变量 ws，用于调用手机服务程序；定义了 Handler 成员变量 handler 及相应的匿名类，主要作用是实现多线程中的界面局部刷新功能；定义了 ServiceConnection 类成员变量 sc 及相应的匿名类，用于实现服务绑定功能。

由于手机服务程序涉及远程访问服务器，因此在 Add 按钮事件响应 onClick()函数中必须加多线程操作。在多线程操作中完成：获取两个加数值 num1、num2，调用手机服务 ws.calc(num1,num2)，结果保存在变量 n 中，并将 n 显示在界面中。但是，在多线程中无法进行界面更新，必须在主线程中进行更新，因此将 n 封装在 Message 对象中，利用 Handler 技术实现消息的发送、异步接收、解析，以及将结果 n 显示在 TextView 组件中。

总之，若手机服务耗时，必须利用多线程技术，而且多线程一般在客户端体现，不要在服务端体现，这一点很重要。

④ 手机服务端程序 WorkService.java。

```
public class WorkService extends Service {
    private final IWorkService.Stub binder = new IWorkService.Stub() {
        public int calc(int num1,int num2) throws RemoteException {
            String result = "", unit = "";
            String req="http://192.168.1.100:8080/myapp/add.jsp?";
            req += "num1="+num1+"&num2="+num2;
            try {
```

```
                URL u = new URL(req);
                BufferedReader in = new BufferedReader(new InputStreamReader
(u.openStream()));
                while ((unit = in.readLine()) != null) {
                    result += unit;
                }
                result = result.trim();
            }catch(Exception e){   }
            return Integr.parseInt(result);
        }
    };
    public IBinder onBind(Intent intent) { return binder;   }
}
```

依据 AIDL 文件，定义了 IWorkService.Stub 成员变量 binder 及相应的匿名类实现，binder 由 onBind()函数返回到手机客户端。

calc()的功能是：将 num1、num2 添加到 Http 请求的参数中，以 GET 方式访问服务器 add.jsp，将服务器返回的结果保存在字符串 result 中，将 result 转换成整型数返回给手机客户端。

⑤ 服务器端：在动态 Web 工程 myapp 中编制 add.jsp，如下所示。

```
<%
    int num1 = Integer.parseInt(request.getParameter("num1"));
    int num2 = Integer.parseInt(request.getParameter("num2"));
    out.print(num1+num2);
%>
```

程序功能比较简单，首先利用 getParameter()获得 num1、num2 对应的字符串值，然后将字符串转化成整数，最后相加后将结果输出到手机服务端。

本示例测试时，要保证手机、服务器在同一网段。先启动 Tomcat 服务器，再执行手机客户端 App。

习题 8

一、简答题

1. 启动 Service 与绑定 Service 有什么不同？

2. 利用 Messenger 技术实现绑定 Service 的优点和缺点分别是什么?

3. 利用 AIDL 绑定 Service 有什么优点?

二、程序题

1. 完善例 8-2 简易的 MP3 播放器,在主界面上增加输入音乐文件名编辑框,当单击"开始播放"按钮时,播放该文件音乐。

2. 在主界面上输入两个加数,利用 AIDL 技术将返回结果显示在主界面上。已知 AIDL 接口定义如下所示。

```
interface IAddService {
    int add(int one, int two);
}
```

第9章

数据存储与共享

本章首先讲解内部存储、外部存储、资源文件存储;接着讲解 SharedPreferences 存储、数据库存储,并对 Content Provider 组件进行详细的介绍。

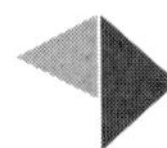

9.1 内部存储

9.1.1 存储目录

内部存储是手机系统自带的存储,一般空间比较小,存储比较快,可以以文件形式存在。存储目录可通过 Context 或 Environment 类获得,如下所示。

- File Environment.getDataDirectory(),获得内部存储根目录 File 对象。
- File Context.getCacheDir(),获得应用缓存目录。
- File Context.getFilesDir(),获得应用文件目录。
- File getFileStreamPath(String fileName),获得 fileName 具体文件的 File 对象。若无 fileName 文件,则返回文件根目录。

注意:上述函数中上下文对象调用的相关函数都与 App 工程的包名相关。

【例 9-1】 验证上述 4 个基本内部存储目录函数的结果。

该示例涉及主界面文件 main.xml 和主应用文件 MainActivity.java。下面一一说明。

① 主界面文件 main.xml。

```
<?xml version="1.0" encoding="utf-8"?>
<LinearLayout xmlns:android="http://schemas.android.com/apk/res/android"
android:layout_width="match_parent"
android:layout_height="match_parent">
    <TextView
        android:id="@+id/mytext"
```

```
        android:layout_width="match_parent"
        android:layout_height="wrap_content" />
</LinearLayout>
```

以上代码定义了一个 id 为 mytext 的 TextView 组件，用以显示 4 个常用基本内部存储目录函数的调用结果。

② 主应用文件 MainActivity.java。

```
public class MainActivity extends AppCompatActivity {
    protected void onCreate(Bundle savedInstanceState) {
        super.onCreate(savedInstanceState);
        setContentView(R.layout.main);
        String direct = Environment.getDataDirectory().toString();
        String cachedir = this.getCacheDir().toString();
        String fileroot = this.getFilesDir().toString();
        String filepath = this.getFileStreamPath("test").toString();
        String s = "";
        s = "内部存储根目录=" + direct +"\n"+
                "工程缓存目录=" + cachedir+"\n"+
                "工程文件目录=" +fileroot+"\n"+
                "工程文件 test 路径="+filepath;
        TextView tv = (TextView)findViewById(R.id.mytext);
        tv.setText(s);
    }
}
```

本示例工程的包名为 com.example.dqjbd.we，运行后在 TextView 组件显示内容，见表 9-1。

表 9-1　TextView 显示结果

内部存储根目录=/data 工程缓存目录=/data/user/0/com.example.dqjbd.we/cache 工程文件目录=/data/user/0/com.example.dqjbd.we/files 工程文件 test 路径=/data/user/0/com.example.dqjbd.we/files/test

9.1.2　存储文件

Android 内部文件存储与计算机 Java IO 是一致的，Java 基础中学过的所有输入输出流都可移植到 Android 中。根本思想是：若想对文件进行读写操作，必须知道文件的绝对路径,在计算机中可方便地写出文件路径,在 Android 中稍微麻烦一些。

根据 9.1.1 节知识可知,若想在工程中对文件 a.txt 进行操作,它的路径代码必须如下所示。

```
String path = Context.getFileStreamPath("a.txt").getPath();
```

知道了 path 值,就可灵活对 a.txt 进行读写。

【例 9-2】 界面上有一输入控件,可输入文本内容,当单击 OK 按钮时,将文本内容存入文件 a.txt。

该示例涉及主界面文件 main.xml 和主应用文件 MainActivity.java,下面一一说明。

① 主界面文件 main.xml。

```
<?xml version="1.0" encoding="utf-8"?>
<LinearLayout xmlns:android="http://schemas.android.com/apk/res/android"
    android:layout_width="match_parent"
    android:layout_height="match_parent"
    android:orientation="vertical">
    <EditText
        android:id="@+id/mytext"
        android:layout_width="match_parent"
        android:layout_height="100dp" />
    <Button
        android:id="@+id/mysave"
        android:layout_width="wrap_content"
        android:layout_height="wrap_content"
        android:text="OK"/>
</LinearLayout>
```

以上代码定义了一个 EditText 控件,id 为 mytext,用于输入文本内容;一个 OK 按钮,id 为 mysave,当单击此按钮时,读取 EditText 控件中的文本内容,并将其存入文件 a.txt。

② 主应用文件 MainActivity.java。

```
public class MainActivity extends AppCompatActivity {
    protected void onCreate(Bundle savedInstanceState) {
        super.onCreate(savedInstanceState);
        setContentView(R.layout.main);
        Button b = (Button)findViewById(R.id.mysave);
        b.setOnClickListener(new View.OnClickListener() {
```

```
            public void onClick(View v) {
                EditText et = (EditText)findViewById(R.id.mytext);
                String s = et.getText().toString();
                try {
                    String path =MainActivity.this.getFileStreamPath("a.txt")
.getPath();
                    FileOutputStream out = new FileOutputStream(path);
                    out.write(s.getBytes("utf-8"));
                    out.flush();
                    out.close();
                }catch(Exception e){ }
            }
        });
    }
}
```

本示例创建了内部存储文件 a.txt。若想读出其内容，利用 FileInputStream 流类即可，关键代码见表 9-2。

表 9-2 读内存文件关键代码

```
String path = this.getFileStreamPath("a.txt").getPath(); //获得文件路径
FileInputStreamin = new FileInputStream(path);          //获得文件输入流
intlen = in.available();                                //获得文件总长度 len
bytebuf[] = new byte[len];                              //开辟字节缓冲区 buf，大小为 len 字节
in.read(buf);                                           //读文件，并将其读到字节缓冲区 buf 中
in.close();                                             //关闭文件
String s =new String(buf, "utf-8");                     //将字节缓冲区变为可视字符串
```

其实，Android 提供了更为方便的输入输出流函数，如下所示。

- FileOutputStream openFileOutput (String name, int mode) throws FileNotFoundException

第 1 个参数 name 是文件名称，该参数不能包含描述路径的斜杠；第 2 个参数 mode 用于指定访问权限，可以为如下值：MODE_PRIVATE，表示文件只能被创建它的工程访问；MODE_APPEND，追加模式，若文件存在，则在文件尾添加内容，否则创建文件；MODE_WORLD_READABLE，表示可被其他应用程序读，但不能写；MODE_WORLD_WRITEABLE，表示可被其他应用程序读和写。

- FileInputStream openFileInput(String name)throws FileNotFoundException name 是文件名称，不是路径。

为了加强理解这两个函数的应用方法，写出与之等价的代码，见表 9-3。

表 9-3 等价输入输出流代码表

功能	代 码	等 价 代 码
文件输入流	String path = this.getFileStreamPath("a.txt").getPath(); FileInputStreamin = new FileInputStream(path);	FileInputStreamin = openFileInput("a.txt");
文件输出流	String path = this.getFileStreamPath("a.txt").getPath(); FileOutputStream in = new FileOutputStream(path);	FileOutputStream in = openFileOutput("a.txt");

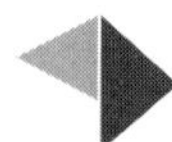

9.2 外部存储

9.2.1 存储目录

每个 Android 设备都支持共享的外部存储用来保存文件，这些外部存储可以是 SD 卡等可以读写的存储介质。把手机与计算机通过 USB 口相连，可以很方便地浏览手机上的外部存储文件。存储目录可通过 Context 或 Environment 类获得，如下所示。

- Environment.getExternalStorageDirectory()，获得外部存储根目录 File 对象。
- Context.getExternalCacheDir()，获得外部存储工程缓存 File 对象。
- Context.getExternalFilesDir(String name)，获得外部存储工程 name 目录 File 对象，当 name 是空串时，返回工程外部存储文件的根目录。

注意：上述函数中上下文对象调用的相关函数都与 App 工程的包名相关。

【例 9-3】 验证上述 3 个基本外部存储目录函数的结果。

该示例涉及主界面文件 main.xml 和主应用文件 MainActivity.java。下面一一说明。

① 主界面文件 main.xml。

```
<?xml version="1.0" encoding="utf-8"?>
<LinearLayout xmlns:android="http://schemas.android.com/apk/res/android"
android:layout_width="match_parent"
android:layout_height="match_parent">
    <TextView
        android:id="@+id/mytext"
        android:layout_width="match_parent"
        android:layout_height="wrap_content" />
</LinearLayout>
```

以上代码定义了一个 id 为 mytext 的 TextView 组件，用以显示 4 个常用基本外部存储目录函数的调用结果。

② 主应用文件 MainActivity.java。

```
public class MainActivity extends AppCompatActivity {
    protected void onCreate(Bundle savedInstanceState) {
        super.onCreate(savedInstanceState);
        setContentView(R.layout.main);
        String direct = Environment.getExternalStorageDirectory().toString();
        String cachedir = this.getExternalCacheDir().toString();
        String fileroot = this.getExternalFilesDir("").toString();
        String filepath = this.getExternalFilesDir("test").toString();
        String s = "";
        s = "外部存储根目录=" + direct +"\n"+
                "工程缓存目录=" + cachedir+"\n"+
                "工程文件根目录=" +fileroot+"\n"+
                "工程文件 test 目录路径="+filepath;
        TextView tv = (TextView)findViewById(R.id.mytext);
        tv.setText(s);
    }
}
```

本示例工程的包名为 com.example.dqjbd.we，运行后在 TextView 组件显示内容，见表 9-4。

表 9-4　TextView 显示外部存储目录结果

外部存储根目录=/storage/emulated/0
工程缓存目录=/storage/emulated/0/Android/data/com.example.dqjbd.we/cache
工程文件根目录=/storage/emulated/0/Android/data/com.example.dqjbd.we/files
工程文件 test 目录路径=/storage/emulated/0/Android/data/com.example.dqjbd.we/files/test

9.2.2　存储文件

【例 9-4】 界面上有一输入控件，可输入文本内容，当单击 OK 按钮时，将文本内容存入外部存储文件 a.txt。

该示例涉及配置文件 Androidmanifest.xml；主界面文件 main.xml；主应用文件 MainActivity.java。下面一一说明。

① 配置文件 Androidmanifest.xml：增加外部存储读写权限。

```
<uses-permission android:name="android.permission.READ_EXTERNAL_STORAGE">
</uses-permission>
<uses-permission android:name="android.permission.WRITE_EXTERNAL_STORAGE">
</uses-permission>
```

② 主界面文件 main.xml。

```
<?xml version="1.0" encoding="utf-8"?>
<LinearLayout xmlns:android="http://schemas.android.com/apk/res/android"
    android:layout_width="match_parent"
    android:layout_height="match_parent"
    android:orientation="vertical">
    <EditText
        android:id="@+id/mytext"
        android:layout_width="match_parent"
        android:layout_height="100dp" />
    <Button
        android:id="@+id/mysave"
        android:layout_width="wrap_content"
        android:layout_height="wrap_content"
        android:text="OK"/>
</LinearLayout>
```

以上代码定义了一个 EditText 控件，id 为 mytext，用于输入文本内容；一个 OK 按钮，id 为 mysave，当单击此按钮时，读取 EditText 控件中的文本内容，将其存入外部文件 a.txt。

③ 主应用文件 MainActivity.java。

```
public class MainActivity extends AppCompatActivity {
    protected void onCreate(Bundle savedInstanceState) {
        super.onCreate(savedInstanceState);
        setContentView(R.layout.main);
        Button b = (Button)findViewById(R.id.mysave);
        b.setOnClickListener(new View.OnClickListener() {
            public void onClick(View v) {
                EditText et = (EdirText)findViewById(R.id.mytext);
                String s = et.getText().toString();
                try {
                    String path =MainActivity.this.getExternalFilesDir("")
```

```
.getPath();
                    File f = new File(path, "a.txt");
                    FileOutputStream out = new FileOutputStream(f);
                    out.write(s.getBytes("utf-8"));
                    out.flush();
                    out.close();
                }catch(Exception e){   }
            }
        });
    }
}
```

可以看出,外部文件的保存过程如下：首先利用 getExternalFilesDir(""),实参为空串,获得工程外部存储根目录 path,然后利用 File f=new File(path,"a.txt")建立 a.txt 文件的 File 对象 f,最后利用 new FileOutputStream(f)建立字节文件输出流 out,保存获得的字符串内容,关闭文件。

9.2.3 共享文件夹

手机上有一些文件夹是系统定义好的、功能明确的目录,如音乐目录、相册目录、下载目录等,获取这些目录利用以下函数即可。

File getExternalStoragePublicDirectory(String type),type 是字符串标识变量,可取如下值。

- Environment.DIRECTORY_MUSIC,音乐目录。
- Environment.DIRECTORY_PICTURES,图片目录。
- Environment.DIRECTORY_MOVIES,电影目录。
- Environment.DIRECTORY_DOWNLOADS,下载目录。
- Environment.DIRECTORY_DCIM,相机拍照或录像文件的存储目录。
- Environment.DIRECTORY_DOCUMENTS,文件文档目录。

【例 9-5】 编程显示图片目录下的内容。

本示例涉及配置文件 Androidmanifest.xml；主界面文件 main.xml；主应用文件 MainActivity.java。下面一一加以说明。

① 配置文件 Androidmanifest.xml：增加外部存储读写权限,与例 9-4 一致,略。

② 主界面文件 main.xml。

```
<?xml version="1.0" encoding="utf-8"?>
```

```
<LinearLayout xmlns:android="http://schemas.android.com/apk/res/android"
    android:layout_width="match_parent"
    android:layout_height="match_parent"
    android:orientation="vertical">
    <TextView
        android:id="@+id/mytext"
        android:layout_width="match_parent"
        android:layout_height="match_parent" />
</LinearLayout>
```

以上代码定义了一个 TextView 组件,id 为 mytext,程序一运行,就将图片目录下的内容显示在 TextView 组件中。

③ 主应用文件 MainActivity.java。

```
public class MainActivity extends AppCompatActivity {
    protected void onCreate(Bundle savedInstanceState) {
        super.onCreate(savedInstanceState);
        setContentView(R.layout.main);
        int check = ActivityCompat.checkSelfPermission(this, Manifest
.permission.READ_EXTERNAL_STORAGE);
        if(check== PackageManager.PERMISSION_DENIED){
            ActivityCompat.requestPermissions(this,new String[]
                  {Manifest.permission.READ_EXTERNAL_STORAGE},1);
        }
        try {
            TextView tv = (TextView) findViewById(R.id.mytext);
            File f =Environment.getExternalStoragePublicDirectory
(Environment.DIRECTORY_PICTURES);
            File fs[] = f.listFiles();
            String s = "";
            for(int i=0; i<fs.length; i++)
                s += fs[i].getName()+"\n";
            tv.setText(s);
        }catch(Exception e){  }
    }
}
```

onCreate()函数中包含动态授权代码,涉及 ActivityCompat 的如下两个静态方法。

• int checkSelfPermission(Context ctx, String mark),检测是否授权函数,ctx 是上

下文对象，mark 是标识字符串，一般是 Manifest.persission 对象中的常量。若已授权，则返回 PackageManager. PERMISSION _ GRANTED，否则返回 PackageManager.PERMISSION_DENIED。

• void requestPermissions(Context ctx，String marks[]，int requestCode)，ctx 是上下文对象，marks 是授权标识数组，一般取自 Manifest.persission 对象中的常量，requestCode 是请求码，一般设置为 1。

onCreate()函数中利用 checkSelfPermission()检查用户是否有 Manifest.Permission.READ_EXTERNAL_STORAGE 读外部存储权限许可，若没有，则用 requestPermissions()函数设置即可。

也许有读者问：在 Androidmanifest.xml 配置文件中已经设置了外部存储读写权限，为什么还要用代码动态设置？这与 Android 版本有关，有些静态设置可以，有些静态设置不行，必须动态设置。为了保险起见，凡是涉及外部存储的，用静态设置＋动态设置最好。

获取图片目录内容的关键步骤是：首先利用 getExternalStoragePublicDirectory()函数获得图片目录的 File 对象 f，然后利用 listFiles()函数获得该目录的 File 对象数组 fs，最后遍历该数组，利用 getName()函数获得每个 File 对象的名称，并将其显示到 TextView 组件中。

【例 9-6】 实现手机外部存储目录遍历功能。

初始时显示手机外部存储根目录下的所有内容。当按某子项时：若它是目录，则显示第 2 级目录下的内容，可级联显示；若它是文件，则不执行任何操作。

实现该功能的核心思想是：①由一个 Activity，而不是多个 Activity 实现。因为可级联显示目录下的内容，所以一定是该 Activity 反复自身调用；②同一 Activity 却可显示不同目录的内容，一定是 Intent 对象传递的参数不同，本示例传递的是相对于外部存储根目录的偏移量。基于以上两点，就可实现外部存储目录遍历功能。

本示例涉及的文件有：配置文件 Androidmanifest.xml；主界面文件 main.xml；主应用文件 MainActivity.java。下面一一加以说明。

① Androidmanifest.xml：增加外部存储读写权限，同例 9-5 一致，略。

② 主界面文件 main.xml。

```
<?xml version="1.0" encoding="utf-8"?>
<LinearLayout xmlns:android="http://schemas.android.com/apk/res/android"
    android:layout_width="match_parent"
    android:layout_height="match_parent"
    android:orientation="vertical">
    <TextView
```

```
        android:id="@+id/mydirect"
        android:layout_width="match_parent"
        android:layout_height="30dp" />
    <ListView
        android:id="@+id/mylist"
        android:layout_width="match_parent"
        android:layout_height="match_parent">
    </ListView>
</LinearLayout>
```

以上代码定义了一个 TextView 组件，id 为 mydirect，用于显示当前正显示的目录；一个 ListView 组件，id 为 mylist，用于显示当前目录下的所有子项内容。

③ 主应用文件 MainActivity.java。

```
public class MainActivity extends AppCompatActivity {
    String direct;
    File fs[] ;
    /* direct 代表当前要显示的目录(相对于外部存储根目录),fs[]是 File 对象数组,用于
保存当前目录下所有子项的 File 对象。 */
    void getDirect(){
        Intent t = this.getIntent();
        direct = t.getStringExtra("direct");
        if(direct==null) direct="";
    }
    /* 通过 Intent 对象获得当前要显示的目录字符串值,若为空,表明是第一次调用该
Activity,显示的是外部存储根目录下的各子项内容 */
    void displayDirect(){
        TextView tv = (TextView)findViewById(R.id.mydirect);
        tv.setText("当前目录:"+direct);
    }
    /* 将带显示的当前目录字符串显示在 TextView 组件中 */
    void setReadPermission(){
        int check = ActivityCompat.checkSelfPermission(this, Manifest
.permission.READ_EXTERNAL_STORAGE);
        if(check== PackageManager.PERMISSION_DENIED){
            ActivityCompat.requestPermissions(this,new String[]
                {Manifest.permission.READ_EXTERNAL_STORAGE},1);
        }
    }
```

```
    /*设置读权限,说明同例 9-5 一致,略*/
    void displayList(){
        ArrayList<String> ary = new ArrayList();
        try {
            File f = Environment.getExternalStorageDirectory();
            File f2 = new File(f, direct);
            fs = f2.listFiles();
            String s = "";
            for(int i=0; i<fs.length; i++)
                ary.add(fs[i].getName());
        }catch(Exception e){  }
        ArrayAdapter<String> ad = new ArrayAdapter<String>(this,android.R
.layout.simple_list_item_1, ary);
        ListView lv = (ListView)findViewById(R.id.mylist);
        lv.setAdapter(ad);
    }
```

/*功能是将当前待显示目录的各子项显示在 ListView 组件中:首先定义 ArrayList 对象 ary;然后根据外部存储根目录对象 f 及当前显示目录值建立 File 对象 f2,利用 listFiles() 函数获得当前目录下的各子项 File 对象,保存在成员变量 fs 中,遍历 fs 数组,将各子项名称保存在 ary 数组对象中;最后将 ListView 组件通过 ArrayAdapter 适配器动态绑定 ary 数组对象,完成当前目录各子项内容的界面显示。*/

```
    protected void onCreate(Bundle savedInstanceState) {
        super.onCreate(savedInstanceState);
        setContentView(R.layout.main);
        getDirect();
        displayDirect();
        setReadPermission();
        displayList();
        ListView lv = (ListView)findViewById(R.id.mylist);
        lv.setOnItemClickListener(new AdapterView.OnItemClickListener() {
            public void onItemClick(AdapterView<?> parent, View view, int pos,
long id) {
                if(fs[pos].isDirectory()){
                    String cur = direct+"/"+fs[pos].getName();
                    Intent t = new Intent(MainActivity.this, MainActivity.class);
                    t.putExtra("direct",cur);
                    startActivity(t);
                }
```

```
            }
        });
    }
}
```

onCreate()按顺序执行了获得目录函数 getDirect()、显示目录函数 displayDirect()、权限设置函数 setReadPermission()、列表显示函数 displayList()，然后为 ListView 组件增加了 OnItemClick 事件监听器。在事件响应函数中，首先利用 fs 对象及响应位置 pos 判定 fs[pos]是否为下一级目录对象，若不是，则什么也不做；若是，则根据当前正显示的目录值 direct 及 fs[pos].getName()合成新的待显示的目录 cur，保存在 Intent 对象中。最后通过 startActivity()函数完成 MainActivity 的自身再启动。

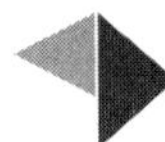

9.3 资源文件存储

在 Android 中，除了可以在内部和外部存储设备上读写文件，还可以访问在工程 res 目录下的 raw 和 xml 子目录下的文件，这两个子目录下的文件都会被映射在 R.java 中，均可通过 R.raw.XXX 或 R.xml.XXX 结合系统函数访问资源文件。本节仅对 RAW 文件下的文本文件进行了解析，同样可扩展至 xml 目录下的文件。

一般来说，我们访问的都是文本文件，下面示例都是结合读取文本文件展开的。

【例 9-7】 在工程 res 目录下建子目录 raw，在其下再建一个文本文件 my.txt 及 XML 文件 my2.xml，内容均相同，如表 9-5 所示。

表 9-5 文本文件示例

res/my.txt	res/my2.xml
user=zhang pwd=123456	<? xml version="1.0" encoding="UTF-8" standalone="no"? > <! DOCTYPE properties SYSTEM "http://java.sun.com/dtd/properties.dtd"> <properties> <entry key="user">zhang</entry> <entry key="pwd">123456</entry> </properties>

Java util 工具包下有一个类 Properties，它非常易于解析某些特定格式的文本文件，格式是“键---值”对应关系。对于 TXT 文本而言，格式是“键=值”；对于 XML 文件而言，要求根节点是 properties，可以包含多个 entry 子标签，键写在 entry 标签的属性中，且必须是 key=“键”，值写在 entry 起始标签与结束标签之间。

其实，在工程中类似 my.txt、my2.xml 结构多存在于自定义配置文件中，而且非常常见，复杂结构的自定义文件其实用得并不多。

解析上述两文件涉及主界面文件 main.xml 及主应用文件 MainActivity.java。下面一一介绍。

① 主界面文件 main.xml，该界面仅是象征意义，空白即可，略。

② 主应用文件 MainActivity.java。

```
public class MainActivity extends AppCompatActivity {
    protected void onCreate(Bundle savedInstanceState) {
        super.onCreate(savedInstanceState);
        setContentView(R.layout.main);
        Properties p = new Properties();
        InputStream in = getResources().openRawResource(R.raw.my);
        try {
            p.load(in);
            in.close();
        }catch(Exception e){   }
        String user = p.getProperty("user");
        String pwd = p.getProperty("pwd");
        Toast.makeText(this, "my.txt:user="+user+"\n"+pwd, Toast.LENGTH_
LONG).show();
        InputStream in2 = getResources().openRawResource(R.raw.my2);
        try {
            p.loadFromXML(in2);
            in2.close();
        }catch(Exception e){   }
        user = p.getProperty("user");
        pwd = p.getProperty("pwd");
        Toast.makeText(this, "my2.xml:user="+user+"\n"+pwd, Toast.LENGTH_
LONG).show();
    }
}
```

onCreate()函数的上半部分是解析文本文件：首先产生 Properties 对象 p；然后利用 getResources( ) 获得系统资源 Resources 对象，再利用 Resources 对象的 openRawResource(R.raw.my)，直接获得 my.txt 的 InputStrem 流对象 in；最后利用 p.load(in)将 p 和 in 绑定，应用 p.getProperty(“user”)，p.getProperty(“pwd”)获得用户名与密码。

onCreate()函数的下半部分是解析 XML 文件,其过程几乎与解析文本文件一致,只是用 loadFromXML()函数代替了 load()函数。

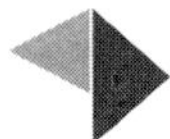

9.4 SharedPreferences 存储

9.4.1 概述

SharedPreferences 是一种轻量级的数据存储方式,其采用的是键值对的存储方式,只能存储少量数据,支持存储的数据类型有 boolean、float、int、long、String。SharedPreferences 将内容存储到一个 XML 文件中,路径在/data/data/<packagename>/shared_prefs/下。

9.4.2 基本用法

SharedPreferences 保存数据的步骤为:首先获取 SharedPreferences 对象;然后获取 Editor 对象并执行相关操作。

① 获取 SharedPreferences 对象函数。

SharedPreferences getSharedPreferences(String name, int mode),通过 Context 调用该方法。它有两个参数:第一个参数 name 指定了 SharedPreferences 存储文件名;第二个参数 mode 指定了操作的模式,一般为 Context.MODE_PRIVATE。采用这种方式创建的文件可以被整个应用中的所有组件使用。

例如,this.getSharedPreferences("my", Context.MODE_PRIVATE)。此行代码表示创建的文件名是 my.xml。

② 获取 Editor 对象并执行相关操作。

```
SharedPreferences sp = this. getSharedPreferences( "my", Context.MODE_
PRIVATE );
Editor et = sp.editor();                        //获取 Editor 对象
et.putString("user", "Li");     //利用 putString()保存键-值对,示例中的键是 user,值是 Li
et.commit();                                    //利用 commit()函数确认提交
```

读取 SharedPreferences 存储的内容步骤为:获取 SharedPreferences 对象,利用 getString()函数获取已知键对应的值。一个简单的示例如下所示。

```
SharedPreferences sp = this. getSharedPreferences( "my", Context.MODE_PRIVATE );
String user = sp.getString("user", "");     //获取键 user 对应的保存值
```

getString()函数有两个参数：第 1 个参数是键字符串；第 2 个参数是默认值。该函数的含义是若键有对应的值，则返回该值，否则返回设置的默认值。

【例 9-8】 两个 Activity 类：MainActivity 和 TwoActivity。在 MainActivity 类中利用 SharedPreferences 保存数据。主界面上有一个按钮，当单击此按钮时，启动 TwoActivity，读取 SharedPreences 对象保存的数据，并利用 Toast 显示。

① 主界面 main.xml。

```
<?xml version="1.0" encoding="utf-8"?>
<LinearLayout xmlns:android="http://schemas.android.com/apk/res/android"
    android:layout_width="match_parent"
    android:layout_height="match_parent"
    android:orientation="vertical">
    <Button
        android:id="@+id/myok"
        android:layout_width="wrap_content"
        android:layout_height="wrap_content"
        android:text="OK"/>
</LinearLayout>
```

以上代码定义了一个 OK 按钮，id 为 myok，当单击此按钮时，启动 TwoActivity。

② 主应用文件 MainActivity.java。

```
public class MainActivity extends AppCompatActivity {
    protected void onCreate(Bundle savedInstanceState) {
        super.onCreate(savedInstanceState);
        setContentView(R.layout.main);
        SharedPreferences sp = this.getSharedPreferences("my",Context.MODE_
PRIVATE);
        SharedPreferences.Editor et = sp.edit();
        et.putString("user","Li");
        et.putInt("age",20);
        et.commit();
        Button b = (Button)findViewById(R.id.myok);
        b.setOnClickListener(new View.OnClickListener() {
            public void onClick(View v) {
                Intent t = new Intent(MainActivity.this,TwoActivity.class);
                startActivity(t);
            }
```

```
        });
    }
}
```

如果程序已启动，就利用 SharedPreferences 对象保存用户的姓名：键是 user，值是 Li；保存用户的年龄值：键是 age，值是整数 20。

按钮响应事件函数定义了启动 TwoActivity 的代码。

③ TwoActivity 界面文件 two.xml：该界面仅是象征意义，空白即可，略。

④ TwoActivity.java 源文件。

```
public class Two extends AppCompatActivity {
    protected void onCreate(Bundle savedInstanceState) {
        super.onCreate(savedInstanceState);
        setContentView(R.layout.activity_two);
        SharedPreferences sp = getSharedPreferences("my", Context.MODE_
PRIVATE);
        String user = sp.getString("user","");
        int age = sp.getInt("age",0);
        Toast.makeText(this,user+":"+age,Toast.LENGTH_LONG).show();
    }
}
```

以上代码利用 SharedPreferences 对象获取用户的姓名及年龄，并利用 Toast 完成显示。

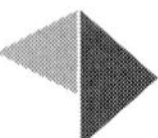

9.5 数据库存储

随着手机技术的不断发展，外部存储空间的不断扩大，智能手机可以实现简单的数据库功能，这为手机与用户信息交互提供了更广阔的拓展空间。本节将对建库、建表、常用的 SQL 操作进行深入的论述。

9.5.1 命令行建库

Android SDK 中 platform-tools 目录下有应用程序 sqlite3.exe，利用它就可以完成数据库的所有常规操作。若读者有 MySQL 数据库操作经验，就会很快掌握 sqlite3 命令的各种用法。命令行操作 MySQL 数据库与操作 Android 数据库非常相似。

首先利用 cmd 命令进入 DOS 窗口，之后进入 Android SDK 中的 platforms-tools 目

录，运行 sqlite3 mydb，这就在当前目录下创建了名为 mydb 的 Android 数据库。进入数据行命令操作如图 9-1 所示。

图 9-1　进入数据行命令操作

可以发现，在当前目录 platform-tools 下生成了 mydb 数据库文件。而且当运行 sqlite3 mydb 命令后，发现盘符发生了变化，变为 sqlite>，这是成功进入 Android 数据库命令行操作的标识。若想在其他目录上生成数据库，只需加上目录(当然得保证该目录存在)，例如 sqlite3 d:/mytest/mydb，表明在 d:/mytest 目录下创建数据库 mydb。

sqlite3 的其他常用命令如下所示。

① 查看数据库文件信息命令(注意，命令前须带字符'.')：sqlite>.database。

② 查看所有表的创建语句：sqlite>.schema。

③ 查看指定表的创建语句：sqlite>.schema table_name。

④ 以 SQLl 语句的形式列出表内容：sqlite>.dump table_name。

⑤ 输出帮助信息：sqlite>.help。

⑥ 退出 sqlite 终端命令：sqlite>.quit 或 sqlite>.exit。

同样，sqlite3 支持命令行操作 SQL 语句。本书仅列出最普通的 SQL 语句的形式，若想查复杂的 SQL 语句，还要查一些专业书籍，如下所示。

① 建立数据表。

```
create table table_name(field1 type1, field2 type1, ...);
```

table_name 是要创建的数据表名称，fieldx 是数据表内的字段名称，typex 则是字段类型。

② 添加数据记录。

```
insert into table_name(field1, field2, ...) values(val1, val2, ...);
```

field 是字段名称，val 为需要存入字段的值。

③ 修改数据记录。

```
update table_name set field1=val1, field2=val2 where expression;
```

where 是 SQL 语句中用于条件判断的命令，expression 为判断表达式。

④ 删除数据记录。

```
delete from table_name [where expression];
```

若不加判断条件，则清空表中的所有数据记录。

⑤ 查询数据记录。

select 指令的基本格式：select col1，col2　from table_name [where expression]；

col 是待查询的表列，expression 是条件表达式。

⑥ 删除数据表或索引。

drop table table_name；　table_name 是表名。

9.5.2　程序建库与操作

Android 提供了利用程序建库的方法，这涉及两个系统类 SQLiteOpenHelper 和 SQLiteDatabase，下面分别介绍。

① SQLiteOpenHelper 类。

该类帮助用户简化对 SQLite 数据库的操作。该类是一个抽象类，必须用户自己实现并重写 onCreate()和 onUpGrade()方法；此外，还必须重写构造方法。构造方法包括上下文、数据库名称、版本等形参。

SQLiteOpenHelper 类常用的方法如下所示。

- SQLiteDatabase getWritableDatabase()，创建或打开可读/写的数据库(通过返回的 SQLiteDatabase 对象进行操作)。
- SQLiteDatabase getReadableDatabase()，创建或打开可读的数据库(通过返回的 SQLiteDatabase 对象进行操作)。
- void onCreate(SQLiteDatabase db)，创建函数，若数据库第 1 次创建，则会调用，即第 1 次运行 getWritableDatabase()或 getReadableDatabase()时调用 onCreate()函数。

- void onUpgrade(SQLiteDatabase db, int oldVersion, int newVersion),重写函数,数据库升级时自动调用。
- close(),关闭数据库。

② SQLiteDatabase 类。

该类封装了数据库增、删、改、查等常用功能,下面分别加以介绍。

- long insert(String table, String mark, ContentValues values),插入函数。table 是表名,mark 一般为 null,values 是 ContentValues 对象。ContentValues 对象也是键值映射对,键是表的列名,不是随意设置的。

ContentValues 的用法示例如下所示。

```
ContentValues obj = new ContentValues();
obj.put("studno",1000);                    //当前学生的学号 1000
obj.put("studname","zhang");               //当前学生的姓名 zhang
```

- int delete(String table,String expression,String[] args),删除函数。table 是表名;expression 是条件表达式(不含 where),表达式中待定的值用"?"代替;args 是字符串数组,依次填充 expression 中的"?"。该函数的返回值代表真正删除的数据库表记录条数。例如:delete("student","studno=?", new String[]("1000")),表示删除学生表中学号为 1000 的记录。
- int update(String table, ContentValues values, String expression, String[] args),更新函数。table 是表名、values 是 ContentValues 对象,ContentValues 对象也是键值映射对,键是表的列名,不是随意设置的;expression 是条件表达式(不含 where),表达式中待定的值用"?"代替;args 是字符串数组,依次填充 expression 中的"?"。该函数的返回值代表真正更新的数据库表记录条数。
- 查询函数:查询是数据库中最复杂、最常用的操作,其常用的函数如下所示。

```
Cursor query(Stringtable,String[]columns,String selection,
       String selectArgs,String groupBy,String havingBy,String orderBy)
Cursor query(Stringtable,String[]columns,String selection,
       String selectArgs,String groupBy,String havingBy,String orderBy,
String limit)
Cursor query(boolean dinstinct,Stringtable,String[]columns,String selection,
       String selectArgs,String groupBy,String havingBy,String orderBy)
Cursor query(boolean dinstinct,Stringtable,String[]columns,String selection,
       String selectArgs,String groupBy,String havingBy,String orderBy,
String limit)
```

上述函数的参数基本是按 SQL 查询语句划分的：查询结果显示的列名＋条件表达式＋分组字段＋分组条件＋排序字段＋限制字段。

查询结果可通过游标 Cursor 对象访问。Cursor 常用函数如表 9-6 所示。

表 9-6　Cursor 常用函数

功　能	函　数
游标移动	boolean isBeforeFirst()，　是否在第 1 条记录前面，若是，返回 true，否则返回 false boolean moveToFirst()，　是否移到第一条记录 boolean moveToNext()，　是否移到下一条记录 boolean moveToPrevious()，　是否移到上一条记录 boolean moveToLast()，　是否移到最后一条记录 boolean isAfterLast()，　是否移到最后一条记录的下一条记录
游标取值	String getString(int col)，取得查询结果第 col 列的字符串值，col 可以为 0。 同理，有 getInt(int col)，getShort(intcol)，getLong(int col)，getFloat(int col)，getDouble(int col)函数
其他	int getColumnCount()，获得查询结果的列数； String getColumnName(int col)，获得查询结果第 col 列字段的名称

- void execSQL(String sql)，上面介绍的增、删、改、查等基本操作只能完成一些比较简单的数据库操作，对于一些复杂的数据库操作(比如多种操作结合的语句)就无能为力了。所以，SQLiteDatabase 提供了 execSQL()方法直接执行 SQL 语句。
- static SQLiteDatabase openDatabase(String path，SQLiteDatabase.CursorFactory factory，int flags)。

使用静态方法 openOrCreateDatabase()打开或创建数据库文件。其中，参数 path 表示要打开或者创建的数据库文件路径；factory 表示一个可选的 factory 类，当开始查询时，可以通过该类实例化 cursor，默认将该参数设置为 null；flags 指定了访问数据库的模式，可以将该参数指定为 OPEN_READWRITE、OPEN_READONLY 和 CREATE_IF_NECESSARY，这 3 个标志分别表示读写方式打开数据库、只读方式打开数据库，以及如果数据库文件不存在，则创建数据库文件。该方法的返回值是打开或者创建的数据库文件 SQliteDataBase 对象。

【例 9-9】 应用 SQLiteOpenHelper、SQLiteDatabase 类创建数据库 school，先在数据库中创建学生表(共两个字段：学号，整型；姓名，字符串)，再向表中存入两条记录。

该示例涉及的文件有：主界面文件 main.xml；主应用文件 MainActivity.java；数据库帮助类文件 MyDbHelper.java。下面一一加以说明。

① 数据库帮助文件类 MyDbHelper.java。

```
public class MyDBHelper extends SQLiteOpenHelper {
    Context ctx;
    public MyDBHelper(Context ctx, String name){
        super(ctx,name,null,1);
        this.ctx = ctx;
    }
    public void onCreate(SQLiteDatabase db) {
        String s = "create table student(studno integer,name char(20))";
        String s2= "insert into student values(1000,'zhang')";
        String s3= "insert into student values(1001,'li')";
        db.execSQL(s);
        db.execSQL(s2); db.execSQL(s3);
        Toast.makeText(ctx, "SUCCESS", Toast.LENGTH_LONG).show();
    }
    public void onUpgrade(SQLiteDatabase db, int oldVersion, int newVersion) {  }
}
```

构造函数 MyDBHelper()的第二个形参 name 表示数据库名称，不包含路径，必须调用 super()父类构造方法，才能完成数据库 name 的创建或打开。本示例中，super()的第四个参数是数据库版本号，本例中默认是 1，正常情况下应由调用方程序传入。数据库版本号很关键，有时它会直接决定 onUpgrade()的编写。

onCreate()函数主要应用 SQLiteDatabase 对象 db 创建数据库表或插入数据，用的函数都是 execSQL()，完成了 student 数据库表的创建及插入了两条学生数据，当 Toast 显示 SUCCESS 时表明数据库创建成功。

理想情况是将创建数据库的所有 SQL 语句保存在某文本文件中。onCreate()函数对文本文件的 SQL 语句操作即可完成数据库的创建，而且 onCreate()中的代码也简洁很多。

② 主界面文件 main.xml。

```
<?xml version="1.0" encoding="utf-8"?>
<LinearLayout xmlns:android="http://schemas.android.com/apk/res/android"
    android:layout_width="match_parent"
    android:layout_height="match_parent"
    android:orientation="vertical">
    <Button
        android:id="@+id/myadd"
```

```
            android:layout_width="wrap_content"
            android:layout_height="wrap_content"
            android:text="Add"/>
        <Button
            android:id="@+id/myshow"
            android:layout_width="wrap_content"
            android:layout_height="wrap_content"
            android:text="Show"/>
        <ListView
            android:id="@+id/mylist"
            android:layout_width="match_parent"
            android:layout_height="match_parent">
        </ListView>
    </LinearLayout>
```

主界面其实是为例 9-10 做准备的，本例中不涉及实质代码。一个 id 为 myadd 的按钮，用于向数据库添加学生信息；一个 id 为 myshow 的按钮，用于启动显示学生信息事件；一个 id 为 mylist 的 ListView 组件，用于显示学生表的所有记录信息。

③ 主应用程序 MainActivity.java。

```
public class MainActivity extends AppCompatActivity {
    MyDBHelper helper = new MyDBHelper(this, "school");
    protected void onCreate(Bundle savedInstanceState) {
        super.onCreate(savedInstanceState);
        setContentView(R.layout.main);
        SQLiteDatabase db = helper.getWritableDatabase();
        db.close();
    }
}
```

以上代码定义了 MyDBHelper 成员变量 helper，且直接初始化完毕，表明要创建的数据库名是 school。

当安装好 App，第一次运行的时候，若运行到 helper.getWritableDatabase()，则触发 MyDBHelper 类中的 onCreate()函数运行，从而完成数据库的创建。当再次重新运行 App 时，则发现不再运行 MyDBHelper 类中的 onCreate()函数了。也就是说，MyDBHelper 类中的 onCreate()函数仅运行一次。这一点非常重要，因为不希望每次运行 App 时都要重新创建数据库，这不符合常规逻辑。

什么时候希望重新初始化数据库？一般来说，当数据库版本变化时，由于 App 每次重新运行时，当第一次运行 getWritableDatabase()或 getReadableDatabase()时，都会运行 MyDBHelper 类中的 onUpgrade()函数，在此处添加代码即可，如下所示。

```
public void onUpgrade(SQLiteDatabase db, int oldVersion, int newVersion) {
    if(newVersion != oldVersion){ //在此处加数据库更新处理代码;}
}
```

由于 SQLiteDatabase 对象占用较多的资源，因此当不用它的时候要用 close()函数关闭，养成良好的编程习惯。

【例 9-10】 例 9-9 主要讲了数据库的创建，本例继续完善，主要完成学生表信息的添加和显示。

例 9-9 界面中有一个添加按钮、一个显示按钮和一个 ListView 组件。直接在 MainActivity.Java 中添加两个按钮的消息响应，代码如下所示。

① 添加功能。

```
Button b = (Button)findViewById(R.id.myadd);
b.setOnClickListener(new View.OnClickListener() {
    public void onClick(View v) {
        int no = (int)(1000 + 1000 * Math.random());
        String name = "Li"+(int)(Math.random() * 100);
        SQLiteDatabase db = helper.getWritableDatabase();
        ContentValues cv = new ContentValues();
        cv.put("studno",no); cv.put("name",name);
        db.insert("student",null,cv);
        db.close();
    }
});
```

为了演示方便，学号、姓名都是随机产生的。学号的范围为[1000,2000]，姓名的范围为["Li0","Li100"]，每次产生一个学号和一个姓名，利用 ContentValues 对象进行封装，最后通过 insert()函数将学生信息添加到 student 表中。

② 查询功能。

```
Button b2 = (Button)findViewById(R.id.myshow);
b2.setOnClickListener(new View.OnClickListener() {
    public void onClick(View v) {
        ArrayList<String> ary = new ArrayList();
        try {
            SQLiteDatabase db = helper.getWritableDatabase();
```

```
            Cursor cs = db.query("student", null, null, null, null, null, null,
null);
            if (cs.moveToFirst()) {
                while (!cs.isAfterLast()) {
                    int no = cs.getInt(0); String name=cs.getString(1);
                    ary.add(""+no+":"+name);
                    cs.moveToNext();
                }
            }
            db.close();
            ArrayAdapter<String> ad = new ArrayAdapter<String>
                (MainActivity.this,android.R.layout.simple_list_item_1,ary);
            ListView lv = (ListView)findViewById(R.id.mylist);
            lv.setAdapter(ad);
        }
        catch(Exception e){Toast.makeText(MainActivity.this,e.getMessage(),
Toast.LENGTH_LONG).show();}
    }
});
```

首先产生 ArrayList 对象 ary，然后获得 SQLiteDatabase 对象 db，调用 query()函数获得 Cursor 游标对象 cs，遍历 cs，获得每一个记录的学号、姓名，将其连接成一个字符串添加到 ary 数组对象中。最后建立 ArrayAdapter 数组适配器对象 ad，并与 id 为 R.id.mylist 的 ListView 组件绑定，完成学生信息的显示。

实验时，可先单击“添加”按钮，每单击一次就向 student 表增加一条记录，然后单击“显示”按钮，在界面上验证添加结果。

【例 9-11】 程序运行后仅利用 SQLiteDatabase 类创建数据库 test.db（不借助 SQLiteOpenHelper），创建 3 个字段个数不相同的表，并分别插入一条数据。界面上有一个 EditText 控件，用于输入表名；一个 OK 按钮，当单击此按钮时，获得表名值，并将该表中的所有记录显示在 TextView 组件中。

① 主界面文件 main.xml。

```
<?xml version="1.0" encoding="utf-8"?>
<LinearLayout xmlns:android="http://schemas.android.com/apk/res/android"
    android:layout_width="match_parent"
    android:layout_height="match_parent"
    android:orientation="vertical">
```

```
    <EditText
        android:id="@+id/mytext"
        android:layout_width="match_parent"
        android:layout_height="wrap_content" />
    <Button
        android:id="@+id/myok"
        android:layout_width="wrap_content"
        android:layout_height="wrap_content"
        android:text="OK"/>
    <TextView
        android:id="@+id/myview"
        android:layout_width="match_parent"
        android:layout_height="match_parent">
    </TextView>
</LinearLayout>
```

主界面中有一个 EditText 组件，id 为 mytext，用于输入表名；一个 Button 按钮，当单击此按钮时，获得在 EditText 组件中输入的表名，并将该表记录显示在 id 为 myview 的 TextView 组件中。

② 主应用文件 MainActivity.java。

```
public class MainActivity extends AppCompatActivity {
    protected void onCreate(Bundle savedInstanceState) {
        super.onCreate(savedInstanceState);
        setContentView(R.layout.main);
        final SQLiteDatabase db = SQLiteDatabase.openDatabase(getFilesDir()
+"/test.db",null,
              SQLiteDatabase.OPEN_READWRITE | SQLiteDatabase.CREATE_IF_
NECESSARY);
        try {
            db.execSQL("create table stud(studno integer,studname char(20))");
            db.execSQL("insert into stud values(1000,'zhang')");
            db.execSQL("create table stud2(studno integer,studname char(20),
sex char(6))");
            db.execSQL("insert into stud2 values(1000,'zhang','boy')");
            db.execSQL("create table stud3(studno integer,studname char(20),
sex char(6),major char(10))");
            db.execSQL("insert into stud3 values(1000,'zhang','boy','math')");
            Toast.makeText(this, "SUCCESS!!!", Toast.LENGTH_LONG).show();
```

```
        }catch(Exception e){     }
        Button b = (Button)findViewById(R.id.myok);
        b.setOnClickListener(new View.OnClickListener() {
            public void onClick(View v) {
                try {
                    EditText et = (EditText) findViewById(R.id.mytext);
                    String table = et.getText().toString();
                    Cursor cs = db.query(table, null, null, null, null, null, null);
                    int count = cs.getColumnCount();
                    String total = "";
                    if(cs.moveToFirst()){
                        while(!cs.isAfterLast()){
                            String unit = "";
                            for(int i=0; i<count; i++){
                                unit += cs.getString(i)+"-";
                            }
                            total += unit+"\n";
                            cs.moveToNext();
                        }
                    }
                    TextView tv = (TextView)findViewById(R.id.myview);
                    tv.setText(total);
                }catch(Exception e){     }
            }
        });
    }
}
```

onCreate()函数主要完成两个功能：创建数据库；为 OK 按钮增加事件响应。

创建数据库的过程如下：直接利用 SQLiteDatabase 的静态函数 openDatabase()创建空数据库 test.db。该函数第 1 个参数实参为：getFilesDir()+"/test.db"，表明要在内部存储中创建数据库 test.db；第 3 个参数实参为：OPEN_READWRITE | CREATE_IF_NECESSARY，表明创建的数据库可读可写。然后利用 exexSQL()函数创建了三个表 stud、stud2、stud3，字段数目分别是 2、3、4，每个表都增加了一条记录。

OK 按钮响应函数功能如下：首先从 EditText 组件中获得待显示的表名，利用 query()函数对该表所有字段查询，结果保存在游标 Cursor 对象 cs 中。我们知道创建的三个表 stud、stud2、stud3 字段数均不相同，如何用相同的代码适应表字段数的变化呢？答案是利用 getColumnCount()函数，通过该函数可获得查询结果的字段个数 count，有了这一数值，就可利用循环获得当前游标各个字段的具体数值。本示例将当前游标各字段值保存在 unit 中，unit 的累加值保存在 total 中，代表了当前显示表的所有记录，最终将 Total 值显示在 id 为 myview 的 TextView 组件中。

数据库内部存储其响应会很快，但若数据库很大，利用内部存储就不恰当了，采用外部存储就是必然的选择，对于本例只需将 OpenDatabase()所在行语句改为如下即可。

```
final SQLiteDatabase db = SQLiteDatabase.openDatabase(getExternalFilesDir
("")+"/test.db",null,
        SQLiteDatabase.OPEN_READWRITE | SQLiteDatabase.CREATE_IF_NECESSARY);
```

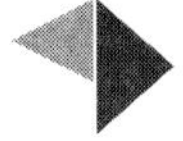

9.6 ContentProvider 组件

9.6.1 简介

ContentProvider，内容提供者，是 Android 的重要跨进程组件。之前学过跨进程的组件有 Activity、BroadcastReceiver、Service，这 3 个组件可以在跨进程中进行调用。那么，ContentProvider 组件跨进程通信与上述 3 个组件有什么不同呢？描述如图 9-2 所示。

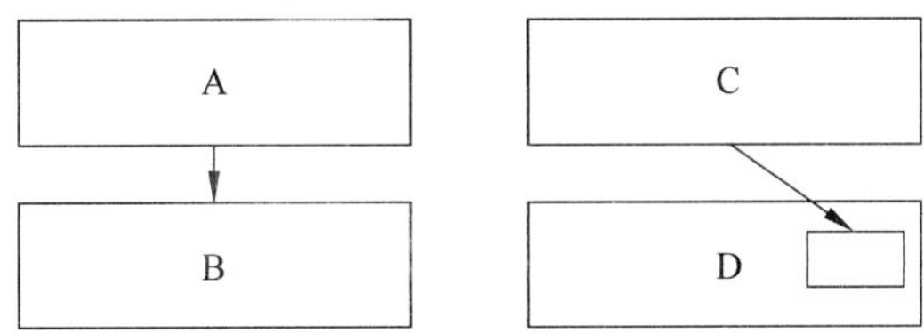

图 9-2 4 种组件跨进程通信简图

A、B 在两个进程中，A 能启动 B，但不能访问 B 的内部，这是 Activity、BroadCastReceiver、Service 相互间跨进程的特点；C、D 是两个进程，C 能进入 D 访问 D 的子单元，说得再通俗一点，C 能跨进程访问 D 的函数，既然能访问 D 的函数，也就说明 C、D 进程共享 D 的数据，这是 ContentProvider 跨进程访问的特点，也是叫作 ContentProvider 名称的原因。

9.6.2 最简单的示例

ContentProvider 涉及较多相关的类，为此先编制一个最简单的 ContentProvider 组件跨进程通信示例，有了一定感性认识后，再加深对相关知识的理解。

示例功能是：产生两个工程，先在工程 1 中实现内部 ContentProvider 调用，再从工程 2 中应用工程 1 中的 ContentProvider 调用。

工程 1 中涉及的文件有：ContentProvider 类文件 MyProvider.java；配置文件 AndroidManifest.xml；主界面文件 main.xml；主应用文件 MainActivity.java。下面一一介绍。

① ContentProvider 类文件 MyProvider.java。

```
public class MyProvider extends ContentProvider {
    public boolean onCreate() {  return true;}
    public Cursor query(Uri uri, String[] projection, String selection, String[]
selectionArgs, String sortOrder) {
        return null;
    }
    public String getType(@NonNull Uri uri) {  return null;  }
    public Uri insert(Uri uri, ContentValues values) { return null;}
    public int delete(Uri uri, String selection, String[] selectionArgs)
{  return 0;  }
    public int update(Uri uri, ContentValues values, String selection, String[]
selectionArgs) {
        return 10;
    }
}
```

该类中的函数是共享函数，必须从 ContentProvider 派生，上述 6 个函数 onCreate()、query()、getType()、insert()、delete()、update()是基类 ContentProvider 中的抽象函数，在子类中必须被重写，跨进程调用一般指 insert()、delete()、update()、query()，即增、删、改、查 4 种功能函数。本示例演示利用 ContentProvider 调用 update()函数，为了简化问题，仅让其返回整数 10。其余 5 个函数的代码均是默认代码，没做任何改动。

② 配置文件 AndroidManifest.xml。

每个自定义 ContentProvider 类都要在配置文件中增加一个＜provider＞标签节点，本示例的 MyProvider 对应的配置文件信息如下所示。

```
<provider
    android:authorities="com.example.dqjbd.we.myprovider"
    android:name=".MyProvider"
    android:exported="true">
```

```
</provider>
```

其中，android：name 用于指明具体的自定义 ContentProvider 类；android：exported 指明是否可被外部进程访问，若为 true，表明可被外部进程访问，否则不能被外部进程访问；android：autoorities，授权字符串，外部进程通过该字符串才能访问 MyProvider 类。

③ 主界面文件 main.xml。

```
<?xml version="1.0" encoding="utf-8"?>
<LinearLayout xmlns:android="http://schemas.android.com/apk/res/android"
android:layout_width="match_parent"
android:layout_height="match_parent">
    <Button
        android:id="@+id/myok"
        android:layout_width="wrap_content"
        android:layout_height="wrap_content"
        android:text="OK"/>
</LinearLayout>
```

以上代码定义了一个 OK 按钮，id 为 myok，用于响应内部进程 ContentProvider 调用功能。

④ 主应用文件 MainActivity.java。

```
public class MainActivity extends AppCompatActivity {
    protected void onCreate(Bundle savedInstanceState) {
        super.onCreate(savedInstanceState);
        setContentView(R.layout.main);
        Button b = (Button) findViewById(R.id.myok);
        b.setOnClickListener(new View.OnClickListener() {
            public void onClick(View v) {
                try {
                    Uri uri = Uri.parse("content://com.example.dqjbd.we
.myprovider/a");
                    ContentResolver resolver = getContentResolver();
                    int n = resolver.update(uri, null, null, null);
                    Toast.makeText(MainActivity.this,"SUCCESS"+n,Toast.LENGTH_
LONG).show();
                }
                catch(Exception e){      }
            }
```

```
            });
        }
    }
```

可以看出，内部 ContentProvider 调用代码在 try 块中，定义 Uri 对象时，有一个重要的字符串前缀"content://com.example.dqjbd.we.myprovider"，content 必不可少，后续的 com.example.dqjbd.we.myprovider 与配置文件 Androidmanifest.xml 中＜provider＞标签中的 android：authoritiess 属性设置一致，这一点至关重要。代码中还涉及 ContentResolver 类，后续还有详细说明，通过 ContentResolver 对象 resolver 调用 update()函数，其实就是调用 MyProvider 类中的 update()函数，将其返回值利用 Toast 显示。

实践证明，try 块中的内部 ContentProvider 调用成功，update 函数成功返回了 10，并利用 Toast 显示在界面中。

既然内部调用 ContentProvider 成功，那是否意味着只将上文 try 块中的代码复制到另外一个进程中就可以使得外部调用 ContentProvider 成功呢？继续做实验，建立工程 2，假设其主应用为 MainActivity2，将上文 try 块中的代码复制到其中，如下所示。

```
public class MainActivity2 extends AppCompatActivity {
    protected void onCreate(Bundle savedInstanceState) {
        super.onCreate(savedInstanceState);
        setContentView(R.layout.main);
        Button b = (Button)findViewById(R.id.myok);
        b.setOnClickListener(new View.OnClickListener() {
            public void onClick(View v) {
                try {
                    Uri uri = Uri.parse("content://com.example.dqjbd.we
.myprovider/a");
                    ContentResolver resolver = getContentResolver();
                    //ContentValues cv = new ContentValues();
                    //cv.put("studno",1000);
                    //int n = resolver.update(uri, cv, null, null);
                    int n = resolver.update(uri, null, null, null);
                    Toast.makeText(MainActivity.this,"SUCCESS"+n,Toast.LENGTH_
LONG).show();
                }
                catch(Exception e){       }
            }
        });
```

```
    }
}
```

也就是说，MainActivity2 与 MainActivity 的代码完全一致，当运行时程序却出现异常，将 try 块中的代码 int n＝resolver.update(uri,null,null,null)用注释中的 3 行语句替换，再运行程序，发现外部调用工程 1 中的 MyProvider 类中的 update()函数成功。也就是说，外部调用和内部调用 ContentProvider 大致相同，细节处稍有不同。对本例外部调用而言，也容易理解：当更新某一数值时，需要把数值传进去，若传入 null，逻辑上是行不通的。

9.6.3 相关类介绍

通过 9.6.2 中的示例，可知 ContentProvider 涉及许多相关的类，下面一一说明。

1. Uri 类

Uri 是统一资源标识符（uniform resource identifier）的简称，本节仅讲述它在 ContentProvider 中的应用。Uri 代表了要操作的数据，主要包含了两部分信息：需要操作的 ContentProvider 以及对 ContentProvider 中的哪些数据进行操作。一个 Uri 由以下几部分组成，如下所示。

```
content://com.example.dqjbd.we.myprovider/path
[scheme]  [   authority权限字符串          ][路径]
```

ContentProvider 的 scheme 已经由 Android 规定为 content://；主机名（或叫 authority 权限字符串）用于唯一标识这个 ContentProvider，外部调用者可以根据这个标识调用。路径（path）可以用来表示要操作的数据，路径的构建应根据业务而定，例如：

- 要操作 student 表中 id 为 5 的记录，可以构建路径/student/5。
- 要操作 student 表中 id 为 5 的记录的 name 字段，可以构建路径 student/5/name。
- 要操作 student 表中的所有记录，可以构建路径/student。

当然，要操作的数据不一定来自数据库，也可以是文件、XML 或网络等其他存储方式。

如果要把一个字符串转换成 Uri，可以使用 Uri 类中的 parse()方法，例如：

```
Uri uri = Uri.parse("content://com.example.dqjbd.we.myprovider ")
```

2. UriMatcher 类

自定义 ContentProvider 类必须重写的 6 个函数中只有 onCreate()函数的参数中没

有 Uri 对象，因此，在其他 5 个函数中很可能根据 Uri 判定执行哪个分支功能。这时使用 UriMatcher 就可以方便地过滤，进行下一步功能，如果不用 UriMatcher，就需要手动过滤字符串，用起来有点麻烦，维护性也不好。

UriMatcher 类常用的方法如下所示。

- UriMatcher(int code)，构造方法，code 是定义的一个整数值，当执行匹配函数 match()时，若没有匹配值，则返回 code。
- void addURI(String authority,String path, int code)，增加 URI 对应的匹配值。当执行匹配函数 match()时，若恰好是 authority+path，则返回 code。
- int match(uri)，返回与 uri 匹配的整数值。

上述 3 个函数的具体应用见表 9-7 所示。

表 9-7 UriMatcher 类应用举例

UriMatcher um =new UriMatcher(0);	um 有值 0
um.addURI("com.test","aaa",1);	um 有值 0,1
um.addURI("com.test","bbb",2);	um 有值 0,1,2
int value = um.match(Uri.parse("content://com.test/aaa"));	由于该 uri 实参值与第 1 个 addURI()的函数参数值匹配，所以返回 1

由上可知，UriMatcher 是一 map 对象，键由 authority+path 字符串组成，值即 code。

3. ContentResolver 类

ContentResolver 是通过 Uri 查询 ContentProvider 中提供的数据的。除了 Uri，还必须知道需要获取的数据段的名称，以及此数据段的数据类型。其常用函数如下所示。

- Uri insert(Uri uri,ContentValues cv)，增加函数。
- int delete(Uri uri,String where, String[] args)，删除函数。
- update(Uri uri,ContentValues cv,String where,String[]args)，修改函数。
- Cursor query(Uri uri,String[]projection,String select, String[]selectargs, String sortOrder)，查询函数。
- String getType(Uri uri);获得类型函数。

4. ContentProvider 类

ContentProvider 类是一个抽象类，它的派生类必须实现 6 个抽象方法。其中，insert()、delete()、update()、query()、getType()函数与 ContentResolver 相应的函数几乎一致，只有 onCreate()函数需要读者加深理解。其实，只要在 9.6.2 节 MyProvider 中的 onCreate()

函数中加一条 Toast 语句，当首次装载并运行 App 时，就会发现 onCreate()函数运行了。这说明 onCreate()函数不是主动调用的，是系统自动调用的。而且不论反复运行多少次 App，onCreate()函数都不再运行。这说明 MyProvider 类中的 onCreate()函数仅运行一次，因此可以得出：该函数的主要功能是在此处创建数据共享对象，为其他进程所共享。

9.6.4 实现 SharedPreferences 共享

读者都有这样的经历，某网站需要输入登录账号及密码，当首次输入并保存账号、密码后，之后再进该网站时保存的账号、密码会显示在界面上，不用再输入了。本节就利用 ContentProvider 类实现 SharedPreferences 内部存储文件共享，并实现不同进程登录信息自动显示的功能。

对每个工程来说，一般包名是不同的，将包名作为键，值为"用户名，密码"合成的字符串，保存到 SharedPreferences 共享文件中即可。

本示例是在 9.6.2 节中示例的基础上改进的，涉及的文件有：配置文件 AndroidManifest.xml；ContentProvider 类文件 MyProvider.java；主界面文件 main.xml；主应用文件 MainActivity.java。下面一一详细介绍。

① 配置文件 AndroidManifest.xml：同 9.6.2 节一致，略。

② ContentProvider 类文件 MyProvider.java。

```
public class MyProvider extends ContentProvider {
    public boolean onCreate() {
        SharedPreferences sp = this.getContext().getSharedPreferences
("myshare.txt",Context.MODE_PRIVATE);
        return true;
    }
    public Cursor query(Uri uri, String[] projection, String selection, String[]
selectionArgs, String sortOrder) {
        return null;
    }
    public String getType(@NonNull Uri uri) {
        SharedPreferences sp = this.getContext().getSharedPreferences
("myshare.txt", Context.MODE_PRIVATE);
        String key = uri.getPath();
        key = key.substring(1);
        return sp.getString(key,null);
```

```
    }
    public Uri insert(@NonNull Uri uri, @Nullable ContentValues values) {return
null;}
    public int delete(Uri uri, String selection, String[] selectionArgs)
{return 0;}
    public int update(Uri uri, ContentValues values, String selection, String[]
selectionArgs) {
        try {
            String package = values.getAsString("package");
            String info = values.getAsString("info");
            SharedPreferences sp = this.getContext().getSharedPreferences
("myshare.txt", Context.MODE_PRIVATE);
            SharedPreferences.Editor et = sp.edit();
            et.putString(package, info);
            et.commit();
        }catch(Exception e){ return 0; }
        return 1;
    }
}
```

insert()、query()、delete()函数保持默认代码,没有做任何改动。

onCreate()函数中利用 SharedPreferences 创建了共享内存文件 myshare.txt,用于保存账号与密码,为各应用程序所共享。

update()函数实现了账号、密码保存功能,getType()实现了账号、密码查询功能。也许有读者会问:这两个函数与我们要求的功能有些词不达意,增加应该是 insert(),查找应该是 query()。但是 query()函数返回的是游标对象 Cursor,这在 SharedPreferences 对象中很难实现;insert()函数的返回值是 Uri,不能返回 String。综合起来,本示例选择了 onupdate()及 getType()两个函数,也有其他办法可以解决读者所说的词不达意问题,增加一个中间类即可,读者可独立完成。

注意 update()函数中参数 ContentValues 对象 values 的定义。values 封装了工程中用户名和密码的信息,如何用相同的代码屏蔽不同工程间的差异?其实很简单,在 values 对象中传过来两个固定键值的信息,本例的两个键值为 package、info。根据键 package,可以获得工程的包名,根据键 info,可以获得用户名及密码的合成信息字符串值,将获得的两个值保存到 SharedPreferences 共享对象中即可。

getType()函数中,要注意 Uri 对象 uri 的作用。由于该函数只有一个参数,因此必须根据该参数确定在 SharedPreferences 对象中查询的键值,也就是说,必须将工程的包名隐藏在传入的 Uri 对象中,因此得该 Uri 对象对应的字符串如下所示。

```
content://授权字符串/包名
```

因此,在 getType()函数中首先要利用 getPath()获得包名字符串的内容,但默认得出的结果形如"/包名",多了一个前缀"/",将其去掉后,就能得到所需的包名,根据该值,就能查询得出 SharedPreferences 共享对象对应的用户名、密码信息了。

③ 主界面文件 main.xml。

```
<?xml version="1.0" encoding="utf-8"?>
<LinearLayout xmlns:android="http://schemas.android.com/apk/res/android"
android:layout_width="match_parent"
android:layout_height="match_parent"
    android:orientation="vertical">
    <EditText
        android:id="@+id/myuser"
        android:layout_width="match_parent"
        android:layout_height="wrap_content" />
    <EditText
        android:id="@+id/mypwd"
        android:layout_width="match_parent"
        android:layout_height="wrap_content"/>
    <Button
        android:id="@+id/myok"
        android:layout_width="wrap_content"
        android:layout_height="wrap_content"
        android:text="OK"/>
</LinearLayout>
```

为了首先在进程内测试 ContentProvider 调用,编制了此界面。主界面中有两个 EditText 组件,id 分别为 myuser、mypwd,用于输入用户名及密码;一个 Button 组件,id 为 myok,用于将用户名、密码信息保存到 SharedPreferences 共享对象中。

④ 主应用文件 MainActivity.java。

```
public class MainActivity extends AppCompatActivity {
    protected void onCreate(Bundle savedInstanceState) {
        super.onCreate(savedInstanceState);
        setContentView(R.layout.main);
        Uri uri = Uri.parse("content://com.example.dqjbd.we.myprovider/com
```

```
.example.dqjbd.we");
        ContentResolver resolver = getContentResolver();
        String s = resolver.getType(uri);
        if(s != null){
            String unit[] = s.split(",");
            EditText et = (EditText)findViewById(R.id.myuser);
            et.setText(unit[0]);
            et = (EditText)findViewById(R.id.mypwd);
            et.setText(unit[1]);
        }
        Button b = (Button)findViewById(R.id.myok);
        b.setOnClickListener(new View.OnClickListener() {
            public void onClick(View v) {
                try {
                    EditText et = (EditText)findViewById(R.id.myuser);
                    String user = et.getText().toString();
                    et = (EditText)findViewById(R.id.mypwd);
                    String pwd = et.getText().toString();
                    Uri uri = Uri.parse("content://com.example.dqjbd.we
.myprovider");
                    ContentResolver resolver = getContentResolver();
                    ContentValues cv = new ContentValues();
                    cv.put("package", "com.example.dqjbd.we");
                    cv.put("info",user+","+pwd);
                    int n = resolver.update(uri, cv, null, null);
                }
                catch(Exception e){    }
            }
        });
    }
}
```

从各 Uri 对象中可看出，授权字符串是 com.example.dqjbd.we.myprovider。由于本示例的工程包名为 com.example.dqjbd.we，因此当 onCreate()函数运行时，将包名附在授权字符串的后面形成 Uri 对象，然后启动 ContentResolver 对象 resolver，调用 getType()函数查找 com.example.dqjbd.we 包名对应的用户名及密码的合成字符串 s。若 s 非空，表明从 SharedPreferences 共享对象中获得了已保存的用户名、密码信息，将 s 按“,”拆分得字符串数组 unit，unit[0]是用户名，unit[1]是密码，将用户名、密码值显示在对应的

EditText 组件中即可；若 s 为空，表明用户名、密码还没有保存到 SharedPreferences 共享对象中。

首先在按钮的消息响应函数中获得用户名 user、密码 pwd；然后创建 Uri 对象，不需要路径部分。创建 ContentResolver 对象 resolver，并创建 ContentValues 对象 cv，其中建立两组键值映射对：第 1 组键是 package，值是包名 com.example.dqjbd.we；第 2 组键是 info，值是用户名和包名的合成串，用“，”分割；最后调用 update()函数完成用户名、包名的保存。

实验时，首次运行本示例 App，输入用户名、密码值，单击 OK 按钮，则将已输入的用户名、密码值保存到 SharedPreferences 对象中。当再次反复重启 App 时，会发现原先输入的用户名、密码值自动显示在界面上，表明我们在进程内利用 ContentProvider 实现了共享对象的存储。

继续做实验，创建第 2 个 App 工程，主界面与第 1 个工程的主界面完全相同，主应用与第一个应用的主应用类 MainActivity 几乎相同，仅两行代码稍有不同，改动后的代码如下所示。

- Uri uri = Uri.parse("content://com.example.dqjbd.we.myprovider/com.example.dqjbd.we2")；

由于第 2 个工程包名为 com.example.dqjbd.we2，因此 Uri 路径部分由此包名替代。

- cv.put("package", "com.example.dqjbd.we2")；

由于第 2 个工程包名为 com.example.dqjbd.we2，因此 put()函数的第 2 个参数由此包名替代。

实验时：保证第 1 个工程已经启动过，运行第 2 个工程，输入用户名、密码值，单击 OK 按钮，则将已输入的用户名、密码值保存到 SharedPreferences 对象中。再次反复重启第 2 个工程 App 的时候，会发现原先输入的用户名、密码值自动显示在界面上，表明我们在进程间利用 ContentProvider 实现了共享对象的存储。

9.6.5 实现数据库共享

尽管利用 ContentProvider 可以实现文本文件、XML 等文件的共享，但最恰当的还是数据库共享，从 query()函数返回 Cursor 游标对象就更能体现出这一点。数据库共享的编程模型与 9.6.4 节论述的 SharedPreferences 共享大同小异。

考虑一个应用：工程 A 仅负责创建空数据库 test.db，其中有两个表，无记录。一个表 stud(包含学号 studno，整型；姓名 studname，字符型)；一个表 teac(包含教师号 teacno，整型；姓名 teacname，字符型)。工程 B 负责操作工程 1 创建的数据库 test.db，并完成学生表 stud 或教师表 teac 数据的添加与显示。

1. 工程 A 包含的文件

① 主界面文件 main.xml：仅起到象征意义，空白即可，略。

② 配置文件 Androidmanifest.xml：配置 ContentProvider 信息。

```
<provider
    android:authorities="com.example.dqjbd.we.myprovider"
    android:name=".MyProvider"
    android:exported="true">
</provider>
```

可以看出，授权字符串是 com.example.dqjbd.we.myprovider，ContentProvider 类是 MyProvider，exported 属性是 true，表明 MyProvider 可以被外部进程访问。

ContentProvider 类文件 MyProvider.java。

```
public class MyProvider extends ContentProvider {
    static String auth="com.example.dqjbd.we.myprovider";
    static UriMatcher um = new UriMatcher(0);
    static{
        um.addURI(auth,"stud",1);
        um.addURI(auth,"teac",2);
    }
    /* 首先定义了静态授权字符串成员变量 auth,然后定义了静态的集合类成员变量 um,可知
um.match()函数返回的值是 0 或 1,或 2:当 Uri 对象包含 auth 和"stud"时,返回 1,表明要对
数据库中的 stud 表进行操作;当 Uri 对象包含 auth 和"teac"时,返回 2,表明要对数据库中的
teac 表进行操作;其余情况返回 0。 */
    public boolean onCreate() {
        Toast.makeText(this.getContext(),"Create",Toast.LENGTH_LONG).show();
        Context ctx = this.getContext();
        SQLiteDatabase db = SQLiteDatabase.openDatabase(ctx.getFilesDir()+"/
test.db",null,
                SQLiteDatabase.OPEN_READWRITE | SQLiteDatabase.CREATE_IF_
NECESSARY);
        try {
            db.execSQL("create table stud(studno integer,studname char(20))");
            db.execSQL("create table teac(teacno integer,teacname char(20))");
        }catch(Exception e){      }
        db.close();
        return true;
```

```
    }
    /* 创建了 test.db 数据库,两个空表 stud 及 teac */
    public Uri insert(@NonNull Uri uri, @Nullable ContentValues values){
        Context ctx = this.getContext();
        SQLiteDatabase db = SQLiteDatabase.openDatabase(ctx.getFilesDir()+"/
test.db",null,
                SQLiteDatabase.OPEN_READWRITE );
        int code = um.match(uri);
        if(code==1) db.insert("stud",null,values);
        if(code==2) db.insert("teac",null,values);
        return null;
    }
    /* 首先利用 openDatabase()函数获得数据库对象 db,然后根据 uri 匹配值决定是对
stud 表还是 teac 表进行添加操作 */
    public Cursor query(Uri uri, String[] projection, String selection, String
[] selectionArgs, String sortOrder) {
        Context ctx = this.getContext();
        SQLiteDatabase db = SQLiteDatabase.openDatabase(ctx.getFilesDir()+"/
test.db",null,
                SQLiteDatabase.OPEN_READWRITE );
        int code = um.match(uri);
           if (code == 1) return db.query ("stud", projection, selection,
selectionArgs,null,null,sortOrder);
           if (code == 2) return db.query ("teac", projection, selection,
selectionArgs,null,null,sortOrder);
        return null;
    }
    /* 首先利用 openDatabase()函数获得数据库对象 db,然后根据 uri 匹配值决定是对
stud 表还是 teac 表进行查询操作 */
    public String getType(@NonNull Uri uri) {return null;}
    public int delete(Uri uri, String selection, String[] selectionArgs)
{return 0;}
    public int update(Uri uri, ContentValues values, String selection, String[]
selectionArgs) {return 0;}
    /* 这 3 个函数保持默认代码,读者可根据需要增加相应功能 */
}
```

③ 主应用文件 MainActivity.java。

```
public class MainActivity extends AppCompatActivity {
    protected void onCreate(Bundle savedInstanceState) {
        super.onCreate(savedInstanceState);
        setContentView(R.layout.main);
    }
}
```

该应用一运行，MyProvider 类的 onCreate()函数就运行，数据库 test.db 创建完毕。

2. 工程 B 包含的文件

为了方便，这里仅编制了学生表 stud 的数据增加与显示代码，读者可仿此代码编制教师表相应的功能。

① 主界面文件 main.xml。

```
<?xml version="1.0" encoding="utf-8"?>
<LinearLayout xmlns:android="http://schemas.android.com/apk/res/android"
    android:layout_width="match_parent"
    android:layout_height="match_parent"
    android:orientation="vertical">
    <EditText
        android:id="@+id/studno"
        android:layout_width="match_parent"
        android:layout_height="wrap_content" />
    <EditText
        android:id="@+id/studname"
        android:layout_width="match_parent"
        android:layout_height="wrap_content" />
    <Button
        android:id="@+id/studadd"
        android:layout_width="wrap_content"
        android:layout_height="wrap_content"
        android:text="添加"/>
    <Button
        android:id="@+id/studshow"
        android:layout_width="wrap_content"
        android:layout_height="wrap_content"
        android:text="显示"/>
    <TextView
```

```
        android:id="@+id/myview"
        android:layout_width="match_parent"
        android:layout_height="match_parent" />
</LinearLayout>
```

以上代码定义了两个 EditText 组件，id 分别为 studno、studname，用于输入学号及姓名；一个 id 为 studadd 的 Button 组件，用于完成学生信息的添加；一个 id 为 studshow 的 Button 组件，用于响应学生信息显示事件；一个 id 为 myview 的 TextView 组件，用于显示学生信息。

② 主应用文件 MainActivity.java。

```
public class MainActivity extends AppCompatActivity {
    protected void onCreate(Bundle savedInstanceState) {
        super.onCreate(savedInstanceState);
        setContentView(R.layout.main);

        Button b = (Button)findViewById(R.id.studadd);
        b.setOnClickListener(new View.OnClickListener() {
            public void onClick(View v) {
                try {
                    EditText et = (EditText) findViewById(R.id.studno);
                    String no = et.getText().toString();
                    et = (EditText) findViewById(R.id.studname);
                    String name = et.getText().toString();
                    ContentValues cv = new ContentValues();
                    cv.put("studno", no);
                    cv.put("studname", name);
                    Uri uri = Uri.parse("content://com.example.dqjbd.we
.myprovider/stud");
                    ContentResolver resolver = getContentResolver();
                    resolver.insert(uri, cv);
                }catch(Exception e){   }
            }
        });
        /* 添加按钮响应函数中，首先获得输入的学号 no、姓名 name，然后将此两值保存至
ContentValues 对象 cv 中，最后建立 Uri 对象，其路径部分用表名 stud 填充，表明要对 stud
表进行添加。启动 ContentResolver 对象 resolver，调用 insert()函数完成跨进程的数据库
学生信息数据的添加。*/
```

```
        Button b2 = (Button)findViewById(R.id.studshow);
        b2.setOnClickListener(new View.OnClickListener() {
            public void onClick(View v) {
                String total = ""; String unit = "";
                try {
                    Uri uri = Uri.parse("content://com.example.dqjbd.we
.myprovider/stud");
                    ContentResolver resolver = getContentResolver();
                    Cursor cs = resolver.query(uri,null,null,null,null);
                    while(cs.moveToNext()){
                        String no = cs.getString(0);
                        String name = cs.getString(1);
                        unit = no+","+name;
                        total += unit+"\n";
                    }
                    TextView tv = (TextView)findViewById(R.id.myview);
                    tv.setText(total);
                }catch(Exception e){     }
            }
        });
    }
}
```

显示按钮响应函数中，首先建立 Uri 对象，路径部分用表名填充，启动 ContentResolver 对象 resolver；然后调用 query()函数获得 stud 表的记录游标 Cursor 对象 cs，遍历 cs，获得每条记录 unit(包含学号、姓名，中间用逗号分隔)，进而累加成所有记录值 total；最后将 total 值显示在 TextView 组件中。

实验时，先运行工程 A，目的是创建空数据库 test.db。然后关闭工程 A，运行工程 B，输入一条学生记录，单击“添加”按钮后，再单击“显示”按钮，发现数据显示在界面上，说明我们利用 ContentProvider 跨进程操作数据库存储成功。

9.6.6 系统数据库共享

手机联系人信息封装在数据库中，手机联系人提供了 ContentProvider 机制，使得外部进程访问相关数据成为可能。本节在自定义工程中实现访问手机联系人的联系电话、邮件等功能。为实现这些功能，必须知道访问联系人、电话、邮件的 Uri 字符串及各字段名称，在系统 ContactsContract 类中均能查到所需信息，见表 9-8。

表 9-8　ContactsContract 类中的相关信息

序号	功能字符串	说　　明
1	ContactsContract.Contacts.CONTENT_URI	联系人 Uri 对象
2	ContactsContract.CommonDataKinds.Phone. CONTENT_URI	电话 Uri 对象
3	ContactsContract.CommonDataKinds.Email. CONTENT_URI	邮件 Uri 对象
4	ContactsContract.Contacts._ID	联系人 id 字段名称
5	ContactsContract.Contacts.DISPLAY_NAME	联系人 name 字段名称
6	ContactsContract.CommonDataKinds.PhoneNUMBER	电话字段名称
7	ContactsContract.CommonDataKinds.Email.DATA	邮件字段名称

实现本功能涉及的文件描述如下所示。

① 配置文件 Androidmanifest.xml：增加读写联系人信息权限，如下所示。

```
<uses-permission android:name="android.permission.READ_CONTACTS"></uses-
permission>
<uses-permission android:name="android.permission.WRITE_CONTACTS"></uses-
permission>
```

② 主界面文件 main.xml。

```
<?xml version="1.0" encoding="utf-8"?>
<LinearLayout xmlns:android="http://schemas.android.com/apk/res/android"
    android:orientation="vertical"
    android:layout_width="match_parent"
    android:layout_height="match_parent">
    <Button
        android:id="@+id/myok"
        android:layout_width="wrap_content"
        android:layout_height="wrap_content"
        android:text="ok"/>
    <TextView
        android:id="@+id/result"
        android:layout_width="match_parent"
        android:layout_height="match_parent"
</LinearLayout>
```

以上代码定义了一个 OK 按钮，id 为 myok，当单击此按钮时，执行获取数据库联系人相关信息操作；一个 TextView 组件，id 为 result，用于显示联系人姓名、电话、邮件等信息。

③ 主应用文件 MainActivity.java。

```
public class MainActivity extends   Activity{
    protected void onCreate(Bundle savedInstanceState) {
        super.onCreate(savedInstanceState);
        setContentView(R.layout.main);
        ActivityCompat.requestPermissions(this,new String[]
                {Manifest.permission.READ_CONTACTS,Manifest.permission.WRITE_
CONTACTS},1);
        Button b = (Button)findViewById(R.id.myok);
        b.setOnClickListener(new View.OnClickListener() {
        public void onClick(View v) {
            try {
                StringBuilder sb = new StringBuilder();
                ContentResolver resolver = getContentResolver();
                Cursor cursor = resolver.query(ContactsContract.Contacts
.CONTENT_URI, null, null, null, null);
                while (cursor.moveToNext()) {
                    int idindex = cursor.getColumnIndex(ContactsContract
.Contacts._ID);
                    int nameindex = cursor.getColumnIndex(ContactsContract
.Contacts.DISPLAY_NAME);
                    int id = cursor.getInt(idindex);
                    String name = cursor.getString(nameindex);
                    Cursor phone =resolver.query(ContactsContract
.CommonDataKinds.Phone.CONTENT_URI,
                    null, ContactsContract.CommonDataKinds.Phone.CONTACT_
ID + "=" + id, null, null);
                    while (phone.moveToNext()) {
                        int phoneindex = phone.getColumnIndex(
                            ContactsContract.CommonDataKinds.Phone.NUMBER);
                        String phonenumber = phone.getString(phoneindex);
                        sb.append(name + ":" + phonenumber + "\n");
                    }
                    phone.close();
                    Cursor email = resolver.query(ContactsContract
.CommonDataKinds.Email.CONTENT_URI,
                        null, ContactsContract.CommonDataKinds.Phone.CONTACT_
ID + "=" + id, null, null);
```

```
                    while (email.moveToNext()) {
                        int emailindex = email.getColumnIndex(
                            ContactsContract.CommonDataKinds.Email.DATA);
                        String emailvalue = email.getString(emailindex);
                        sb.append(name + ":" + emailvalue + "\n");
                    }
                    email.close();
                }
                    cursor.close();
                    TextView tv = (TextView) findViewById(R.id.result);
                    tv.setText(sb.toString());
                }catch(Exception e){   }
            }
        });
    }
}
```

onCreate()函数的开始部分增加了动态添加读写联系人权限代码，主要是屏蔽不同 Android 版本间的差异，因此本示例既有静态权限设置，又有动态权限设置，保证在不同 Android 版本下都能正确运行。

为了便于理解 onCreate()代码中关于查询联系人姓名、电话、邮箱的具体过程，可参考表 9-9。

表 9-9　数据库信息查询流程描述

```
获得联系人 Uri 对应的查询 Cursor 游标对象 cursor
while(有联系人记录)
    根据联系人关键字、姓名字段名称获得对应的索引位置 idindex、nameindex
    根据 idindex、nameindex 获得联系人的具体 id 值及姓名 name
    根据电话 Uri 对象查询联系人为 id 的电话信息，获得结果集 Cursor 游标对象 phone
    while(有电话信息)
        根据电话字段信息，获得电话索引位置 phoneindex
        根据 phoneindex，获得具体的电话号码 phonenumber
    end while
    根据邮件 Uri 对象查询联系人为 id 的邮件信息，获得结果集 Cursor 游标对象 email
    while(有邮件信息)
        根据邮件字段信息，获得邮件索引位置 emailindex
        根据 emailindex，获得具体邮件名称 emailvalue
    end while
end while
```

习题 9

一、简答题

1. 什么是内部存储和外部存储?
2. 什么是 SharedPreferences 存储?
3. SQLiteOpenHelper 类的作用是什么?
4. ContentProvider 的作用是什么?

二、程序题

1. 界面上有一 EditText 控件,当单击“确定”按钮时,将 EditText 控件中输入的内容保存在外部文件 a.txt 中。

2. 创建 SQLite 外部数据库文件 mydb 及学生表 student,在主界面中输入学生信息(学号、姓名、专业、学院),单击“确定”按钮后,将学生信息保存到数据库表 student 中。

第 10 章

图形与动画

本章首先讲解 Android 2D 绘制基本图形、文字、位图的方法，接着对 path 路径绘图进行详细的描述，最后讲解帧动画、补间动画、属性动画的技术与应用。

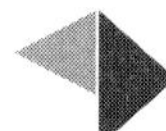 10.1 2D 绘图

10.1.1 最简单的绘图

Android 2D 绘图首先要定义 View 的派生类，要重写绘图函数 onDraw(Canvas cv)。下面先编制一个最简单的实例：画一条坐标(100,100)到坐标(300,300)的线段，其涉及的文件如下所示。

① 自定义 View 派生类 MyView.java。

```
public class MyView extends View {
    public MyView(Context ctx, AttributeSet attr){
        super(ctx, attr);
    }
    protected void onDraw(Canvas canvas) {
        super.onDraw(canvas);
        Paint p = new Paint();
        p.setColor(Color.RED);
        p.setStrokeWidth(3);
        canvas.drawLine(100,100,300,300,p);
    }
}
```

自定义 View 类绘图一定要包含一个构造方法，其形参有两个，分别是上下文

Context 对象及 AttributeSet 属性集合对象。在构造方法内直接利用 super()调用基类包含上述两个形参的构造方法。Android 系统创建 MyView 对象时，调用了包含上述两个形参的构造方法。若 MyView 类中没有上述构造方法，则程序运行时会出现异常。

图形绘制功能是在重写的 onDraw()函数中完成的，可以看出，Paint 是定制画笔及设置画笔属性的，示例中设置画笔颜色为红，线宽为 3。canvas 是画布的意思，所有绘制都是在其上完成的。

② 主界面文件 main.xml。

```
<?xml version="1.0" encoding="utf-8"?>
<LinearLayout xmlns:android="http://schemas.android.com/apk/res/android"
android:layout_width="match_parent"
android:layout_height="match_parent">
    <com.example.dqjbd.we.MyView
        android:layout_width="match_parent"
        android:layout_height="match_parent" />
</LinearLayout>
```

采用垂直线性布局，仅包含一个 MyView 组件，在写该组件节点时，一定要写上全包路径。

③ 主应用文件 MainActivity.java。

```
public class MainActivity extends AppCompatActivity {
    protected void onCreate(Bundle savedInstanceState) {
        super.onCreate(savedInstanceState);
        setContentView(R.layout.main);
    }
}
```

最简单的主程序，当运行时，会看见在手机窗口中出现一条红色的线段。

继续做实验，在 main.xml 中，在<com.example.dqjbd.we.MyView>节点中增加背景属性设置：android:background="#000000"，当再重新运行 App 时，会发现手机黑色背景中有一条红色线段。也就是说，该属性值一定是在创建 MyView 对象时经构造方法传进去的，即经 MyView 构造方法的第二个形参 AttributeSet 对象传入的，因此该构造方法必须写，而且不能写错。

10.1.2 相关类简介

可以看出，Android 2D 绘图涉及两个主要的类 Paint 与 Cancas，下面一一详细介绍。

1. Paint 类

Paint 类也称为画笔，它在画图的过程中定义各种参数，如颜色、线条样式、图案样式等。其常用函数如下所示。

- void setColor(int c)，设置颜色。
- void setStrokeWidth(float width)，设置画笔宽度。
- void setAntiAlias(bolean mark)，若 mark 为 true，则抗锯齿功能打开，否则关闭。抗锯齿是指在图像中，物体边缘总会或多或少地呈现三角形的锯齿。抗锯齿就是指对图像边缘进行柔化处理，使图像边缘看起来更平滑，更接近实物的物体。
- void setAlpha(int value)，value 的范围为[0,255]，数值越小越透明，反之越不透明。
- void setARGB(int a, int r, int g, int b)，设置透明度和红、绿、蓝颜色值。
- void setStyle(Style style)，画笔样式有 3 种：Paint.Style.FILL，填充内部；Paint.Style.STROKE，描边；Paint.Style.FILL_AND_STROKE，填充内部和描边。
- void setStrokeJoin(Join join)，设置线条连接处样式。当绘图样式为 STROKE / FILL_AND_STROKE 时，该方法用于指定线条连接拐角样式。常用的样式有 3 种：Paint.Join.MITER、Paint.Join.ROUND、Paint.Join.BEVEL，形状如图 10-1 所示。

图 10-1 线条连接拐角样式

- void setStrokeCap(Cap cap)，设置箭头样式。常用的 3 种形式：Paint.Cap.BUTT，平头；Paint.Cap.ROUND，圆头；Paint.Cap.SQUARE，方头。
- void setShader(Shader sh)，颜色渐变函数，包括线性渐变、径向渐变和扫描渐变等常用渐变。指定两种或两种以上颜色，根据颜色过渡算法自动计算中间的过渡颜色，从而形成渐变效果，对于开发人员来说，无需关注中间渐变颜色。
- void setTextSize(float size)，设置文本字体的大小。
- void setTextAlign(Align align)，设置文本的对齐方式，常用的参数值为 Align.CENTER、Align.LEFT、Align.RIGHT。

2. Canvas 类

可以把 Canvas 理解成系统提供给用户的一块内存区域，它提供了一整套对这个内存区域进行操作的方法，其常用方法如下所示。

- void drawRect(Rect rect, Paint paint)，绘制一个矩形。
- void drawPath(Path path, Paint paint)，绘制一条路径。
- void drawBitmap(Bitmap bitmap, Rect src, Rect dst, Paint paint)，贴图函数。参数 1 是常规的 Bitmap 对象，参数 2 是源区域(这里是 bitmap)，参数 3 是目标区域(在 canvas 中的位置和大小)，参数 4 是 Paint 对象。因为用到缩放和拉伸的功能，所以当原始 Rect 不等于目标 Rect 时，性能将会大幅下降。
- void drawLine(float startX, float startY, float stopX, float stopY, Paint paint)，画线函数。参数 1 是起始点的 x 轴坐标，参数 2 是起始点的 y 轴坐标，参数 3 是终点的 x 轴坐标，参数 4 是终点的 y 轴坐标，最后一个参数 Paint 为对象。
- void drawPoint(float x, float y, Paint paint)，画点函数。参数 1 为水平的 x 轴坐标，参数 2 为垂直的 y 轴坐标，参数 3 为 Paint 对象。
- void drawText(String text, float x, float y, Paint paint)，画文本函数。参数 1 是 String 类型的文本，参数 2 是 x 轴坐标，参数 3 是 y 轴坐标，参数 4 是 Paint 对象。
- void drawOval(Rect oval, Paint paint)，画椭圆函数。参数 1 是矩形区域，参数 2 为 Paint 对象。
- void drawCircle(float cx, float cy, float radius,Paint paint)，绘制圆函数。参数 1 是中心点的 x 轴坐标，参数 2 是中心点的 y 轴坐标，参数 3 是半径大小，参数 4 是 Paint 对象。
- void drawArc(RectF oval, float startAngle, float sweepAngle, boolean useCenter, Paint paint)，画弧函数。参数 1 是矩形区域对象，参数 2 是起始角度，参数 3 是扫描角度(顺时针旋转)，参数 4 决定着是封闭曲线还是弧线(如果是 true，则包括椭圆中心，将形成一个封闭曲线；如果是 false，则将是一条弧线)，参数 5 是 Paint 对象。

10.1.3 图像变换

Android 2D 图像变换主要包括平移、缩放、旋转等，下面一一介绍。

1. 平移

主要函数是 void translate(float dx, float dy)，即坐标原点相对当前坐标原点在 x 轴、y 轴上移动的距离，dx、dy 可正，可负。其示例如图 10-2 所示。

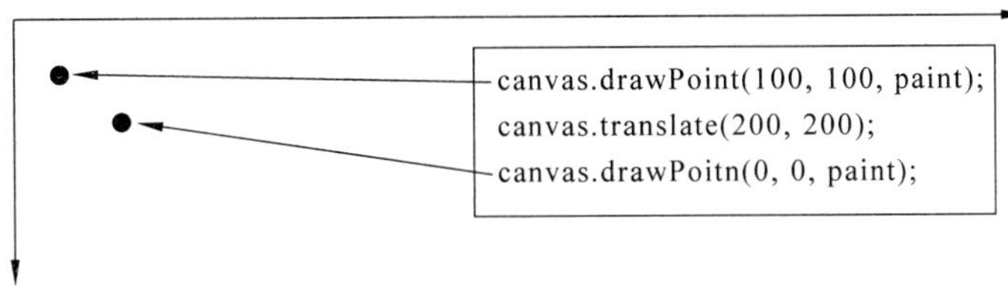

图 10-2　平移示意图

2. 缩放

主要函数有如下两个。

- void scale(float sx, float sy)：sx 为横向的缩放倍数，默认为 1，若小于 1 则缩小，若大于 1 则放大；sy 为纵向的缩放倍数，默认为 1，若小于 1 则缩小，若大于 1 则放大。sx、sy 均大于 0，其示例如图 10-3 所示。
- void scale(float sx, float sy,float dx, float dy)：sx、sy 解释同上；dx、dy 与坐标原点偏移有关。该函数可以等价理解为如下的描述：

```
void scale(float sx, float sy,float dx, float dy){
    scale(sx,sy);
    translate(dx,dy);
}
```

也就是说，该函数的功能可等价理解为首先执行一个具有两个参数的 scale()缩放函数，然后执行坐标平移函数 translate()。

缩放函数示例如图 10-3 所示。

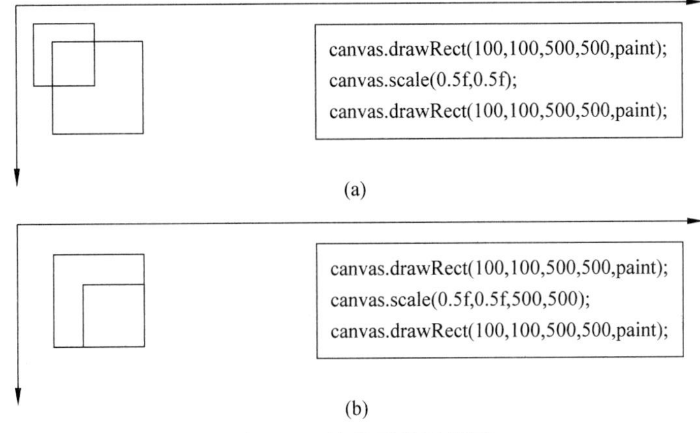

图 10-3　缩放函数示例图

图 10-3(a)比较容易理解,用的是具有两个参数的 scale()函数,横纵缩放比都是 0.5,对(100,100,500,500)矩形而言,即简单的坐标缩放,结果矩形为(50,50,250,250)。

对图 10-3(b)而言,要仔细理解,它用的是具有 4 个参数的 scale()函数,代码中缩小矩形的代码分析如表 10-1 所示。

表 10-1 具有 4 个参数的 scale()函数缩放分析

源 代 码	等 价 代 码

canvas.scale(0.5f,0.5f,500,500);
canvas.drawRect(100,100,500,500,paint);
	canvas.scale(0.5f,0.5f);
	canvas.translate(500,500)
	canvas.drawRect(100,100,500,500,paint);

根据等价代码可知,矩形横、纵缩放比都是 0.5;由 translate(500,500)知,坐标原点移至(250,250);由 drawRect(100,100,500,500,paint)及缩放比均为 0.5,可得画的矩形范围(50,50,250,250),由于是以原点(250,250)为基准画的矩形,因此缩小后矩形真实坐标是(300,300,500,500)。

3. 旋转

主要函数有如下两个。

- void rotate(float degrees): degrees 是旋转的角度,正值为顺时针,负值为逆时针。
- void rotate(float degrees, float px, float py): degrees 是旋转的角度,正值为顺时针,负值为逆时针,px、py 为旋转中心的 x 坐标、y 坐标,如果不指定旋转中心,则默认为点(0,0)。

旋转函数示例如图 10-4 所示。

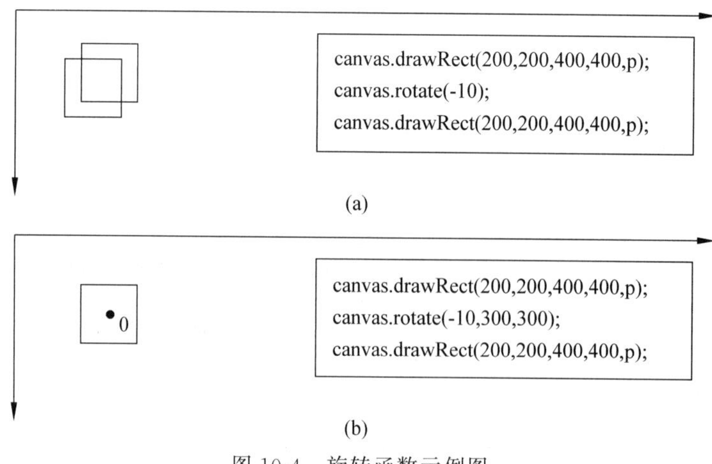

图 10-4 旋转函数示例图

图 10-4(a)用的是单参数的 rotate()函数，旋转中心的坐标是(0,0)；图 10-4(b)用的是具有 3 个参数的 rotate()函数，后两个参数指明了旋转中心，旋转中心的坐标是(300，300)，随着旋转中心的不同，得到的图形也不同。

10.1.4 Path 应用

Path 封装了由直线和曲线(二次、三次贝塞尔曲线)构成的几何路径。可以用 Canvas 中的 drawPath 把这条路径画出来，也可以用剪裁画布根据路径绘制文字。有时会用 Path 描述一个图像的轮廓，所以也会称为轮廓线(轮廓线仅是 Path 的一种使用方法，两者并不等价)。Path 相关函数可分为以下几种，下面一一说明。

1. 线段函数

- void moveTo(float x, float y)：设置操作的起始点位置。
- void lineTo(float x, float y)：添加上一个点到当前点之间的直线到 Path 对象。
- void close()：Path 对象中最后一个点与第一个点连接，形成一个闭合区域。
- void setLastPoint(float dx,float dy)：重置当前 Path 中最后一个点的位置，如果在绘制之前调用，则效果和 moveTo 相同。

线段函数应用示例见表 10-2。

表 10-2　线段函数应用示例

画 A->B->C	画 A->B->C->A	体会 setLastPoint()代码

Path pa = new Path();
pa.moveTo(X_A,Y_A);
pa.lineTo(X_B,Y_B);
pa.lineTo(X_C,Y_C);
canvas.drawPath(pa,paint);
	Path pa = new Path();
	pa.moveTo(X_A,Y_A);
	pa. lineTo (X_B, Y_B); pa. lineTo (X_C,Y_C);
	pa.close();
	canvas.drawPath(pa,paint);
		Path pa = new Path();
		pa.moveTo(X_A,Y_A);
		pa. lineTo (X_B, Y_B); pa. lineTo (X_C,Y_C);
		pa.setLastPoint(X_D,Y_D);
		canvas.drawPath(pa,paint);

	利用 close()函数形成封闭曲线	Path 路径经过点 A、B、C，此时最后一个点是 C，但由于 setLastPoint()设置成最后一个点是 D，所以最终画出的线段是 A->B->D，与 C 无关

注意：利用 Path 绘制直线一定要将画笔设置为 Paint.Style.STROKE 描边模式，否则是画不出直线的，一个关键代码示例如下所示。

```
Paint p = new Paint();
```

```
p.setStrokeWidth(10);                        //设置画笔宽度
p.setStyle(Paint.Style.STROKE);              //设置为描边模式,此行是画出 Path 定义的
                                               直线的关键
p.setColor(Color.RED);                       //设置画笔为红色
Path pa = new Path();                        //创建 Path
pa.moveTo(100,100);pa.lineTo(200,300);       //定义(100,100)- (200,300)的线段
canvas.drawPath(pa, p);                      //画出线段
```

2. 基本图形函数

- void addCircle (float x, float y, float radius, Path.Direction dir):添加圆形。
- void addOval (RectF oval, Path.Direction dir):添加椭圆。
- void addRect (float left,float top,float right,float bottom,Path.Direction dir):添加矩形。
- void addRect (RectF rect, Path.Direction dir):添加矩形。
- void addRoundRect (RectF rect, float[] radii, Path.Direction dir):添加圆角矩形。
- void addRoundRect (RectF rect, float rx, float ry, Path.Direction dir):添加圆角矩形。
- void addArc (RectF oval, float startAngle, float sweepAngle):添加一段圆弧。
- void arcTo (RectF oval, float startAngle, float sweepAngle):添加一段圆弧,如果圆弧的起点和上次最后一个坐标点不相同,就连接两个点。

可以看出,上述前 6 个函数都有一个形参 Direction 对象 dir,用来表明是顺时针画还是逆时针画,其值为以下静态常量:Path.Direction.CW,表明顺时针画;Path.Direction.CCW,表明逆时针画。

可能有读者问:画圆及画方时,顺时针画和逆时针画结果不一样吗?这两个参数有什么作用?主要是在一些特殊的情况下,利用 setLastPoint()修改最后一点的位置,从而影响图形的整体体现。下面以画方为例,如图 10-5 所示。

图 10-5(a)的含义是:顺时针添加矩形 ABCD,由于 setLastPoint()重新设置了最后一个点为 E,因此 Path 中包含的节点是 ABCE,当真实执行画矩形函数时,画的是 AB、BC、CE、EA 线段。也就是说,画矩形函数画出的有可能不是矩形。

图 10-5(b)的含义是:逆时针添加矩形 ADCB,由于 setLastPoint()重新设置了最后一个点为 D,因此 Path 包含的节点是 ADCE,当真实执行画矩形函数时,画的是 AD、DC、CE、EA 线段。

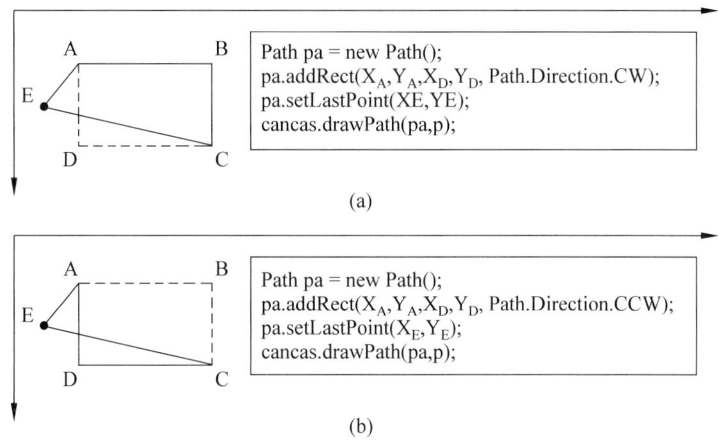

图 10-5　顺时针、逆时针影响图

10.1.5　贝塞尔曲线

1. 基本原理

贝塞尔曲线是计算机图形的重要曲线，它是由一个起始点、一个结束点及多个控制点组成的。贝塞尔曲线可绘制丰富的图形。下面简要论述它的原理。

(1) 二阶贝塞尔曲线原理，如图 10-6 所示。

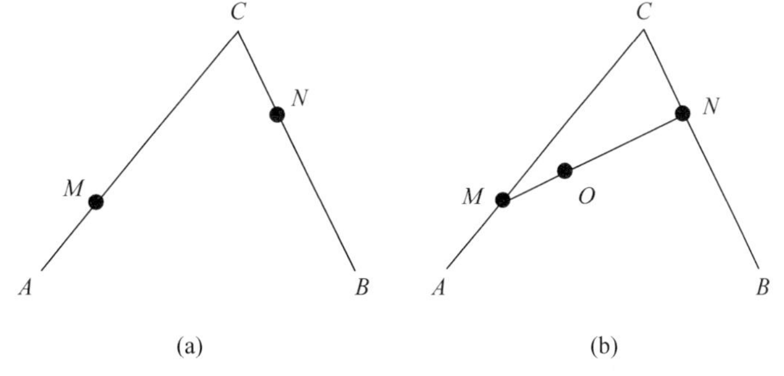

图 10-6　二阶贝塞尔曲线原理简图

A 是贝塞尔曲线的起始点，B 是结束点，C 是控制点。如图 10-6(a)：在 AC 上取一点 M，在 CB 上取一点 N，满足 $AM:MC=CN:NB$；如图 10-6(b)：连接 MN，在 MN 上取一点 O，满足 $MO:ON=AM:MC$，则 O 点的集合就形成了二阶光滑的贝塞尔曲线。

(2) 三阶贝塞尔曲线原理。

明白了二阶贝塞尔曲线原理,三阶贝塞尔曲线原理也就容易理解了,如图 10-7 所示。

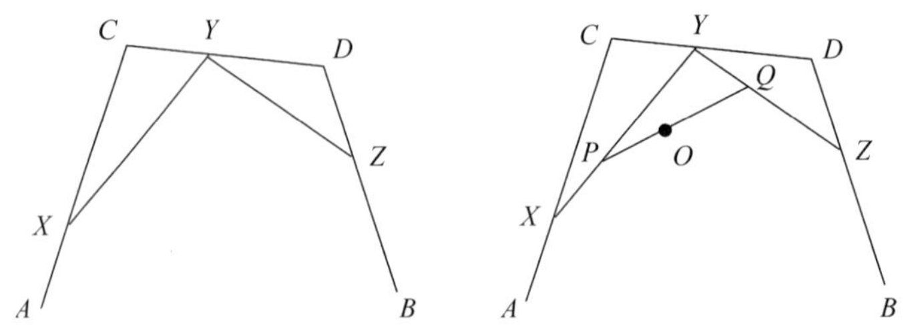

图 10-7 三阶贝塞尔曲线

A 是贝塞尔曲线的起始点,B 是结束点,C、D 是控制点。如图 10-7(a):在 AC、CD、DB 上分别取点 X、Y、Z,$AX : XC = CY : YD = DZ : ZB = t$;如图 10-7(b):在 XY、YZ 上分别取点 P、Q,$XP : PY = YQ : QZ = t$,在 PQ 上取点 O,满足 $PO : OQ = t$,则 O 点的集合就形成了三阶贝塞尔曲线。

很明显,求 A、B、C、D 四点三阶的贝塞尔曲线转换成了求 X、Y、Z 三点二阶的贝塞尔曲线,进而转变成求 P、Q 两点一阶的贝塞尔曲线。也就是说,多阶的贝塞尔曲线都可转化为求一阶的贝塞尔曲线。

2. 基本函数

Path 类提供了添加二阶、三阶贝塞尔曲线的方法,如下所示。

void quadTo(float x1,float y1,float x2, float y2):x1、y1 是二阶贝塞尔曲线中控制点的坐标值,x2、y2 是结束点的坐标值。

void cubicTo(float x1,float y1,float x2, float y2,float x3,float y3):x1、y1 是三阶贝塞尔曲线中控制点 1 的坐标值,x2、y2 是控制点 2 的坐标值,x3、y3 是结束点的坐标值。

可能有读者会问:上述两个函数中为什么没有贝塞尔曲线起始点的坐标?其实,起始点坐标由 moveTo()函数确定,例如,下述代码画出了起始点为(x1,y1),结束点为(x2,y2),控制点为(x3,y3)的二阶贝塞尔曲线。

```
Path pa = new Path();
pa.moveTo(x1,y1);
pa.quadTo(x2,y2,x3,y3);
canvas.drawPath(pa,p);
```

【例 10-1】 绘制二阶贝塞尔曲线:起始点、结束点确定,坐标分别为(100,500),(700,500)。控制点由手触摸事件决定。本示例涉及的文件如下所示。

① 主界面文件 main.xml。

```
<?xml version="1.0" encoding="utf-8"?>
<LinearLayout xmlns:android="http://schemas.android.com/apk/res/android"
    android:orientation="vertical"
    android:layout_width="match_parent"
    android:layout_height="match_parent">
    <com.example.dqjbd.we2.MyView
        android:id="@+id/mytestview"
        android:layout_width="match_parent"
        android:layout_height="match_parent">
    </com.example.dqjbd.we2.MyView>
</LinearLayout>
```

垂直线性布局，仅包含自定义组件 MyView，id 为 mytestview。

② 自定义 View 类文件 MyView.java。

```
public class MyView extends View implements View.OnTouchListener{
    PointF start = new PointF(100,500);
    PointF end = new PointF(700,500);
    PointF control = new PointF(400,300);
    public MyView(Context ctx, AttributeSet attr){
        super(ctx, attr);
    }
    protected void onDraw(Canvas canvas) {
        super.onDraw(canvas);
        Paint p = new Paint();
        p.setStrokeWidth(10);
        p.setStyle(Paint.Style.STROKE);
        p.setColor(Color.RED);
        Path pa = new Path();
        pa.moveTo(start.x,start.y);
        pa.quadTo(control.x,control.y, end.x,end.y);
        canvas.drawPath(pa, p);
    }
    public boolean onTouch(View v, MotionEvent event) {
        control.x = event.getX();
        control.y = event.getY();
        invalidate();
```

```
            return false;
        }
    }
```

一方面，MyView从View派生；另一方面，由于要在该类中处理手触摸事件，因此必须实现OnTouchListener接口。

成员变量start是二阶贝塞尔曲线的起始坐标，end是结束坐标，control是控制点坐标。onDraw()函数实现了绘制二阶贝塞尔曲线的全过程。onTouch()函数是手触摸事件响应函数，根据触摸位置，改变控制点control的坐标值，进而通过invalidate()函数通知系统要重新更新MyView显示界面，启动onDraw()函数完成二阶贝塞尔函数的重新绘制。

③ 主应用文件MainActivity.java。

```
public class MainActivity extends  Activity{
    protected void onCreate(Bundle savedInstanceState) {
        super.onCreate(savedInstanceState);
        setContentView(R.layout.main);
        MyView view = (MyView) findViewById(R.id.mytestview);
        view.setOnTouchListener(view);
    }
}
```

主应用文件主要是注册触摸侦听事件TouchListener，设置其响应函数在自定义View中。

10.1.6 位图操作

Bitmap对象本质上是一张图片的内容在内存中的表达形式。它将图片的内容看作由存储数据的有限个像素点组成；每个像素点存储该像素点位置的ARGB值。每个像素点的ARGB值确定下来，这张图片的内容就相应地确定下来了。与位图相关的类如下所示。

1. BitmapFactory

BitmapFactory是一个工具类，它提供了大量的方法，用于从不同的数据源解析、创建Bitmap对象。常用函数如下所示。

- Bitmap decodeFile(String pathName)：从pathName指定的文件中解析、创建Bitmap对象。

- Bitmap decodeFileDescriptor（FileDescriptor fd，Rect outPadding）：用于从 FileDescriptor 对应的文件中解析、创建 Bitmap 对象。
- Bitmap decodeResource(Resources res，int id)：用于给定的资源 ID 从指定资源中解析、创建 Bitmap 对象。
- Bitmap decodeStream(InputStream is，Rect outPadding)：用于从指定输入流中解析、创建 Bitmap 对象。

例如，在 App 工程中的 res\drawable\目录下创建图像资源文件 my.jpg，则其对应的资源号为 R.drawable.my，将其显示在自定义 View 屏幕上的关键代码如下所示。

```
protected void onDraw(Canvas canvas) {
    super.onDraw(canvas) *
    Bitmap bmp = BitmapFactory.decodeResource(getResources(),R.drawable.my);
    Paint p = new Paint();
    canvas.drawBitmap(bmp,0,0,p);
}
```

2. Bitmap

Bitmap 是一个位图类，它提供了许多方法，如下所示。

- Bitmap createBitmap(Bitmap source)：创建与原位图 source 等同的位图。
- Bitmap createBitmap(Bitmap source,int x,int y,int width,int height)：创建的位图与原位图 source 中从起始位置(x,y)开始，宽 width，高 height 的图像一致。
- Bitmap createBitmap(int width，int height,Config config)：创建一个宽 width、高 height 的位图。config 是一个枚举型，它代表了 Bitmap 可能的配置情况。一个配置描述的是这些像素信息是如何存储的，影响图片的质量和透明度。其常用的枚举值如下所示。

ALPHA_8，每个像素信息只存储 alpha 这一项信息。

ARGB_4444，该值在 level 13 时就不建议使用了，是一个低质量的配置。

ARGB_8888，每个像素信息占 4 字节(即 32 个二进制位)的存储空间，alpha、red、green、blue 各占 8 个二进制位，配置非常灵活，提供了最好的质量。

RGB_565，每个像素信息占 2 字节(即 16 个二进制位)，并且仅存储 RGB 信息，没有 alpha 信息，其中 Red 5 位，Green 6 位，Blue 5 位，该配置项在不需要提供透明度的情况下更有用。

- boolean compress(Bitmap.CompressFormat format，int quality，OutputStream stream)：用于将 Bitmap 对象压缩为指定格式并保存到指定的文件输出流中。

format 可以是 Bitmap.CompressFormat.PNG、Bitmap.CompressFormat.JPEG、Bitmap.CompressFormat.WEBP。quality 能体现压缩比，取值范围为[0,100]。例如，quality＝30，表示压缩率为 70%；quality＝100，表示压缩率为 0%，即不压缩。

【例 10-2】 界面上有一 save 按钮及自定义 View，当在 View 中用手触摸时，设触摸点的坐标为(x,y)，画一条(0,0)-(x,y)的线段，因此可以在 View 中画许多线段。当单击 save 按钮时，将所画内容保存成位图文件 a.jpg。本示例涉及的文件如下所示。

① 主界面文件 main.xml。

```
<?xml version="1.0" encoding="utf-8"?>
<LinearLayout xmlns:android="http://schemas.android.com/apk/res/android"
    android:orientation="vertical"
    android:layout_width="match_parent"
    android:layout_height="match_parent">
    <Button
        android:id = "@+id/mysave"
        android:layout_width="wrap_content"
        android:layout_height="wrap_content"
        android:text="save"/>
    <com.example.dqjbd.we2.MyView
        android:id="@+id/mytestview"
        android:layout_width="match_parent"
        android:layout_height="match_parent">
    </com.example.dqjbd.we2.MyView>
</LinearLayout>
```

以上代码定义了一个 save 按钮，id 为 mysave；一个自定义 MyView 组件，id 为 mytestview。

② 自定义 View 类文件 MyView.java。

```
public class MyView extends View  implements View.OnTouchListener{
    Bitmap b;
    Canvas mycanvas;
    Point point = new Point();
    /*成员变量 b 是要保存的位图对象,point 用于保存手触摸屏幕的坐标,mycanvas 是画布
对象,与保存的位图 b 对象关联,如何关联见下文描述。*/
    public MyView(Context ctx,AttributeSet attr){
        super(ctx, attr);
```

```
        b = Bitmap.createBitmap(400,400, Bitmap.Config.ARGB_8888);
        mycanvas = new Canvas(b);
    }
    /* 将位图对象 b 与 mycancas 画布关联起来,在 mycanvas 上绘制图形就是在位图 b 对象
中绘制图形 */
    protected void onDraw(Canvas canvas) {
        super.onDraw(canvas);
        Paint p = new Paint();
        canvas.drawBitmap(b,0,0,null);
        p.setColor(Color.BLUE);
        p.setStrokeWidth(10);
        mycanvas.drawLine(0,0,point.x,point.y,p);
        canvas.drawBitmap(b,0,0,null);
    }
    /* 要仔细理解 mycanvas 与 canvas 的关系。mycanvas 是内存画布,canvas 是界面画布。
核心思想是:首先在 mycanvas 内存画布中绘制图形,再利用 canvas 中的 drawBitmap()函数
将 mycanvas 画布中的内容显示在界面中。 */
    public boolean onTouch(View v, MotionEvent event) {
        point.x = (int)event.getX(); point.y = (int)event.getY();
        invalidate();
        return false;
    }
    /* 手触摸事件响应函数的功能:首先获得触摸点的坐标,然后利用 invalidate()函数通知
系统要进行界面更新。 */
    public void save(){
        try {
            String path =getContext().getExternalFilesDir("").getPath();
            File f = new File(path, "a.jpg");
            FileOutputStream out = new FileOutputStream(f);
            b.compress(Bitmap.CompressFormat.JPEG,100,out);
            out.close();
        }catch(Exception e){   }
    }
}
```

保存位图 b 对象,将其保存在该 App 工程的外部存储目录中,一般保存在 Android/data/工程包名/files/目录下。

③ 主应用文件 MainActivity.java。

```
public class MainActivity extends   Activity{
    protected void onCreate(Bundle savedInstanceState) {
        super.onCreate(savedInstanceState);
        setContentView(R.layout.main);
        final MyView view = (MyView)findViewById(R.id.mytestview);
        view.setOnTouchListener(view);
        Button b = (Button)findViewById(R.id.mysave);
        b.setOnClickListener(new View.OnClickListener() {
            public void onClick(View v) {   view.save();   }
        });
    }
}
```

主应用文件主要是为 MyView 注册 TouchListener 侦听事件，为 save 按钮增加 ClickListener 侦听事件，在其响应函数中调用 MyView 中的 save()函数完成位图对象的保存。

10.1.7 绘制文字

1. 确定位置

绘制文字常用的函数是 drawText(String s, float x,float y,Paint p)，其中(x,y)坐标并不是绘制字符串的左上角坐标，它是指绘制文本基线最左边的位置，如图 10-8 所示。

图 10-8 字符基线与字符示例图

可知，字符基线的位置非常重要，箭头所指坐标值(x,y)才是 drawText()函数中的 x、y 值。

一般来说，水平起始位置 x 比较易算，但有时需要知道字符串长度，例如，若让字符串居中显示，需要用到 measureText()函数，示例代码如下所示。

```
Paint p = new Paint();p.setTextSize(100);
String s = "Hello, world";         //居中显示的字符串 s
```

```
float len = p.measureText(s);    //获得字符串实际像素长度 len
float x =( getWidth()+len)/2;   //居中显示 x 坐标=(View 视窗宽度+字符串像素长度)/2
canvas.drawText(s, x, 100, p);   //水平居中显示字符串,但字符基线坐标 y 是固定的,
                                  是 100
```

计算垂直位置有时稍复杂一些,需要读者首先掌握 FontMetrics 类。FontMetrics 是字体的度量类,是指对于指定字号的某种字体,在度量方面的各种属性,其参数如下所示。

baseline,字符基线。

ascent,字符最高点到 baseline 的推荐距离。

top,字符最高点到 baseline 的最大距离。

descent,字符最低点到 baseline 的推荐距离。

bottom,字符最低点到 baseline 的最大距离。

leading,行间距,即前一行的 descent 与下一行的 ascent 之间的距离。

其参数位置示意图如图 10-9 所示。

图 10-9　FontMetrics 字体度量图

例如,让字符串在自定义 View 视窗中水平、垂直居中的代码示例如下所示。

```
Paint p = new Paint();
p.setTextSize(100);
String s = "Hello, world";
float len = p.measureText(s);    //获得字符串实际像素长度 len
float x =( getWidth()+len)/2;   //水平居中 x 坐标=(View 视窗宽度+字符串像素长度)/2
Paint.FontMetrics fm = p.getFontMetrics();       //获得字体度量对象
float height =  fm.ascent + fm.descent;          //计算字的高度
float y = ( getHeight()+height)/2;               //垂直居中 y 坐标
canvas.drawText(s, x, y, p);
```

10.2 动画

10.2.1 帧动画

帧动画就是简单地将 N 张静态图片收集起来，然后通过控制依次显示这些图片，人眼"视觉残留"的原因，会给我们造成动画的"错觉"，与放电影的原理一样。

与帧动画相关的类是 AnimationDrawable，其主要函数如下所示。

- AnimationDrawable()：构造方法。
- void addFrame(Drawable draw, int duration)：draw 是加入的 Drawable 对象，duration 是该帧图像显示的时间，单位是毫秒。
- void setOneShot(boolean oneshot)：若 oneshot 为 true，则播放一次，否则循环播放。
- void start()：动画开始播放。
- void stop()：动画停止播放。
- int getDuration(int i)：获得第 i(0 基)帧图像播放的时间，单位为毫秒。
- Drawable getFrame(int i)：获得第 i(0 基)帧图像的 Drawable 对象。
- int getNumberOfFrames()：获得动画包含的总帧数。

【例 10-3】 将 3 张图片放到工程目录 res/drawable 下，利用帧动画循环播放这 3 张图片。相关文件如下所示。

① 主界面文件 main.xml。

```
<?xml version="1.0" encoding="utf-8"?>
<LinearLayout xmlns:android="http://schemas.android.com/apk/res/android"
android:layout_width="match_parent"
android:layout_height="match_parent"
android:orientation="vertical">
    <Button
        android:id="@+id/mystart"
        android:layout_width="wrap_content"
        android:layout_height="wrap_content"
        android:text="帧动画"/>
    <ImageView
        android:id="@+id/myview"
        android:layout_width="match_parent"
```

```
        android:layout_height="match_parent" />
</LinearLayout>
```

以上代码定义了一个帧动画按钮,id 为 mystart,用于启动帧动画;一个 ImageView 组件,id 为 myview,作为播放帧动画的载体。注意:对帧动画载体,一般选择 ImageView,其他组件是不行的。

② 主界面文件 MainActivity.java。

```
public class MainActivity extends AppCompatActivity implements View
.OnClickListener{
    public void onCreate(Bundle savedInstanceState) {
        super.onCreate(savedInstanceState);
        setContentView(R.layout.main);
        Button b = (Button)findViewById(R.id.mystart);
        b.setOnClickListener(this);
    }
    public void onClick(View v) {
        AnimationDrawable frm = new AnimationDrawable();  //定义动画框架对象
        frm.addFrame(getResources().getDrawable(R.drawable.a),100);
                                                //增加 a.jpg 帧,播放 100ms
        frm.addFrame(getResources().getDrawable(R.drawable.b),100);
                                                //增加 b.jpg 帧,播放 100ms
        frm.addFrame(getResources().getDrawable(R.drawable.c),100);
                                                //增加 c.jpg 帧,播放 100ms
        frm.setOneShot(false);                  //false 表明是循环播放,非一次 oneshot
        ImageView iv = (ImageView)findViewById(R.id.myview);
        iv.setImageDrawable(frm);               //将 Animation 播放对象放到 ImageView
                                                  载体中
        frm.start();                            //开始播放帧动画
    }
}
```

运行本示例,单击帧动画按钮后,会看到 a.jpg、b.jpg、c.jpg 不停地循环播放。

Android 还提供了将帧图像资源封装在配置文件中,从而启动动画的方法。对例 10-3 实现相同的功能,仅做如下改动即可。

① 在工程目录 res/drawable 下增加自定义文件 myanimation.xml,内容如下所示。

```
<?xml version="1.0" encoding="utf-8"?>
<animation-list xmlns:android="http://schemas.android.com/apk/res/android"
```

```
    android:oneshot="false">
    <item android:drawable="@drawable/a" android:duration="100" />
    <item android:drawable="@drawable/b" android:duration="100" />
    <item android:drawable="@drawable/c" android:duration="100" />
</animation-list>
```

该 XML 文件必须定义在 drawable 目录下，若定义在其他目录，则会出现异常。根节点必须是<animation-list>，android:oneshot="false"，表明是循环播放。每个<item>子节点定义每帧的图片文件及设置播放的毫秒数。

② 修改帧动画按钮响应函数，如下所示。

```
public void onClick(View v) {
    ImageView iv = (ImageView)findViewById(R.id.myview);
    iv.setImageResource(R.drawable.myanimation);
                                    //设置 ImageView 为 myanimation.xml 中的资源图像
    AnimationDrawable frm = (AnimationDrawable)iv.getDrawable();
    frm.start();
}
```

10.2.2 补间动画

帧动画是通过连续播放图片模拟动画效果，而补间动画只需指定动画开始和动画结束的状态值，动画变化的中间帧则由系统计算并补齐。所有补间动画类均从 Animation 抽象类派生，其载体可以是 ImageView，也可以是 View 的其他派生类。Animation 类常用的函数如下所示。

- void setDuration(int millisecond)；设置播放的毫秒数。
- void setFillAfter(boolean mark)；mark 为 true 表示动画结束后停留在结束画面，mark 为 false 表示动画结束后恢复到开始画面。
- void setRepeatMode()；设置重播模式。Animation.RESTART 表示从头开始，Animation.REVERSE 表示倒过来开始，默认为 Animation.RESTART。
- void setRepeatCount(int times)；设置重播次数。

上述函数为所有补间动画所共有。按特点划分，补间动画包含灰度动画、平移动画、缩放动画、旋转动画、集合动画等，分别对应不同的 Animation 子类，其构造方法如下所示。

- AlphaAnimation(float from,float to)；灰度动画构造方法，表明动画的初始透明度是 from，结束透明度是 to，取值范围均为[0,1]。0 表示完全不透明，1 表示完全透明。

- TranslateAnimation(float xfrom, float xto, float yfrom, float yto);平移动画构造方法,表明动画左上角的初始坐标是(xfrom,yfrom),结束坐标是(xto,yto)。所有坐标值都是相对于动画左上角的值。
- ScaleAnimation(float xfrom,float xto,float yfrom, float yto);缩放动画构造方法,参数均是倍数值,表明动画左上角横坐标初始时缩放 xFrom 倍,结束时缩放 xto 倍;动画左上角纵坐标初始时缩放 yFrom 倍,结束时缩放 yto 倍。
- RotateAnimation(float from, float to, float ox, float oy);旋转动画构造方法,表明以(ox,oy)为圆心旋转角度,初始角度为 from,结束角度为 to,角度范围为[0,360]。所有坐标值都是相对于动画左上角的值。

【例 10-4】 将一张图片放到工程目录 res/drawable 下,验证灰度动画,所涉及的文件如下所示。

① 主界面文件 main.xml。

```
<?xml version="1.0" encoding="utf-8"?>
<LinearLayout xmlns:android="http://schemas.android.com/apk/res/android"
android:layout_width="match_parent"
android:layout_height="match_parent"
android:orientation="vertical">
    <Button
        android:id="@+id/mystart"
        android:layout_width="wrap_content"
        android:layout_height="wrap_content"
        android:text="灰度动画"/>
    <ImageView
        android:id="@+id/myview"
        android:layout_width="wrap_content"
        android:layout_height="wrap_content"
        android:src="@drawable/a"/>
</LinearLayout>
```

以上代码定义了一个 Button 组件,id 为 mystart,用以启动灰度动画;一个 ImageView 组件,id 为 myview,其图片源设置为 a.jpg,即仅用一张图片就能演示灰度动画过程。

② 主应用文件 MainActivity.java。

```
public class MainActivity extends AppCompatActivity implements View
```

```
.OnClickListener{
    public void onCreate(Bundle savedInstanceState) {
        super.onCreate(savedInstanceState);
        setContentView(R.layout.main);
        Button b = (Button)findViewById(R.id.mystart);
        b.setOnClickListener(this);
    }
    public void onClick(View v) {
        ImageView iv = (ImageView)findViewById(R.id.myview);
        Animation am = new AlphaAnimation(1.0f,0.1f);
                                            //透明度初态 1.0,结束态 0.1
        am.setDuration(3000);               //由"1.0"到"0.1"播放 3 秒
        am.setFillAfter(true);              //动画结束后停留在结束态
        iv.startAnimation(am);              //在 ImageView 组件中播放灰度动画
    }
}
```

若想简单验证其他类型补间动画效果,仅将 Animation am＝new AlphaAnimation(1.0f,0.1f)分别用表 10-3 中的代码替换,其他代码不变即可。

表 10-3　其他补间动画验证方法

分别替换的代码	说　　明
Animation am ＝new TranslateAnimation(0,400,0,500);	平移动画,左上角初始坐标为(0,0),末态坐标为(400,500)。坐标值均相对于左上角
Animation am ＝ new ScaleAnimation(1.0f,2.0f,1.0f,3.0f);	缩放动画,左上角的横坐标初始缩放 1.0 倍,纵坐标缩放 1.0 倍,结束时左上角的横坐标缩放 2.0 倍,纵坐标缩放 3.0 倍
Animation am ＝new RotateAnimation(0,360,200,200);	旋转动画,旋转中心的坐标为(200,200),初始角度为 0,结束角度为 360°。坐标值均相对于左上角

Android 还提供了补间动画状态信息封装在配置文件中,从而启动动画的方法。对例 10-4 实现相同的功能仅做如下改动即可。

① 在工程目录 res 下创建 anim 子目录(必须是该目录,若有,则不用创建),并在其下创建灰度补间动画配置文件 myalpha.xml,如下所示。

```
<?xml version="1.0" encoding="utf-8"?>
<alpha xmlns:android="http://schemas.android.com/apk/res/android"
    android:interpolator="@android:anim/accelerate_decelerate_interpolator"
```

```
    android:fromAlpha="1.0"
    android:toAlpha="0.1"
    android:duration="3000" />
```

同理,若是其他类型补间动画,则根节点必须是 translate,平移动画;scale,缩放动画;rotate,旋转动画。

② 将 MainActivity 类中的 onClick()函数替换成如下内容。

```
public void onClick(View v) {
    ImageView iv = (ImageView)findViewById(R.id.myview);
    Animation am = AnimationUtils.loadAnimation(this, R.anim.myalpha);
                                                    //由配置文件生成动画对象
    iv.startAnimation(am);
}
```

【例 10-5】 将一张图片放到工程目录 res/drawable 下,验证集合动画。

例 10-4 中分别讲解了灰度、平移、缩放、旋转动画,但有时需要图像平移、旋转、缩放等动画同时运行,怎么办? 这就是集成动画,Android 为我们提供了 AnimationSet 类来解决这一问题。本示例所需的文件如下所示。

① 主界面文件 main.xml。

```
<?xml version="1.0" encoding="utf-8"?>
<LinearLayout xmlns:android="http://schemas.android.com/apk/res/android"
android:layout_width="match_parent"
android:layout_height="match_parent"
android:orientation="vertical">
    <Button
        android:id="@+id/mystart"
        android:layout_width="wrap_content"
        android:layout_height="wrap_content"
        android:text="集成动画"/>
    <ImageView
        android:id="@+id/myview"
        android:layout_width="wrap_content"
        android:layout_height="wrap_content"
        android:src="@drawable/a"/>
</LinearLayout>
```

以上代码定义了一个 Button 组件,id 为 mystart,用以启动集成动画;一个

ImageView 组件,id 为 myview,其图片源设置为 a.jpg,即仅用一张图片就能演示集成动画过程。

② 主应用文件 MainActivity.java。

```
public class MainActivity extends AppCompatActivity implements View
.OnClickListener{
    public void onCreate(Bundle savedInstanceState) {
        super.onCreate(savedInstanceState);
        setContentView(R.layout.main);
        Button b = (Button)findViewById(R.id.mystart);
        b.setOnClickListener(this);
    }
    public void onClick(View v) {
        ImageView iv = (ImageView)findViewById(R.id.myview);
        Animation am1 = new ScaleAnimation(1.0f,2.0f,1.0f,2.0f);
        Animation am2 = new TranslateAnimation(0,300,0,500);
        Animation am3 = new RotateAnimation(0,360,iv.getWidth()/2, iv
.getHeight()/2);
        AnimationSet aset = new AnimationSet(this, null);
        aset.addAnimation(am1);
        aset.addAnimation(am2);
        aset.addAnimation(am3);
        aset.setDuration(5000);
        iv.startAnimation(aset);
    }
}
```

onClick()函数中:首先分别定义缩放、平移、旋转对象 am1、am2、am3;然后创建 AnimationSet 对象 aset,利用 addAnimation()函数将 am1、am2、am3 对象添加到 aset 集合中,设置播放时间为 5s;最后利用 startAnimation()函数启动集成动画。

当然,也可以利用配置文件+Java 代码方式启动集成动画。对例 10-5 实现相同的功能仅做如下改动即可。

① 在工程目录 res 下创建 anim 子目录(必须是该目录,若有,则不用创建),并在其下创建灰度补间动画配置文件 myset.xml,根节点必须是 set,如下所示。

```
<?xml version="1.0" encoding="utf-8"?>
<set xmlns:android="http://schemas.android.com/apk/res/android"
    android:duration="5000">
```

```
    <scale
        android:fromXScale="1.0"
        android:fromYScale="1.0"
        android:toXScale="2.0"
        android:toYScale="2.0"/>
    <translate
        android:fromXDelta="0"
        android:fromYDelta="0"
        android:toXDelta="300"
        android:toYDelta="500" />
    <rotate
        android:fromDegrees="0"
        android:toDegrees="360"
        android:pivotX="0.5"
        android:pivotY="0.5"/>
</set>
```

② 将 MainActivity 类中的 onClick()函数替换成如下内容。

```
public void onClick(View v) {
    ImageView iv = (ImageView)findViewById(R.id.myview);
    AnimationSet aset = (AnimationSet) AnimationUtils.loadAnimation(this, R.
anim.myset);
    iv.startAnimation(aset);
}
```

调整配置文件中旋转、平移、缩放的前后顺序，会发现程序运行后所得的动画结果是不同的，集合对象与添加单一 Animation 对象的顺序是相关的。因此，采取“配置文件＋Java 代码”的方式可能更方便，因为不必修改代码，只要修改配置文件就能看到不同的动画效果。

10.2.3 属性动画

补间动画包括灰度、平移、缩放、旋转 4 种类型。进一步思考：若希望展现字体由小到大的过程，或者图像背景变化的动画过程，该如何实现？也就是说，对 View 的每个属性都可能实现一个动画过程，而不仅仅局限于灰度、平移、缩放、旋转 4 种属性，这就是所说的动态属性。

属性动画涉及的类是 ObjectAnimator，其相关函数如下所示。

- ObjectAnimator ofInt(Object target,String propertyName,int … value)：静态方法，定义整型属性的属性动画，target 一般指待动画的视窗 View 对象，propertyName 指属性名称字符串，value 是整数值变长参数类型，表明整型属性值在动画过程中的变化情况。
- ObjectAnimator ofFloat(Object target,String propertyName,int … value)：静态方法，定义浮点属性的属性动画，target 一般指待动画的视窗 View 对象，propertyName 指属性名称字符串，value 是整数值变长参数类型，表明浮点属性值在动画过程中的变化情况。
- ObjectAnimator ofFloat(Object target,String propertyName,int … value)：静态方法，定义浮点属性的属性动画，target 一般指待动画的视窗 View 对象，propertyName 指属性名称字符串，value 是整数值变长参数类型，表明浮点属性值在动画过程中的变化情况。
- ObjectAnimator ofArgb(Object target,String propertyName,int … value)：静态方法，定义颜色属性的属性动画。

【例 10-6】 展示字符串“Hello”字体由小到大的变化过程，涉及的文件如下所示。

① 主界面文件 main.xml。

```
<?xml version="1.0" encoding="utf-8"?>
<LinearLayout xmlns:android="http://schemas.android.com/apk/res/android"
android:layout_width="match_parent"
android:layout_height="match_parent"
android:orientation="vertical">
    <Button
        android:id="@+id/mystart"
        android:layout_width="wrap_content"
        android:layout_height="wrap_content"
        android:text="属性动画"/>
    <TextView
        android:id="@+id/mytext"
        android:layout_width="match_parent"
        android:layout_height="match_parent"
        android:text="HELLO" />
</LinearLayout>
```

以上代码定义了一个 Button 组件，id 为 mystart，用于启动属性动画；一个 TextView 组件，id 为 mytext，直接将待动画的字符串“HELLO”显示在界面上。

② 主应用文件 MainActivity.java。

```
public class MainActivity extends AppCompatActivity implements View.
OnClickListener{
    public void onCreate(Bundle savedInstanceState) {
        super.onCreate(savedInstanceState);
        setContentView(R.layout.main);
        Button b = (Button)findViewById(R.id.mystart);
        b.setOnClickListener(this);
    }
    public void onClick(View v) {
        TextView tv= (TextView)findViewById(R.id.mytext);
        ObjectAnimator anim = ObjectAnimator.ofFloat(tv,"textSize",10,200);
        anim.setDuration(5000);
        anim.start();
    }
}
```

主要理解代码行 ObjectAnimator anim = ObjectAnimator.ofFloat(tv,"textSize",10,200),其含义是:对 TextView 对象 tv 来说,设置其 textSize 文字大小的动画初值时是 10,结束时是 200。属性名称 textSize 属于 TextView 组件的固有内容,不是随便写的,必须保证正确。由于 ofFloat()函数最后的形参是变长类型,因此除了定义字体大小的初值、末值,还可定义字体大小的中间过渡值。

当然,也可以利用“配置文件+Java 代码”的方式实现本示例功能,改动如下所示。

① 在工程目录 res 下创建 animator 子目录(必须是该目录,若有,则不用创建),并在其下创建配置文件 myproperty.xml,根节点是 objectAnimator,如下所示。

```
<?xml version="1.0" encoding="utf-8"?>
<objectAnimator xmlns:android="http://schemas.android.com/apk/res/android"
    android:valueFrom="10"
    android:valueTo="100"
    android:valueType="floatType"
    android:propertyName="textSize"/>
```

② 将 MainActivity 类中的 onClick()函数替换成如下内容。

```
public void onClick(View v) {
    TextView tv= (TextView)findViewById(R.id.mytext);
    Animator anim = AnimatorInflater.loadAnimator(this, R.animator
```

```
.myproperty);
    anim.setTarget(tv);
    anim.setDuration(5000);
    anim.start();
}
```

【例 10-7】 属性集合动画。展示字符串"Hello"字体由小到大，同时平移的功能，涉及的文件如下所示。

① 主界面文件 main.xml。

```
<?xml version="1.0" encoding="utf-8"?>
<LinearLayout xmlns:android="http://schemas.android.com/apk/res/android"
android:layout_width="match_parent"
android:layout_height="match_parent"
android:orientation="vertical">
    <Button
        android:id="@+id/mystart"
        android:layout_width="wrap_content"
        android:layout_height="wrap_content"
        android:text="属性动画"/>
    <TextView
        android:id="@+id/mytext"
        android:layout_width="match_parent"
        android:layout_height="match_parent"
        android:text="HELLO" />
</LinearLayout>
```

以上代码定义了一个 Button 组件，id 为 mystart，用于启动属性动画；一个 TextView 组件，id 为 mytext，直接将待动画的字符串"HELLO"显示在界面上。

② 主应用文件 MainActivity.java。

```
public class MainActivity extends AppCompatActivity implements View
.OnClickListener{
    public void onCreate(Bundle savedInstanceState) {
        super.onCreate(savedInstanceState);
        setContentView(R.layout.main);
        Button b = (Button)findViewById(R.id.mystart);
        b.setOnClickListener(this);
    }
```

```
    public void onClick(View v) {
        TextView tv= (TextView)findViewById(R.id.mytext);
        ObjectAnimator anim1 = ObjectAnimator.ofFloat(tv,"translationX",0, 500);
        ObjectAnimator anim2 = ObjectAnimator.ofFloat(tv,"textSize",10, 100);
        AnimatorSet aset = new AnimatorSet();
        aset.setDuration(5000);
        aset.play(anim1).with(anim2);
        aset.start();
    }
}
```

水平平移、文字大小涉及 TextView 的 translationX、textSize 属性，因此对这两个属性分别创建动画对象 anim1、anim2。其中，在 anm1 水平移动对象中，设置的 X 轴初始、结束坐标值均是相对于 TextView 组件左上角的值。

在属性动画中，将多个动画形成集合的类是 AnimatorSet（而在补间动画中形成动画集合的类是 Animation，它们是不同的类）。该类提供了一个 play()方法，如果向这个方法中传入一个 Animator 对象（ValueAnimator 或 ObjectAnimator），将会返回一个 AnimatorSet.Builder 的实例。AnimatorSet.Builder 中包括以下 4 个方法，其常用函数如下所示。

- Builder void after(Animator anim)：将现有动画插入传入的动画之后执行。
- Builder void after(long delay)：将现有动画延迟指定毫秒后执行。
- Builder void before(Animator anim)：将现有动画插入传入的动画之前执行。
- void with(Animator anim)：同时执行现有动画和传入的动画。

根据以上说明，着重理解代码行 aset.play(anim1).with(anim2)，其含义是在播放 anm1 动画的同时，也播放 anim2 对象。

当然，也可以利用“配置文件+Java 代码”的方式实现本示例功能，改动如下所示。

① 在工程目录 res 下创建 animator 子目录（必须是该目录，若有，则不用创建），并在其下创建配置文件 mymerge.xml，根节点是 set，如下所示。

```
<?xml version="1.0" encoding="utf-8"?>
<set xmlns:android="http://schemas.android.com/apk/res/android"
    android:ordering="sequentially">
    <objectAnimator android:propertyName="translationX"
        android:valueFrom="0"
        android:valueTo="500"
        android:valueType="floatType"/>
```

```
    <objectAnimator android:propertyName="textSize"
        android:valueFrom="10"
        android:valueTo="100"
        android:valueType="floatType"/>
</set>
```

android:ordering 可取 sequentially，表明多个动画依次播放；或者取 together，表明多个动画同时播放。

② 将 MainActivity 类中的 onClick()函数替换成如下内容。

```
public void onClick(View v) {
    TextView tv= (TextView)findViewById(R.id.mytext);
    Animator animator = AnimatorInflater.loadAnimator(this, R.animator
.mymerge);
    animator.setTarget(tv);
    animator.start();
}
```

10.2.4 实用动画技术

利用 Android 系统提供的技术，可以很方便地实现帧动画、补间动画以及属性动画。但是，有时还要知道这些动画的实现原理。也就是说，不用已有的动画知识，而是自己实现帧动画、补间动画以及属性动画，这对深入研究动画技术非常有必要。本节利用延时重绘技术实现帧动画，以及利用设置状态参数实现属性动画。下面一一论述。

1. 延时重绘技术

考虑这样一个应用：给定两幅图像，2s 显示一幅图像，循环显示，涉及的文件如下所示。

① 主界面文件 main.xml。

```
<?xml version="1.0" encoding="utf-8"?>
<LinearLayout xmlns:android="http://schemas.android.com/apk/res/android"
android:layout_width="match_parent"
android:layout_height="match_parent"
android:orientation="vertical">
    <Button
        android:id="@+id/mystart"
        android:layout_width="wrap_content"
```

```
        android:layout_height="wrap_content"
        android:text="开始"/>
    <Button
        android:id="@+id/mystop"
        android:layout_width="wrap_content"
        android:layout_height="wrap_content"
        android:text="停止"/>
    <ImageView
        android:id="@+id/myimg"
        android:layout_width="wrap_content"
        android:layout_height="wrap_content" />
</LinearLayout>
```

以上代码定义了一个开始按钮，id 为 mystart，用于启动帧动画；一个停止按钮，id 为 mystop，用于停止帧动画；一个 ImageView 组件，id 为 myimg，作为帧动画的载体。

② 主应用文件 MainActivity.java。

```
public class MainActivity extends AppCompatActivity implements View.
OnClickListener{
    int index = 0;                                          //图像数组下标索引
    boolean mark = true;                                    //线程运行标识
    Drawable draw[] = new Drawable[2];                      //图像数组
    Handler handler = new Handler(){
        public void handleMessage(Message msg) {
            super.handleMessage(msg);
            int pos = msg.arg1;
            ImageView iv = (ImageView)findViewById(R.id.myimg);
            iv.setImageDrawable(draw[index]);
        }
    };
    /* 利用 Handler 技术加载帧图像,msg 中的 arg1 是待加载图像数组 draw 的数组下标 */
    Runnable fresh = new Runnable() {
        public void run() {
            if(mark==true){
                Message msg = Message.obtain();
                msg.arg1 = index;
                handler.sendMessage(msg);
                index ++ ;
                if(index==2) index = 0;
```

```
                handler.postDelayed(this, 2000);
            }
        }
    };
    /*多线程运行函数,mark为true时运行线程函数体:首先将图像索引index保存在Message对象msg中,并利用Handler技术将消息发送出去;然后控制index索引值,保证其小于2(因为仅用两幅图像做实验);最后调用postDelayed(this,2000),表明2s后再调用一次线程体run()函数。*/
    public void onCreate(Bundle savedInstanceState) {
        super.onCreate(savedInstanceState);
        setContentView(R.layout.main);
        draw[0] = getResources().getDrawable(R.drawable.a);
        draw[1] = getResources().getDrawable(R.drawable.b);
        Button b = (Button)findViewById(R.id.mystart);
        Button b2 = (Button)findViewById(R.id.mystop);
        b.setOnClickListener(this);
        b2.setOnClickListener(this);
    }
    /*首先加载a.jpg、b.jpg两幅图像,生成draw[]数组对象,然后注册开始、停止按钮的OnClickListener帧听事件*/
    public void onClick(View v) {
        int id = v.getId();
        if(id==R.id.mystart){
            new Thread(fresh).start();
        }
        if(id==R.id.mystop)
            mark = false;
    }
}
```

当单击“开始”按钮时,启动帧动画显示线程;当单击“停止”按钮时,置线程停止标识mark为false,从而停止线程运行。

2. 设置状态参数

考虑这样一个应用:演示字符串“HELLO”字体由小到大显示的动画过程,涉及的文件如下所示。

① 主界面文件 main.xml。

```
<?xml version="1.0" encoding="utf-8"?>
<LinearLayout xmlns:android="http://schemas.android.com/apk/res/android"
android:layout_width="match_parent"
android:layout_height="match_parent"
android:orientation="vertical">
    <Button
        android:id="@+id/mystart"
        android:layout_width="wrap_content"
        android:layout_height="wrap_content"
        android:text="开始"/>
    <TextView
        android:id ="@+id/mytext"
        android:layout_width="wrap_content"
        android:layout_height="wrap_content" />
</LinearLayout>
```

以上代码定义了一个“开始”按钮，id 为 mystart，用于启动字体属性动画；一个 TextView 组件，id 为 mytext，用于显示字体不同大小的“Hello”字符串。

② 主应用文件 MainActivity.java。

```
public class MainActivity extends AppCompatActivity implements View.
OnClickListener{
    int startSize = 10;            //定义初始状态下的字体大小
    int endSize = 110;             //定义结束状态下的字体大小
    int curSize = 10;              //当前显示的字体大小
    int inc = 5;                   //字体大小步长值为 5
    Handler handler = new Handler(){
                                   //Handler 技术显示当前字体下的 Hello 字符串
        public void handleMessage(Message msg) {
            super.handleMessage(msg);
            TextView tv = (TextView)findViewById(R.id.mytext);
            tv.setTextSize(msg.arg1);
            tv.setText("HELLO");
        }
    };
    Runnable fresh = new Runnable() {
        public void run() {
            if(curSize <= endSize){
```

```
                Message msg = Message.obtain();
                msg.arg1 = curSize;
                handler.sendMessage(msg);
                curSize += inc;
                handler.postDelayed(this, 200);  //每 200ms 字体大小增加 5
            }
        }
    };
    public void onCreate(Bundle savedInstanceState) {
        super.onCreate(savedInstanceState);
        setContentView(R.layout.main);
        Button b = (Button)findViewById(R.id.mystart);
        b.setOnClickListener(this);
    }
    public void onClick(View v) {
            new Thread(fresh).start();
    }
}
```

习题 10

一、简答题

1. 简述 Canvas 在绘图中的作用。
2. 利用 Path 绘图的特点是什么？
3. 什么是贝塞尔曲线？
4. 什么是帧动画？
5. 什么是补间动画？

二、程序题

1. 在界面上输入两个点的坐标(x1,y1)、(x2,y2)。单击“确定”按钮后，画出两点间的线段。

2. 在 Canvas 中输出字体大小为 30sp 的“HELLO”字符串，其位置依次为顶部水平居中、左侧垂直居中、底部水平居中、水平垂直居中。

3. 5 张图片的帧动画演示。要求：设置每张图片的播放时间，设置播放方式(单次播、循环播)，有播放、暂停、停止播放按钮等。

第 11 章

设 备 操 作

本章首先讲解麦克风、摄像头常规操作，接着描述传感器的应用方法，最后对手机定位技术进行深入的讨论。

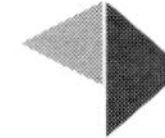

11.1 麦克风

与麦克风相关的类有滑动条(SeekBar)类、音量控制(AudioManager)类、录音(MediaRecorder)类、放音(MediaPlayer)类。下面一一介绍。

11.1.1 SeekBar 类

SeekBar 类是滑动条类，用于进度控制，其主要属性如下所示。

- max，指定滑动条的最大值。
- progress，指定滑动条的当前值。

通过 setOnSeekBarChangeListener(OnSeekBarChangeListener)的回调方法实现 SeekBar 的拖动事件。OnSeekBarChangeListener 接口定义了下面 3 个方法。

- void onProgressChanged(SeekBar seekBar, int progress, boolean fromUser)：进度发生改变时触发。
- void onStartTrackingTouch(SeekBar seekBar)：按住 SeekBar 时触发。
- void onStopTrackingTouch(SeekBar seekBar)：放开 SeekBar 时触发。

11.1.2 AudioManager 类

AudioManager 类提供了访问音量和振铃器控制。通常使用 Context.getSystemService(Context.AUDIO_SERVICE)得到这个类的一个实例。其常用方法如下所示。

- void adjustStreamVolume(int streamType,int direction, int flags)：调整手机指定类型的声音。其中,第一个参数 streamType 指定声音类型,该参数可接受如下几个值。

STREAM_ALARM,手机闹铃的声音。

STREAM_DTMF,DTMF 音调的声音。

STREAM_MUSIC,手机音乐(音频、视频、游戏)的声音。

STREAM_NOTIFICATION,系统提示的声音。

STREAM_RING,电话铃声的声音。

STREAM_SYSTEM,手机系统的声音。

STREAM_VOICE_CALL,语音电话的声音。

第二个参数 direction 指定对声音进行增大还是减小,该参数可接受如下几个值。

ADJUST_LOWER,降低音量。

ADJUST_RAISE,升高音量。

ADJUST_SAME,保持不变,这个值主要用于向用户展示当前的音量。

ADJUST_MUTE,静音。

第三个参数 flags 是调整声音时的标志,例如指定 FLAG_SHOW_UI,则指定调整声音时显示音量进度条。

- void setStreamVolume (int streamType, int value, int flags)：直接设置手机的指定类型的音量值。其中参数 streamType 与 adjustStreamVolume()方法中的第一个参数的意义相同。value 是设置的音量值,flags 是调整声音时的标志,例如指定 FLAG_SHOW_UI,则指定调整声音时显示音量进度条。
- int getStreamVolume (int streamType)：获取指定类型铃声的当前音量。
- int getStreamMaxVolume (int streamType)：获取指定类型铃声的最大音量。

11.1.3 MediaRecorder 录音类

Android 系统是通过使用 MediaRecorder 实现录音以及录像功能的,其主要函数如下所示。

- void prepar()：准备录制。
- void start()：开始录制。
- void stop()：停止录制。
- void reset()：复位 MediaRecorder。
- void release()：释放 MediaRecorder 占用的资源。
- void setAudioEncoder(int encoder)：设置音频记录的编码格式,其取值如下

所示。

AMR_NB：窄带编码。

AMR_WB：宽带编码。

AAC：低复杂度的高级编码。

HE_AAC：高效率的高级编码。

AAC_ELD：高效率的高级编码。

- void setAudioSource(int source)：设置音频记录的音频源。一般设置为麦克风 AudioSource.MIC。
- void setOutputFormat(int format)：设置记录的媒体文件的输出转换格式。输出格式取值说明见表 11-1。

表 11-1　输出格式取值说明

输出格式值	分类	扩展名	格 式 说 明
AMR_NB	音频	.amr	窄带格式
AMR_WB	音频	.amr	宽带格式
AAC_ADTS	音频	.aac	高级的音频传输流格式
MPEG_4	视频	.mp4	MPEG4 格式
THREE_GPP	视频	.3gp	3GP 格式

- void setOutputFile(String strFile)：设置输出文件。

【例 11-1】 实现实时录音功能：界面上有一个 start 按钮、一个 stop 按钮、一个秒表显示器。当单击 start 按钮时开始录音，秒表显示器显示录制的时间；当单击 stop 按钮时，停止录音，并保存相应的录音文件。该程序所涉及的文件如下所示。

① 配置文件 Androidmanifest.xml：增加录音授权，如下所示。

```
<uses-permission android:name="android.permission.RECORD_AUDIO"></uses-
permission>
```

② 主界面文件 main.xml。

```
<?xml version="1.0" encoding="utf-8"?>
<LinearLayout xmlns:android="http://schemas.android.com/apk/res/android"
android:layout_width="match_parent"
android:layout_height="match_parent"
android:orientation="vertical">
    <TextView
```

```
        android:id="@+id/mytime"
        android:layout_width="wrap_content"
        android:layout_height="wrap_content" />
    <Button
        android:id="@+id/mystart"
        android:layout_width="wrap_content"
        android:layout_height="wrap_content"
        android:text="start3"/>
    <Button
        android:id="@+id/mystop"
        android:layout_width="wrap_content"
        android:layout_height="wrap_content"
        android:text="stop"/>
</LinearLayout>
```

以上代码定义了一个 TextView 组件，id 为 mytime，用于显示录音的秒数；一个 id 为 mystart 的 Button 组件，用于启动录音功能；一个 id 为 mystop 的 Button 组件，用于停止录音。

③ 主应用文件 MainActivity.java。

```
public class MainActivity extends AppCompatActivity {
    int count = 0;                                          //秒表计数器值
    Timer timer = new Timer();                              //定时器对象
    MediaRecorder mr = new MediaRecorder();                 //录音对象
    private static final int GET_RECODE_AUDIO = 1;
    private static String[] PERMISSION_AUDIO = {            //权限
            Manifest.permission.RECORD_AUDIO
    };
    void checkGrant(){
        int permission = ActivityCompat.checkSelfPermission(this,
                Manifest.permission.RECORD_AUDIO);
        if (permission != PackageManager.PERMISSION_GRANTED) {
            ActivityCompat.requestPermissions(this, PERMISSION_AUDIO,
                    GET_RECODE_AUDIO);
        }
    }
    /* 检验权限函数，检查是否有录音权限，若没有，则申请录音权限。在配置文件
    Androidmanifest.xml 中已增加了静态录音授权，为什么还要增加动态授权？这是因为随着
    Android 版本的提高，有些权限必须代码动态授权，抛弃静态授权。本例中，静态、动态授权都有，
```

```
主要是为了屏蔽不同 Android 版本的差异。 * /
    Handler handler = new Handler(){
        public void handleMessage(Message msg) {
            super.handleMessage(msg);
            TextView tv = (TextView)findViewById(R.id.mytime);
            tv.setText(""+msg.arg1);
        }
    };
    class MyTimerTask extends TimerTask{
        public void run() {
            try {
                count++;
                Message msg = Message.obtain();
                msg.arg1 = count;
                handler.sendMessage(msg);
            }catch(Exception e){    }
        }
    }
    /* 上文的 handler 与 MyTimerTask 类是相互配合的,主要是为了完成界面中的录音总时
间显示。MyTimerTask 是一个定时器类,是一个多线程类。在其 run()函数中不能刷新界面,因
此本示例中采取了 Handler 技术。每运行一次 run()函数,秒表值 count 都加 1,表示时间又过
了 1s,将其封装成 Message 对象 msg,利用 handler.sendMessage(msg)将信息传送出去。然
后,handler 中重写的 handlerMessage()函数会异步接收 msg,加以解析,获得录制的时间值,
并将其显示在 id 为 mytime 的 TextView 组件中。 * /
    void start(){
        try {
            mr.setAudioSource(MediaRecorder.AudioSource.MIC);
                                                  //设置声音源为麦克风
            mr.setOutputFormat(MediaRecorder.OutputFormat.AMR_NB);
                                                  //设置格式为窄带格式
            mr.setAudioEncoder(MediaRecorder.AudioEncoder.AMR_NB);
                                                  //设置编码为窄带编码
            String path = getExternalFilesDir("").getPath();
                                                  //获得工程外部存储目录
            File f = new File(path, "myfirst.amr");
                                                  //录制文件名为 myfirst.amr
            mr.setOutputFile(f.getPath());
            mr.prepare();                         //准备录制
```

```
            mr.start();                                         //开始录制
        }
        catch(Exception e){    }
    }
    /* 开始录制功能:首先进行录音设置,包括声音源、输出格式、编码格式、输出音频文件路径,
然后调用 prepare()完成录制准备工作,最后调用 start()函数开始录制。 */
    void stop() { mr.stop();}                                   //停止录制
    protected void onCreate(Bundle savedInstanceState) {
        super.onCreate(savedInstanceState);
        setContentView(R.layout.main);
        checkGrant();
        Button btn = (Button)findViewById(R.id.mystart);
        btn.setOnClickListener(new View.OnClickListener() {
            public void onClick(View v) {
                start();
                timer.schedule(new MyTimerTask(), 0, 1000);
            }
        });
        Button btn2 = (Button)findViewById(R.id.mystop);
        btn2.setOnClickListener(new View.OnClickListener() {
            public void onClick(View v) {
                try {
                    stop();
                    timer.cancel();
                }catch(Exception e){
                    TextView tv = (TextView)findViewById(R.id.mytime);
                    tv.setText(e.getMessage());
                }
            }
        });
    }
}
```

首先利用 checkGrant()函数进行录音授权检验,然后对 start、stop 两个按钮进行 OnClickListener 事件注册。在 start 按钮事件响应函数中,启动 MyTimer 定时器,1s 运行一次,同时启动录音功能;在 stop 按钮事件响应函数中停止计时器及停止录音。

11.1.4 MediaPlayer 类

MediaPlayer 类在 8.2.3 节中已经描述过,本节主要完成一个稍复杂的本地音乐播放

器，其仿真界面如图 11-1 所示。

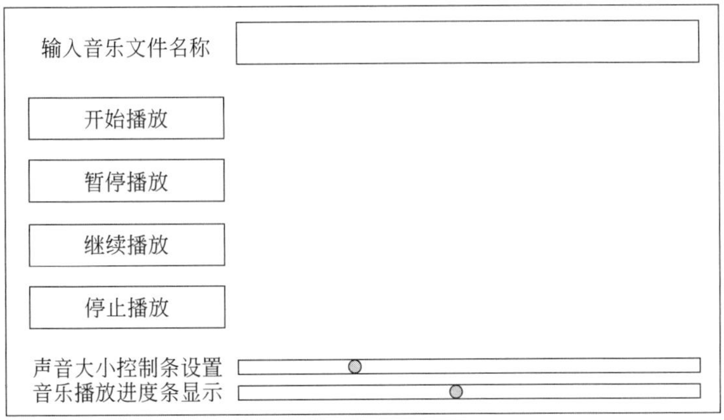

输入音乐文件名称

开始播放

暂停播放

继续播放

停止播放

声音大小控制条设置

音乐播放进度条显示

图 11-1　本地音乐播放器仿真界面图

本示例限定播放手机 music 目录下的音乐文件。输入音乐文件名称，若单击“开始播放”按钮，则开始播放音乐，同时音乐播放进度条显示进度；若单击“暂停播放”按钮，则停止播放；若单击“继续播放”按钮，则音乐从暂停处继续播放音乐；若单击“停止播放”按钮，则音乐播放停止。可以通过调整声音大小控制条控制音量大小。

本示例涉及的文件如下所示。

① 配置文件 Androidmanifest.xml：增加外部存储读写权限。

```
<uses-permission android:name="android.permission.READ_EXTERNAL_STORAGE">
</uses-permission>
<uses-permission android:name="android.permission.WRITE_EXTERNAL_STORAGE">
</uses-permission>
```

② 主界面文件 main.xml。

```
<?xml version="1.0" encoding="utf-8"?>
<LinearLayout xmlns:android="http://schemas.android.com/apk/res/android"
android:layout_width="match_parent"
android:layout_height="match_parent"
android:orientation="vertical">
    <EditText
        android:id="@+id/mypath"
        android:layout_width="match_parent"
        android:layout_height="50dp" />
    <Button
```

```
            android:id="@+id/mystart"
            android:layout_width="wrap_content"
            android:layout_height="wrap_content"
            android:text="开始播放"/>
        <Button
            android:id="@+id/mycontinue"
            android:layout_width="wrap_content"
            android:layout_height="wrap_content"
            android:text="继续播放"/>
        <Button
            android:id="@+id/mypause"
            android:layout_width="wrap_content"
            android:layout_height="wrap_content"
            android:text="暂停播放"/>
        <Button
            android:id="@+id/mystop"
            android:layout_width="wrap_content"
            android:layout_height="wrap_content"
            android:text="停止播放"/>
        <SeekBar
            android:id="@+id/myvolume"
            android:layout_width="match_parent"
            android:layout_height="wrap_content" />
        <SeekBar
            android:id="@+id/myrecord"
            android:layout_width="match_parent"
            android:layout_height="wrap_content" />
    </LinearLayout>
```

该界面与图 11-1 相似，但某些组件界面上缺少一些文字说明，主要是为了简化代码。以上代码定义了一个 EditText 控件，用于输入音乐文件名称；四个 Button 组件，id 分别为 mystart、mycontinue、mypause、mystop，对应开始播放、继续播放、暂停播放、停止播放功能；两个 SeekBar 组件，一个 id 为 myvolume，用于调节音量大小；一个 id 为 myrecord，用于显示音乐的播放进度。

③ 主应用文件 MainActivity.java。

```
public class MainActivity extends AppCompatActivity {
    MediaPlayer mp = new MediaPlayer();      //产生播放器对象
```

```
    Timer timer = new Timer();                //产生定时器对象
    public class MyTimerTask extends TimerTask{
        public void run() {
            int len = mp.getCurrentPosition();
            SeekBar sbar = (SeekBar) findViewById(R.id.myrecord);
            sbar.setProgress(len);
        }
    }
    /* 音乐播放进度显示更新类,定时运行 run()函数:获得播放到此时的毫秒数 len,用该值
设置音乐播放进度条的位置。*/
    void myprepare(){
        try {
            EditText et = (EditText)findViewById(R.id.mypath);
            File root = Environment.getExternalStoragePublicDirectory
(Environment.DIRECTORY_MUSIC);
            File file = new File(root.getPath() + "/" +et.getText().toString());
            mp.reset();
            mp.setDataSource(file.getPath());
            mp.prepare();
            int len = mp.getDuration();
            SeekBar progress = (SeekBar) findViewById(R.id.myrecord);
            progress.setMax(len);
            progress.setProgress(0);

            SeekBar volume = (SeekBar)findViewById(R.id.myvolume);
            final AudioManager am = (AudioManager)getSystemService(Context
.AUDIO_SERVICE);
            int maxvalue = am.getStreamMaxVolume(AudioManager.STREAM_MUSIC);
            volume.setMax(maxvalue);
            volume.setProgress(maxvalue/2);
            am.setStreamVolume(AudioManager.STREAM_MUSIC,maxvalue/2,0);
            volume.setOnSeekBarChangeListener(new SeekBar
.OnSeekBarChangeListener() {
                public void onProgressChanged(SeekBar seekBar, int progress,
boolean fromUser) {
                    am.setStreamVolume(AudioManager.STREAM_MUSIC,progress,0);
                }
                public void onStartTrackingTouch(SeekBar seekBar) { }
```

```
                public void onStopTrackingTouch(SeekBar seekBar) {   }
            });
            timer.schedule(new MyTimerTask(),0,1000);
        }catch(Exception e) {   }
    }
```

/＊该函数实现了音乐播放前的相应准备工作，具体包含如下功能。

- 确定播放的文件。通过 EditText 输入控件获得该文件名称，为了方便，本例假设音乐文件存放在 Music 目录下。
- 设定播放源及获得相关参数。播放器对象 mp 执行 reset()复位操作，相当于将 mp 对象恢复到初始状态。然后利用 setDataSource()函数设定数据源，调用 prepare()函数完成播放前的系统准备工作。
- 设定播放进度条 progress 对象的最大值及初始值。获得播放源所需总毫秒值 len，利用 setMax()函数设置播放进度条的最大值为 len，利用 setProgress()函数设定初始值为 0，表明该音乐还没有播放。
- 设定声音控制进度条 volume 的最大值为 maxvalue，利用 setStreamVolume()函数设定声音的默认值为 maxvalue/2，同时利用 setProgress()函数设置声音控制进度条的值也为 maxvalue/2。最后对该进度条增加 OnSeekBarChangeListener 事件侦听及响应。
- 利用 timer 对象中的 schedule()函数每间隔 1s 运行一次 MyTimerTask 类中的 run()方法，音乐播放进度条相应地更新显示。

```
    */
    protected void onCreate(Bundle savedInstanceState) {
        super.onCreate(savedInstanceState);
        setContentView(R.layout.main);
        int check = ActivityCompat.checkSelfPermission(this, Manifest
.permission.READ_EXTERNAL_STORAGE);
        if(check== PackageManager.PERMISSION_DENIED){
        ActivityCompat.requestPermissions(this,new String[]{
        Manifest.permission.READ_EXTERNAL_STORAGE,Manifest.permission.WRITE_
EXTERNAL_STORAGE},1);
        }
        Button btn0=(Button)findViewById(R.id.mystart);
        btn0.setOnClickListener(new View.OnClickListener() {
            public void onClick(View v) {
                myprepare();
```

```
                mp.start();
            }
        });
        Button btn = (Button)findViewById(R.id.mycontinue);
        btn.setOnClickListener(new View.OnClickListener() {
            public void onClick(View v) { mp.start();}
        });
        Button btn2=(Button)findViewById(R.id.mypause);
        btn2.setOnClickListener(new View.OnClickListener() {
            public void onClick(View v) {mp.pause();}
        });
        Button btn3 = (Button)findViewById(R.id.mystop);
        btn3.setOnClickListener(new View.OnClickListener() {
            public void onClick(View v) {mp.stop();}
        });
    }
}
```

OnCreate()函数主要完成动态设置外部存储权限及对开始播放、继续播放、暂停播放、停止播放 4 个按钮完成 OnClickListener 事件侦听及响应。

11.2　摄像头

手机的摄像功能很强，能照出精美的电子相片，也能连续不断地录影。本节主要讲解照相、录影的基本知识。

11.2.1　相关类简介

1. Camera 类

Camera 类常用函数如下所示。

- Camera open( )，静态方法，返回能打开的 Camera 对象。
- void setDisplayOrientation(int degrees)，设置相机与水平方向的角度，范围为[0，360]。
- void setPreviewDisplay(SurfaceHolder holder)，在 SurfaceHolder 上显示画面。
- void setPreviewCallback(PreviewCallback)，设置回调函数，将视频源数据传递到 onPreviewFrame()方法。

- Parameters getParameters()，获取相机相关参数。
- void stopPreview()，停止预览。
- void release()，释放相机资源。

2. SurfaceView 类

View 是通过刷新重绘视图的，刷新的时间间隔是 16 ms。如果可以在 16 ms 内完成绘制工作，则没有任何问题；如果绘制过程的逻辑很复杂，并且界面更新还非常频繁，这时就会造成界面的卡顿，影响用户体验，为此 Android 提供了 SurfaceView 类来解决这一问题，以满足照相和录影的需求。

此时需要自定义一个类派生自 SurfaceView，并重写 SurfaceHolder.Callback 接口函数，示例如下所示。

```
public class CameraPreview extends SurfaceView implements SurfaceHolder
.Callback {
    public CameraPreview(Context context) {
        super(context);
    }
    public void surfaceCreated(SurfaceHolder holder) {}
    public void surfaceDestroyed(SurfaceHolder holder) { }
    public void surfaceChanged(SurfaceHolder holder, int format, int w, int h) { }
}
```

voidsurfaceCreated(SurfaceHolder holder)：当 SurfaceView 第一次创建后，会立即调用该函数。程序可以在该函数中做一些和绘制界面相关的初始化工作，一般情况下是在另外的线程绘制界面，所以不要在这个函数中绘制 Surface。

void surfaceChanged(SurfaceHolder holder, int format, int width,int height)：当 Surface 的状态(大小和格式)发生变化时会调用该函数，在 surfaceCreated 调用后该函数至少会被调用一次。

void surfaceDestroyed(SurfaceHolder holder)：当 Surface 被摧毁前会调用该函数，该函数被调用后就不能继续使用 Surface，一般在该函数中清理使用过的资源。

11.2.2 照相预览功能

首先编制一个最简单的手机照相预览功能，涉及的文件如下所示。

① 配置文件 Androidmanifest.xml：增加 Camera 允许权限，并增加外部存储读写权限(为后续的功能扩展作准备)。

```
<uses-permission android:name="android.permission.CAMERA"></uses-
permission>
<uses-permission android:name="android.permission.READ_EXTERNAL_STORAGE">
</uses-permission>
<uses-permission android:name="android.permission.WRITE_EXTERNAL_STORAGE"></
uses-permission>
```

② 主界面文件 main.xml。

```
<?xml version="1.0" encoding="utf-8"?>
<LinearLayout xmlns:android="http://schemas.android.com/apk/res/android"
android:layout_width="match_parent"
android:layout_height="match_parent"
android:orientation="vertical">
    <FrameLayout
        android:id="@+id/camera_preview"
        android:layout_width="fill_parent"
        android:layout_height="fill_parent"
        android:layout_weight="1"
        />
    <Button
        android:id="@+id/btn_preview"
        android:text="开始预览"
        android:layout_width="wrap_content"
        android:layout_height="wrap_content"
        android:layout_gravity="center"
        />
</LinearLayout>
```

以上代码定义了一个 FrameLayout 组件，id 为 camera_preview，用于照相预览；一个 Button 组件，id 为 btn_preview，用于启动照相预览功能。

③ 自定义 SurfaceView 类 CameraPreview。

```
public class CameraPreview extends SurfaceView implements SurfaceHolder
.Callback {
    private SurfaceHolder mHolder;
    private Camera mCamera;
    public CameraPreview(Context context, Camera camera) {
        super(context);
        mCamera = camera;
```

```
        mHolder = getHolder();
        mHolder.addCallback(this);
    }
    public void surfaceCreated(SurfaceHolder holder) {
        try {
            mCamera.setPreviewDisplay(holder);
            mCamera.startPreview();
        } catch (Exception e) {  }
    }
    public void surfaceDestroyed(SurfaceHolder holder) {mCamera.release();}
    public void surfaceChanged(SurfaceHolder holder, int format, int w, int h) {}
}
```

CameraPreview 中定义了 SurfaceHolder 成员变量 mHolder，Camera 成员变量 mCamera。在 CameraPreview 构造方法中获得 mHolder 对象值，并设置其回调函数在本类中（由于 addCallback（）函数中的实参是 this）。surfaceCreated（）函数中，通过 setPreviewDisplay()函数设置在 holder 对象中显示，并通过 startPreview()函数启动预览功能。surfaceDestroy()函数中，通过 release()函数释放所有的照相资源。

④ 主应用文件 MainActivity.java。

```
public class MainActivity extends AppCompatActivity {
    private Camera mCamera;                //定义 Camera 成员变量 mCamera
    private CameraPreview mPreview;        //定义预览窗口成员变量 mPreview
    public Camera getCameraInstance(){
        Camera c = null;
        try {  c = Camera.open();  }
        catch (Exception e){  }
        return c;
    }
    /* 通过运行 Camera 类中的静态方法 open()获得 Camera 对象的一个实例，若没有实例，则
返回 null 对象。 */
    public void onCreate(Bundle savedInstanceState) {
        super.onCreate(savedInstanceState);
        setContentView(R.layout.main);
        if (ContextCompat.checkSelfPermission(this, Manifest.permission
.CAMERA)
                == PackageManager.PERMISSION_GRANTED) {
        } else {
```

```
            ActivityCompat.requestPermissions(this,new String[]{Manifest
    .permission.CAMERA,
       Manifest.permission.READ_EXTERNAL_STORAGE,Manifest.permission.WRITE_
    EXTERNAL_STORAGE}, 1);
        }
        Button btn = (Button) findViewById(R.id.btn_preview);
        btn.setOnClickListener(new View.OnClickListener() {
            public void onClick(View v) {
                if(mCamera != null)
                    mCamera.release();
                mCamera = getCameraInstance();
                mCamera.setDisplayOrientation(90);
                mPreview = new CameraPreview(MainActivity.this, mCamera);
                FrameLayout preview = (FrameLayout) findViewById(R.id.camera_
    preview);
                preview.addView(mPreview);
            }
        });
    }
}
```

onCreate()函数中,首先动态检查本示例是否有 Camera 操作权限,若没有,则通过 requestPermissions()函数添加。然后注册“开始预览”按钮的 OnClickListener 事件侦听。在响应函数中,首先通过 getCameraInstance()函数获得 Camera 对象 mCamera,然后通过 setDisplayOrientation()函数设置预览角度为 90°。当设置为 0°时,预览对象与手机宽边平行;当设置为 90°时,预览对象与手机窄边平行。最后通过 mPreview = new CameraPreview(MainActivity.this, mCamera)表明 mCamera 的摄像内容将显示在 mPreview 视窗中,随后将 mPreview 添加在主界面中的 id 为 camera_preview 的 FrameLayout 中,当运行 addView()函数时,CameraPreview 中的 surfaceCreated()、surfaceChanged()函数运行,实现真实的手机预览功能。

可能有读者有如下疑问:完全没有必要在 FrameLayout 中动态添加 CameraPreview 子窗口,在 main.xml 主界面文件中直接定义 CameraPreview 代替 FrameLayout 不更好吗?本示例中是不行的。这是因为若直接将 CameraPreview 添加在 main.xml 中,当程序加载主页面时,Camera 操作有可能还没有授权完毕,势必会出现异常,因此是错误的。

讨论:本示例 App 运行后,单击“开始预览”按钮,可在手机界面中看见预览的影像。若想加一“停止预览”按钮,单击此按钮,则手机画面静止不动,预览功能停止。当再单击

“开始预览”按钮,手机中又出现活动的预览图像,如何实现?具体分以下两步。

- 在 main.xml 中增加“停止预览”按钮节点,如下所示。

```
<Button
    android:id="@+id/stop_preview"
    android:layout_width="wrap_content"
    android:layout_height="wrap_content"
    android:layout_gravity="center"
    android:text="停止预览"/>
```

- 在 MainActivity 类的 onCreate()函数中增加 id 为 stop_preview 的按钮事件响应。

```
Button btn2 = (Button)findViewById(R.id.stop_preview);
btn2.setOnClickListener(new View.OnClickListener() {
    public void onClick(View v) {
        mCamera.stopPreview();
    }
});
```

11.2.3 拍照功能

在 11.2.2 节实现手机预览的前提下捕获一帧图像,将其保存为图像文件,即手机的拍照功能。具体步骤如下所示。

① 在 main.xml 中增加“拍照”按钮节点,如下所示。

```
<Button
    android:id="@+id/btn_capture"
    android:layout_width="wrap_content"
    android:layout_height="wrap_content"
    android:layout_gravity="center"
    android:text="拍照"/>
```

② 在 MainActivity 类的 onCreate()函数中增加 id 为 btn_capture 的按钮事件响应。

```
Button captureButton = (Button) findViewById(R.id.btn_capture);
captureButton.setOnClickListener(
        new View.OnClickListener() {
            public void onClick(View v) {
                mCamera.takePicture(null, null, mPicture);
```

```
            }
        }
);
```

可以看出,拍照功能主要是由 Camera 中的 takePicture()函数完成的,其函数原型如下所示。

```
void takePicture(ShutterCallback shutter, PictureCallback raw, Camera
.PictureCallback jpeg);
```

shutter 是快门回调对象,raw 是源数据流回调对象,jpeg 是 JPEG 格式数据回调对象。前两个参数可设置为 null,假设保存的是最常见的 JPEG 格式图像,所以仅定义第 3 个参数即可,本示例中是 mPicture,它是什么内容？请看下文。

③ JPEG 数据流回调对象定义：在 MainActivity 类中定义即可,如下所示。

```
private Camera.PictureCallback mPicture = new Camera.PictureCallback() {
    public void onPictureTaken(byte[] data, Camera camera) {

        File pictureFile = getOutputMediaFile(MEDIA_TYPE_IMAGE);
        if (pictureFile == null){
            return;
        }
        try {
            FileOutputStream fos = new FileOutputStream(pictureFile);
            fos.write(data);
            fos.close();
        } catch (Exception e) {   }
    }
};
```

可以看出,JPEG 数据流主要是回调 mPicture 对象中的 onPictureTaken(byte[] data, Camera camera)函数,即将截获的数据 data[]保存到文件中。该回调函数首先调用 getOutputMediaFile()函数获得保存文件的 File 对象,然后在 try 块中将数据 data[]保存到 File 对象指向的文件中。

getOutputMediaFile()函数也定义在 MainActivity 类中,如下所示。

```
private static File getOutputMediaFile(int type){
    File mediaStorageDir = new File(Environment
.getExternalStoragePublicDirectory(
            Environment.DIRECTORY_PICTURES), "MyCameraApp");
```

```
    if (! mediaStorageDir.exists()){
        if (! mediaStorageDir.mkdirs()){
            Log.d("MyCameraApp", "failed to create directory");
            return null;
        }
    }
    String timeStamp = new SimpleDateFormat("yyyyMMdd_HHmmss").format(new
Date());
    File mediaFile;
    if (type == MEDIA_TYPE_IMAGE){
        mediaFile = new File(mediaStorageDir.getPath() + File.separator +
                "IMG_"+ timeStamp + ".jpg");
    } else if(type == MEDIA_TYPE_VIDEO) {
        mediaFile = new File(mediaStorageDir.getPath() + File.separator +
                "VID_"+ timeStamp + ".mp4");
    } else {
        return null;
    }
    return mediaFile;
}
```

可以看出，在手机系统 pictures 目录下建立了 MyCameraApp 目录，所有拍照文件均保存在此目录下。拍照文件的命名格式为"IMG_时间戳+.jpg"，录影文件的命名格式为"VID_时间戳+.mp4"。

11.2.4 录影功能

与拍照功能相比，手机录影功能稍显复杂，需要协调好 Camera 对象与 MediaRecorder 对象的关系，其实主要是手机录影的准备阶段需要设置的参数较多。录影流程见表 11-2。

表 11-2 录影流程

序号	功能	说　明
1	打开相机	运用 Camera.open()函数
2	连接预览	通过 Camera.setPreviewDisplay()将 SurfaceView 连接到摄像机
3	开始预览	调用 Camera.startPreview()开始显示实时摄像机图像

续表

序号	功能	说　明
4	录影准备	(1) 解锁相机:调用 Camera.unlock()解锁相机,以供 MediaRecorder 使用 (2) 配置 MediaRecorder setCamera():设置用于视频捕获的摄像机 setAudioSource():设置音频源 setVideoSource():设置视频源 setProfile():设置视频输出格式和编码 (3) 调用 MediaRecorder.prepare()为提供的配置准备 MediaRecorder
5	录影阶段	调用 MediaRecorder 对象的 start()函数
6	停录阶段	调用 MediaRecorder 对象的 stop()函数
7	释放资源	主要调用 MediaRecorder 的 release()函数
8	释放 Camera	主要调用 Camera 对象的 releae()函数

至此,表 11-2 中的 1～3,8 功能均已实现,在此基础之上,只实现表 11-2 中的 4-7 功能即可,所需完善内容如下所示。

① 在主界面文件 main.xml 中增加“录影”按钮节点。

```
<Button
    android:id="@+id/btn_movie"
    android:layout_width="wrap_content"
    android:layout_height="wrap_content"
    android:text="录影"
    android:layout_gravity="center"/>
```

“录影”按钮的 id 为 btn_movie。当单击此按钮时,开始录制视频;当再单击此按钮时,停止录制视频。

② 在 MainActivity 类中增加“录影”按钮的事件响应。

```
Button btn3 = (Button) findViewById(R.id.btn_movie);
btn3.setOnClickListener(
        new View.OnClickListener() {
            public void onClick(View v) {
                if (isRecording) {
                    mMediaRecorder.stop();
                    releaseMediaRecorder();
                    mCamera.lock(); isRecording = false;
```

```
                } else {
                    if (prepareVideoRecorder()) {
                        mMediaRecorder.start();
                        isRecording = true;
                    } else {
                        releaseMediaRecorder();
                    }
                }
            }
        }
    );
```

isRecording 是 MainActivity 类中的布尔成员变量，初始为 false。也就是说，该录影功能响应函数有两个分支，由 isRecording 控制。当首次单击“录影”按钮时，由于 isRecording 为 false，所以走第二个分支：调用录制准备函数 prepareVideoRecorder()，若已准备好，则调用 start()函数开始录影，同时置 isRecording 为 true；再次单击“录影”按钮，由于 isRecording 为 true，所以走第一个分支：首先调用 stop()函数停止录影，然后调用 releaseMediaRecorder()函数释放 MediaRecorder 对象资源。

本函数中涉及 prepareVideoRecorder()及 releaseMediaRecorder()两个函数，它们均是 MainActivity 类中的成员方法，代码如下所示。

```
    private boolean prepareVideoRecorder(){
        try {
            mCamera = getCameraInstance();
            mMediaRecorder = new MediaRecorder();
            //步骤 1:解锁,将 Camera 与 MediaRecorder 关联
            mCamera.unlock();
            mMediaRecorder.setCamera(mCamera);
            //步骤 2:设置音频源、视频源
            mMediaRecorder.setAudioSource(MediaRecorder.AudioSource.CAMCORDER);
            mMediaRecorder.setVideoSource(MediaRecorder.VideoSource.CAMERA);
            //步骤 3:设置输出文件的格式及编码
            mMediaRecorder.setProfile(CamcorderProfile.get(CamcorderProfile
.QUALITY_HIGH));
            //步骤 4:设置输出文件
            mMediaRecorder.setOutputFile(getOutputMediaFile(MEDIA_TYPE_VIDEO)
.toString());
            //步骤 5:设置预览输出
```

```
        mMediaRecorder.setPreviewDisplay(mPreview.getHolder().getSurface());
    }
    catch(Exception e){  }
    try {
        mMediaRecorder.prepare();                    //步骤 6:完成步骤 1~5 的参数准备
    } catch (Exception e) {
        releaseMediaRecorder();
        return false;
    }
    return true;
}
```

本函数主要进行表 11-2 所述的各种设置，见代码中的注释。最后调用 prepare()函数完成真实的参数准备工作，若成功，则返回 true，否则返回 false。该函数中涉及的 getOutputMediaFile()函数与 11.2.3 节中的一致，略。

```
private void releaseMediaRecorder(){
    if (mMediaRecorder != null) {
        mMediaRecorder.reset();
        mMediaRecorder.release();
        mMediaRecorder = null;
        mCamera.lock();
    }
}
```

该函数主要是释放 MediaRecorder 对象资源，最后锁住 Camera 对象，为将来所用。

③ 动态增加录音权限 Manifest.permission.RECORD_AUDIO：用下述代码代替 MainActivity 类 onCreate()函数中相应位置的代码。

```
ActivityCompat.requestPermissions(this,
        new String[]{Manifest.permission.CAMERA,
            Manifest.permission.READ_EXTERNAL_STORAGE,
            Manifest.permission.WRITE_EXTERNAL_STORAGE,
            Manifest.permission.RECORD_AUDIO}, 1);
```

本示例录影后生成的文件形如“VID_时间戳＋.mp4”，保存在手机目录 pictures/MyCameraApp 下。

11.2.5 放映功能

视频放映功能也是由 MediaPlayer 完成的，与 11.1.4 节播放音频的过程几乎一致。

下面编制一个较简单的视频播放功能，涉及的文件如下所示。

① 增加外部存储器读写权限。

```
<uses-permission android:name="android.permission.READ_EXTERNAL_STORAGE">
</uses-permission>
<uses-permission android:name="android.permission.WRITE_EXTERNAL_STORAGE">
</uses-permission>
```

② 自定义 SurfaceView 类 CameraPreview。

```
public class CameraPreview extends SurfaceView {
    public CameraPreview(Context context, AttributeSet attr) {
        super(context, attr);
    }
}
```

该类的功能非常简单，只将视频显示在其中即可。

③ 主界面文件 main.xml。

```
<?xml version="1.0" encoding="utf-8"?>
<LinearLayout xmlns:android="http://schemas.android.com/apk/res/android"
android:layout_width="match_parent"
android:layout_height="match_parent"
android:orientation="vertical">
    <com.example.dqjbd.we.CameraPreview
        android:id="@+id/myview"
        android:layout_width="match_parent"
        android:layout_height="match_parent"
        android:layout_weight="1"/>
    <Button
        android:id="@+id/btn_play"
        android:text="放映"
        android:layout_width="wrap_content"
        android:layout_height="wrap_content"
        android:layout_gravity="center"/>
</LinearLayout>
```

以上代码定义了一个 id 为 myview 的 CameraPreview 节点，用于显示视频图像；一个 id 为 btn_play 的 Button 组件，用于启动放映视频。

④ 主应用程序 MainActivity.java。

```
public class MainActivity extends AppCompatActivity {
    MediaPlayer mp = new MediaPlayer();
    public void onCreate(Bundle savedInstanceState) {
        super.onCreate(savedInstanceState);
        setContentView(R.layout.main);
        ActivityCompat.requestPermissions(this,
            new String[]{Manifest.permission.READ_EXTERNAL_STORAGE,
            Manifest.permission.WRITE_EXTERNAL_STORAGE}, 1);
        Button btn = (Button)findViewById(R.id.btn_play);
        btn.setOnClickListener(new View.OnClickListener() {
            public void onClick(View v) {
                try {
                    File root = Environment.getExternalStoragePublicDirectory
                            (Environment.DIRECTORY_PICTURES);
                    File file = new File(root.getPath() + "/MyCameraApp/VID_
20210507_071440.mp4");
                    mp.reset();
                    mp.setDataSource(file.getPath());
                    CameraPreview cp = (CameraPreview)findViewById(R.id.myview);
                    mp.setDisplay(cp.getHolder());
                    mp.prepare();
                    mp.start();
                }
                catch(Exception e){    }
            }
        });
    }
}
```

以上代码定义了 MediaPlayer 成员变量 mp,直接调用 MediaPlayer()构造方法完成了初始化。onCreate()函数中,首先利用代码动态添加允许外部存储器读写权限,然后注册放映按钮 OnClickListener 侦听事件。在其响应函数中看出是对一个固定视频文件 VID_20210507_071440.mp4 进行播放(该文件是在 11.2.4 节中产生的,当然,也可以是其他文件)。其中最关键的一行语句是 mp.setDisplay(cp.getHolder()),表明将播放对象与显示对象关联起来。当调用 prepare()完成准备工作后,调用 start()函数开始播放视频。本例几乎是最简单的播放视频代码,可在此基础之上增加播放进度滚动条、声音大小控制条等。

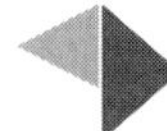

11.3 传感器

11.3.1 简介

手机虽小，但包括了多种多样的传感器，学好传感器基本知识对手机开发至关重要。传感器的种类和数量随智能手机不同而不同，Android 中与传感器相关的类是 Sensor。常用的传感器信息见表 11-3。

表 11-3 常用的传感器信息

序号	类　型	名　称
1	TYPE_ACCELEROMETER	加速度
2	TYPE_MAGNETIC_FIELD	磁场
3	TYPE_ORIENTATION	方位
4	TYPE_LIGHT	光线
5	TYPE_PROXIMITY	距离
6	TYPE_GRAVITY	重力
7	TYPE_LINEAR_ACCELERATION	线性加速度
8	TYPE_GAME_ROTATION_VECTOR	无标定旋转矢量
9	TYPE_MAGNETIC_FIELD_UNCALIBRATED	无标定磁场
10	TYPE_SIGNIFICANT_MOTION	特殊动作
11	TYPE_COUNTER	步行计数

若想获得手机中支持的传感器信息，可参考下述关键代码，进而改进即可。

```
SensorManager sm = (SensorManager)getSystemService(Context.SENSOR_SERVICE);
List<Sensor> list = sm.getSensorList(Sensor.TYPE_ALL); //获得所有传感器对象集合
for(int i=0; i<list.size(); i++){                      //遍历传感器集合
    Sensor ss = list.get(i);
    String name = ss.getName();                        //获得传感器名称
    String type = ss.getStringType();                  //获得传感器类型
}
```

11.3.2 编程步骤

对所有 Android 传感器而言，编程步骤几乎是一致的，一般包含 4 步，如下所示。

① 获得传感器服务。

```
SensrorManager sm= (SensorManager)getSystemService(Context.SENSOR_SERVICE);
```

② 获得具体的传感器对象。

```
Sensor ss =  getDefaultSensor(Sensor.TYPE_XXX);
```

TYPE_XXX 可以是 Sensor 类中前缀为 TYPE 的静态常量，表 11-3 所示仅是部分数据。

③ 注册监听器 SensorEventListener。

注册函数属于 SensorManager 类，不属于 Sensor。其函数原型如下所示。

void registerListener(SensorEventListener se, Sensor ss, int samplerate)；se 是注册的 SensorEventListener 对象，ss 是传感器对象，samplerate 是传感器数据采样时间，单位是 μs。若要接收 N 个传感器对象数据，就要调用 N 遍 registerListener()函数。

④ 实现监听器 SensorEventListener 对象的回调函数 onSensorChanged()、onAccuracyChanged()。当传感器数据发生变化时，就会调用 onSensorChanged()函数；当传感器精度发生变化时，就会调用 onAccuracyChanged()函数。因此，一般在 onSensorChanged()函数中对接收的数据进行处理。

11.3.3 加速度传感器

本节讲述 3 个问题：首先编制加速度传感器数据显示程序，为读者增加感性认识；然后讲解加速度传感器数据的物理含义、加速度传感器的实际应用；最后讲解传感器消息注册优化问题。下面一一说明。

1. 编制加速度传感器数据显示程序

涉及的文件如下所示。

```
//主界面文件 main.xml
<?xml version="1.0" encoding="utf-8"?>
<LinearLayout xmlns:android="http://schemas.android.com/apk/res/android"
android:layout_width="match_parent"
android:layout_height="match_parent"
```

```
android:orientation="vertical">
    <TextView
        android:id="@+id/myinfo"
        android:layout_width="match_parent"
        android:layout_height="wrap_content" />
</LinearLayout>
```

以上代码定义了一个 id 为 myinfo 的 TextView 组件，用于显示加速度数据。

```
//主应用程序 MainActivity.java
public class MainActivity extends AppCompatActivity implements
SensorEventListener{
    public void onCreate(Bundle savedInstanceState) {
        super.onCreate(savedInstanceState);
        setContentView(R.layout.main);
        SensorManager sm = (SensorManager)getSystemService(Context.SENSOR_
SERVICE);
        Sensor ss = sm.getDefaultSensor(Sensor.TYPE_ACCELEROMETER);
        sm.registerListener(this, ss, 1000000);
    }
    public void onSensorChanged(SensorEvent event) {
        String str = "";
        float f[]=event.values;
        str = "X 轴:" +f[0]+"\n";
        str += "Y 轴:" +f[1]+"\n";
        str += "Z 轴:" +f[2]+"\n";
        str += "合成:" + Math.sqrt(f[0] * f[0]+f[1] * f[1]+f[2] * f[2]);
        TextView tv = (TextView)findViewById(R.id.myinfo);
        tv.setText(str);
    }
    public void onAccuracyChanged(Sensor sensor, int accuracy) {    }
}
```

MainActivity 实现了 SensorEventListener 接口，表明要在该类内重写 onSensorChanged()、onAccuracyChange()函数。在 onCreate()函数中，首先获得 SensorManager 对象 sm，进而获得加速度传感器对象 ss，然后利用 sm.registerListener(this, ss, 1000000)完成对加速度传感器对象的 SensorEventListener 事件侦听，第 3 个参数值为 1000000，表明传感器数据采样时间间隔是 1s。

在 onSensorChanged()函数中完成了对加速度传感器数据的处理，加速度数值存在 3

个互相垂直的分量，数值保存在 event.values 中，它是一个浮点数组，赋给了代码中的 f，f[0]、f[1]、f[2]，分别表示 X、Y、Z 轴上的分量值。本示例除了显示 X、Y、Z 轴的分加速度外，还显示了它们的合成加速度。

2. 加速度的数据含义

本示例运行后，手机变换不同角度，待稳定后，在界面中可看出 X、Y、Z 轴的分加速度大小随手机角度变化而变化，而合成加速度的大小是 $9.8\mathrm{m/s^2}$ 左右。可知，手机静态时测量的就是重力加速度的大小。

X、Y、Z 轴是如何标定的？其实它是相对手机平面的，如图 11-2 所示。

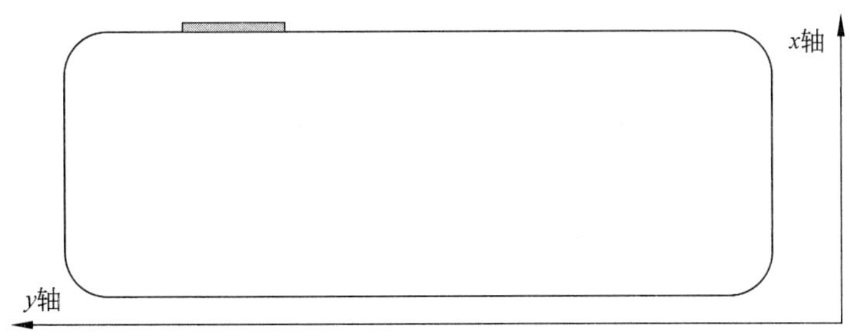

图 11-2 X、Y、Z 轴说明图

X、Y 轴如图 11-2 所示，Z 轴垂直纸面向外，因此当手机位于不同角度时，X、Y、Z 轴也随时变化，重力加速度的 3 个分量值是变化的。例如，若手机的 X 轴与手机重力加速度的方向在一条直线上，则 X 轴的加速度分量值在 $9.8m/s^2$ 左右，Y、Z 轴上的加速度分量值接近 0；若手机的 Y 轴与手机重力加速度的方向在一条直线上，则 Y 轴的加速度分量值在 $9.8m/s^2$ 左右，X、Z 轴上的加速度分量值接近 0。

3. 加速度应用：手机倾斜角度的测定

手机倾斜角度即手机窄边或宽边与水平面的夹角。根据前文知，可获得手机坐标下的 3 个正交方向的加速度值，假设为 X、Y、Z，那么重力加速度向量可表示为

$$\boldsymbol{A}_{\mathrm{g}}=\boldsymbol{X}_{\mathrm{i}}+\boldsymbol{Y}_{\mathrm{j}}+\boldsymbol{Z}_{\mathrm{k}}$$

由于重力加速度的方向是垂直向下，是固定的，因此只要计算在 X、Y、Z 轴方向上的加速度向量 $\boldsymbol{A}_{\mathrm{x}}$、$\boldsymbol{A}_{\mathrm{y}}$、$\boldsymbol{A}_{\mathrm{z}}$ 与 $\boldsymbol{A}_{\mathrm{g}}$ 的夹角，就能计算出手机 X、Y、Z 轴与水平面的夹角。而 $\boldsymbol{A}_{\mathrm{x}}=\boldsymbol{X}_{\mathrm{i}}$、$\boldsymbol{A}_{\mathrm{y}}=\boldsymbol{Y}_{\mathrm{j}}$、$\boldsymbol{A}_{\mathrm{z}}=\boldsymbol{Z}_{\mathrm{k}}$ 是已知的，所以利用高中学习的空间向量知识即可求出两个向量的夹角。例如，求 $\boldsymbol{A}_x$ 与 $\boldsymbol{A}_g$ 的公式如下所示。

$$\mathrm{value}=\boldsymbol{A}_{\mathrm{g}}\cdot\boldsymbol{A}_{\mathrm{x}}/(|\boldsymbol{A}_{\mathrm{g}}|\times|\boldsymbol{A}_{\mathrm{x}}|)$$

value 是 $\boldsymbol{A}_g$、$\boldsymbol{A}_x$ 两向量的夹角余弦值，由于 $\boldsymbol{A}_g$ 向量与水平面垂直，因此 90-value 即是向量 $\boldsymbol{A}_x$ 与水平面的夹角。

上述分析过程写在 11.3.2 节的 onSensorChanged()函数中即可，在界面中就可实时显示手机窄边与水平面的夹角，代码如下所示。

```
public void onSensorChanged(SensorEvent event) {
    String str = "";
    float f[]=event.values;
    float zi = f[0] * f[0];
    float mu = (float)(Math.abs(f[0]) * Math.sqrt(f[0] * f[0]+f[1] * f[1]+f[2] *
f[2]));
    float value = (float)Math.acos(zi/mu);
    float angle = (float)(value * 360/2/Math.PI);
    TextView tv = (TextView)findViewById(R.id.myinfo);
    tv.setText("angle=" + (90-angle));
}
```

在实际应用中，可将求手机倾斜角度转变成求斜面倾角等实际问题，要灵活掌握。

4. 传感器消息注册优化问题

本示例中，传感器消息是在 onCreate()中注册的，程序运行时，onSensorChanged()函数不断地接收传感器的数据，即使切换到其他应用程序，onSensorChanged()也在不停地接收数据，有时这是不必要的。我们希望当本示例是运行焦点时，onSensorChanged()接收数据，当切换到其他应用程序时，onSensorChanged()不接收数据，尽量节省手机内存资源。其实，根据 Activity 生命周期的特点，在 onResume()函数中完成传感器消息注册，在 onStop()函数中撤销传感器消息注册就可以了，关键代码如下所示。

```
protected void onResume() {
    super.onResume();
    SensorManager sm = (SensorManager)getSystemService(Context.SENSOR_
SERVICE);
    Sensor ss = sm.getDefaultSensor(Sensor.TYPE_ACCELEROMETER);
    sm.registerListener(this, ss, 1000000);
}
protected void onStop() {
    super.onStop();
    SensorManager sm = (SensorManager)getSystemService(Context.SENSOR_
SERVICE);
```

```
        Sensor ss = sm.getDefaultSensor(Sensor.TYPE_ACCELEROMETER);
        sm.unregisterListener(this,ss);
    }
```

11.3.4 磁场传感器

Android 磁场传感器可读取磁场的变化，该传感器读取的数据是手机空间坐标系 3 个方向的磁场值，其数据单位为 μT，即微特斯拉。其简单数据获取与 11.3.3 节中讲解的加速度传感器数据获取是一致的。本节讲解稍微复杂一点的应用，即利用磁场传感器、加速度传感器实现简单的指南针功能。其中，涉及的算法均被 Android 系统屏蔽了，只简单地按要求获取数据就可以，涉及的文件如下所示。

① 自定义罗盘显示类 MyView.java。

```
public class MyView extends View {
    float radian;
    boolean mark;
    public MyView(Context ctx, AttributeSet attr){
        super(ctx, attr);
    }
    protected void onDraw(Canvas canvas) {
        super.onDraw(canvas);
        Paint p = new Paint();
        p.setStrokeWidth(30);
        p.setStyle(Paint.Style.STROKE);
        canvas.drawRect(100,100,600,600,p);
        if(mark==true) {
            int Ox = 350, Oy = 350;
            int x = (int)(350 + 250 * Math.sin(radian));
            int y = (int)(350 - 250 * Math.cos(radian));
            canvas.drawLine(Ox, Oy, x, y, p);
            mark = false;
        }

    }
    public void process(float radian){
        this.radian = radian;
        mark = true;
        invalidate();
```

```
    }
}
```

成员变量 radian 表示相对于正北方向的弧度大小，布尔变量 mark 起标识作用。onDraw()函数用于画简单的罗盘以及表针的指向。为了简单，罗盘用正方形模拟。当 mark 为 true 时，才进行表针指向绘制。什么时候 mark 为 true? 当外部类对象调用本类中的 process()函数时，表明有新的数据，这时将新数据赋值给成员变量 radian，mark 置为 true，利用 invalidate()函数通知 MyView 需要重新绘制表针指向。

② 主界面文件 main.xml。

```
<?xml version="1.0" encoding="utf-8"?>
<LinearLayout xmlns:android="http://schemas.android.com/apk/res/android"
android:layout_width="match_parent"
android:layout_height="match_parent"
android:orientation="vertical">
    <TextView
        android:id="@+id/myinfo"
        android:layout_width="match_parent"
        android:layout_height="50dp" />
    <com.example.dqjbd.we.MyView
        android:id="@+id/myview"
        android:layout_width="match_parent"
        android:layout_height="match_parent" />
</LinearLayout>
```

以上代码定义了一个 id 为 myinfo 的 TextView 组件，用于显示与正北方向的夹角；一个 id 为 myview 的自定义 MyView 组件，用于显示指南针的罗盘。

③ 主应用文件 MainActivity.java。

```
public class MainActivity extends AppCompatActivity implements
SensorEventListener{
    SensorManager sm ;                              //传感器管理者对象
    Sensor acceobj;                                 //加速器传感器对象
    Sensor magnobj;                                 //磁场传感器对象
    float accedata[];                               //加速度传感器数据
    float magndata[]; ;                             //磁场传感器数据
    public void onCreate(Bundle savedInstanceState) {
        super.onCreate(savedInstanceState);
        setContentView(R.layout.main);
```

```
        sm = (SensorManager)getSystemService(Context.SENSOR_SERVICE);
                                                        //获得传感器管理者对象
        acceobj = sm.getDefaultSensor(Sensor.TYPE_ACCELEROMETER);
                                                        //获得加速度传感器对象
        magnobj = sm.getDefaultSensor(Sensor.TYPE_MAGNETIC_FIELD);
                                                        //获得磁场传感器对象
    }
    protected void onResume() {
        super.onResume();
        sm.registerListener(this, acceobj, 500000);         //注册加速度传感器事件
        sm.registerListener(this, magnobj, 500000);         //注册磁场传感器事件
    }
    protected void onPause() {
        super.onPause();
        sm.unregisterListener(this,acceobj);                //撤销加速度传感器事件
        sm.unregisterListener(this,magnobj);                //撤销磁场传感器事件
    }
    public void onSensorChanged(SensorEvent event) {
        Sensor ss = event.sensor;
        int type = ss.getType();
        if(type==Sensor.TYPE_ACCELEROMETER)
            accedata = event.values;
        if(type==Sensor.TYPE_MAGNETIC_FIELD)
            magndata = event.values;
        if(accedata!=null && magndata!=null)
            calcProcess();
    }
    public void onAccuracyChanged(Sensor sensor, int accuracy) {    }
    void calcProcess(){
        float matrix[] = new float[9];
        float value[] = new float[3];
        SensorManager.getRotationMatrix(matrix, null, accedata,magndata);
        SensorManager.getOrientation(matrix, value);
        float angle = (float)Math.toDegrees(value[0]);
        TextView tv = (TextView)findViewById(R.id.myinfo);
        tv.setText("angle====" + angle);
        MyView mv = (MyView)findViewById(R.id.myview);
        mv.process(value[0]);
```

```
        }
    }
```

由于在 onSensorChanged()函数中要接收两个传感器数据，因此首先要获得具体的传感器对象 ss，进而获得其特征类型整数值 type。若 type 是加速度传感器标识，则将 event.values 传感器数据赋值给成员变量 accedata；若 type 是磁场传感器标识，则将 event.values 传感器数据赋值给成员变量 magndata。当 accedata、magndata 都已有有效数据时，则调用 calcProcess()函数计算真实的指南针角度值。

在 calcProcess()函数中，首先定义一维数组 matrix，大小为 9，用以表示 3×3 的二维旋转矩阵，并定义一维数组 value，大小为 3；然后利用 SensorManager 中的静态方法 getRotationMatrix()结合已经获得的加速度 accedata、磁场传感器数据 magndata，获得旋转矩阵 matrix 的具体值。利用 SendorManager 中的静态方法 getRotation()，结合 matrix 获得 value 数组的值，value[0]即偏离正北方向的弧度值。最后，一方面将 value[0]转变成角度显示在 TextView 组件中；另一方面将 value[0]传到 MyView 对象中，重画指南针的罗盘，用图形反映指南针偏转的角度。

value 数组包含获得的 3 个有物理含义的值，具体描述如下所示。

value[0]：方位角(绕 *Z* 轴旋转的角度)。这是设备当前罗盘方向与磁北极之间的角度。如果器件的上边缘朝向北方，则方位角为 0°；如果器件的顶部边缘朝南，则方位角为 180°。类似地，如果器件的顶部边缘朝向东方，则方位角为 90°；如果器件的顶部边缘朝向西方，则方位角为 270°。

value[1]：间距(绕 *X* 轴旋转的角度)。这是平行于设备屏幕的平面与平行于地面的平面之间的角度。如果将设备平行于地面，底边最靠近您，并将设备的顶边向地面倾斜，则俯仰角变为正。向相反方向倾斜，将设备的顶部边缘移离地面，导致俯仰角变为负。间距值的范围是－180°～180°。

value[2]：滚动(绕 *Y* 轴旋转的角度)。这是垂直于设备屏幕的平面与垂直于地面的平面之间的角度。如果将设备平行于地面，底边最靠近读者，并将设备的左边缘向地面倾斜，则滚动角度变为正值。向相反方向倾斜，将设备的右边缘朝向地面移动，导致滚动角度变为负值。滚动值的范围为－90°～90°。

11.3.5 计步传感器

这种类型的传感器返回用户自上次重新激活以来的总步数。该值作为浮点数返回(小数部分设置为零)，仅在系统重新引导时才将其重置为零，事件的时间戳设置为采取该事件的最后一步的时间。微信运动就是计步传感器的具体应用，其特点是：进入微信运动,单击“步数排行榜”就能看自己或他(她)人的总步数。本节仅实现微信部分功能:主界面上有“开始”和“查看”两个按钮。单击“开始”按钮,从此时开始计步数;单击“查看”按钮,显示走的总步数。所需相关文件如下所示。

① 主界面文件 main.xml。

```
<?xml version="1.0" encoding="utf-8"?>
<LinearLayout xmlns:android="http://schemas.android.com/apk/res/android"
android:layout_width="match_parent"
android:layout_height="match_parent"
android:orientation="vertical">
    <Button
        android:id="@+id/mystart"
        android:layout_width="wrap_content"
        android:layout_height="wrap_content"
        android:text="开始计数"/>
    <Button
        android:id="@+id/myfind"
        android:layout_width="wrap_content"
        android:layout_height="wrap_content"
        android:text="查询计数"/>
    <TextView
        android:id="@+id/mystep"
        android:layout_width="match_parent"
        android:layout_height="50dp" />
</LinearLayout>
```

以上代码定义了“开始计数”“查询计数”按钮组件;一个 TextView 组件,用于显示步数信息。

② 主应用文件 MainActivity.java。

```
public class MainActivity extends AppCompatActivity implements
SensorEventListener,View.OnClickListener{
    SensorManager sm ;
    Sensor stepobj;
    Button objStart;                              //对应“开始计数”按钮对象
    Button objFind;                               //对应“查询计数”按钮对象
    boolean mark;                                 //mark 为 true 时获得初始步数
    int begin;                                    //初始步数
    public void onCreate(Bundle savedInstanceState) {
```

```
        super.onCreate(savedInstanceState);
        setContentView(R.layout.main);
        sm = (SensorManager) getSystemService(Context.SENSOR_SERVICE);
        stepobj = sm.getDefaultSensor(Sensor.TYPE_STEP_COUNTER);
        objStart = (Button)findViewById(R.id.mystart);
        objStart.setOnClickListener(this) ;
        objFind = (Button)findViewById(R.id.myfind);
        objFind.setOnClickListener(this) ;
    }
    public void onSensorChanged(SensorEvent event) {
        if(mark == true)
            begin = (int)event.values[0];
        else{
            int total = (int)(event.values[0]-begin);
            TextView tv = (TextView)findViewById(R.id.mystep);
            tv.setText("begin="+begin+"\ncur="+(int)event.values[0]+"\n 总步
数:"+total);
        }
        sm.unregisterListener(this, stepobj);
    }
    public void onAccuracyChanged(Sensor sensor, int accuracy) {    }
    public void onClick(View v) {
        Button b = (Button)v;
        if(b==objStart)
            mark = true;
        else
            mark = false;
        sm.registerListener(this, stepobj,1000000);
    }
}
```

MainActivity 实现了 SensorEventListener、OnClickListener，表明传感器消息响应函数、按钮事件响应函数均在本类中。

计步传感器不论程序操作与否，均在工作。因此，只在需要时注册 SensorEventListener 事件即可。所以，当单击“开始计数”“查询计数”按钮时注册一次，当 OnSensorChanged()获取到所需的数据后，直接撤销对 SensorEventistener 事件的侦听。因此，什么时候注册、撤销传感器事件不是一成不变的，和需求分析相关，请灵活掌握。

本示例的计步原理比较简单：当单击“开始计数”按钮时，onSensorChanged()运行，此时获得的 event.values[0]即初始步数，赋值给成员变量 begin；当单击“查询计数”按钮时，onSensorChanged()运行，此时获得的 event.values[0]-begin 即从计数开始到现在走的总的步数。

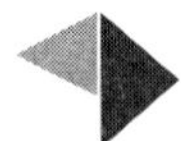

11.4　手机定位

11.4.1　定位原理

手机定位在生活中应用很广，如高德地图，其原理一般有两种情况，如下所示。

- 基站定位：一般服务于手机用户，每个基站都有一个标识符，手机可以搜集周围所有收到信号的基站和它们的标识符，通过互联网发送到云端服务器，再由服务器根据这些基站的位置信息查询并计算出当前位置，然后返回给手机。因为基站信号辐射范围大，所以误差也大，一般在 500 米到几千米。
- GPS 定位：其原理是利用天上的卫星不断地广播信号，地面的 GPS 接收设备收到信号后，通过分析多个卫星信号，就可以计算出地球坐标。GPS 保证全球任何一个地方都可以同时收到至少 4 个卫星的信号，从而可以准确确定所处的经纬度以及海拔位置。

11.4.2　相关类介绍

与手机定位相关的类有 LocationManager、Criteria、Location，相关的接口有 LocationListener。下面一一介绍。

1. LocationManager 类

LocationManager 类提供连接本地位置的服务，这些本地位置允许应用程序获取定期更新设备的地理位置。LocationManager 类不能实例化，但可以通过 Context.getSystemService(Context.LOCATION_SERVICE)方法获得 LocationManager 的实例。其常用函数如下所示。

- List<String> getAllProviders()，返回 List 集合，包含所有位置服务提供者的名称。
- String getBestProvider(Criteria criteria, boolean enabledOnly)，返回最符合标准的位置服务提供者的名称。

- String getLastKnownLocation(String provider)，指定位置服务提供者名称，返回该位置服务提供的位置信息。
- LocationProvider getProvider(String name)，指定服务提供者名称的返回位置提供者实例。
- void requestLocationUpdates(String provider, long minTime, float minDistance, LocationListener listener)，通过指定的位置服务提供者请求位置更新服务。
- void removeUpdates(LocationListener listener)，移除所有指定 LocationListener 的位置更改信息。

2. Criteria 类

该类表示应用程序选择位置服务的标准(或条件)，Criteria 给用户提供了多种因素的标准设置，LocationManager 可以根据这个设定好的标准，自动选择所需的 LocationProvider。该类的常用函数如下所示。

- void setAccuracy(int accuracy)，设置经纬度的精准度。可选的参数有 ACCURACY_FINE，准确；ACCURACY_COARSE，粗略。
- void setAltitudeRequired(boolean altitudeRequired)，设置是否需要获取海拔数据。
- void setBearingAccuracy(int accuracy)，设置方向的精确性。可选的参数有 ACCURACY_LOW，低；ACCURACY_HIGH，高；NO_REQUIREMENT，没有要求。
- void setBearingRequired(boolean bearingRequired)，设置是否需要获得方向信息。
- void setCostAllowed(boolean costAllowed)，设置是否允许在定位过程中产生资费，比如流量等。
- void setHorizontalAccuracy(int accuracy)，获取水平方向经纬度的精准度。可选的参数有 ACCURACY_LOW，低；ACCURACY_MEDIUM，中；ACCURACY_HIGH，高；NO_REQUIREMENT，无要求。
- void setPowerRequirement(int level)，设置耗电量的级别。
- void setSpeedAccuracy(int accuracy)，设置速度的精确度。
- void setSpeedRequired(boolean speedRequired)，设置是否提供速度的要求。
- void setVerticalAccuracy(int accuracy)，设置垂直距离的海拔高度的精度。

3. Location 类

通过该类可获得相关位置的信息，具体函数如下所示。

- double getAltitude()，获得海拔高度。
- double getLatitude()，获得纬度。
- double getLongitude()，获得经度。

4. LocationListener

该接口定义了 4 个抽象函数，LocationManager 对象的回调函数如下所示。

public void onLocationChanged(Location location)。

public void onStatusChanged(String provider, int status, Bundle extras)。

public void onProviderEnabled(String provider)。

public void onProviderDisabled(String provider)。

主要重写 onLocationChanged()函数，在此函数内获得精度、纬度等位置信息，进行相关操作。

考虑这样一个应用：利用手机定位功能随时获得所处的经度、纬度值。其涉及的文件如下所示。

① 配置文件 Androidmanifest.xml：增加定位粗糙、精确权限。

```
<uses-permission android:name="android.permission.ACCESS_COARSE_LOCATION" />
<uses-permission android:name="android.permission.ACCESS_FINE_LOCATION" />
```

② 主界面文件 main.xml。

```
<?xml version="1.0" encoding="utf-8"?>
<LinearLayout xmlns:android="http://schemas.android.com/apk/res/android"
android:layout_width="match_parent"
android:layout_height="match_parent"
android:orientation="vertical">
    <Button
        android:id="@+id/mystart"
        android:layout_width="wrap_content"
        android:layout_height="wrap_content"
        android:text="启动定位"/>
    <TextView
        android:id="@+id/myinfo"
        android:layout_width="match_parent"
```

```
    android:layout_height="200dp" />
</LinearLayout>
```

以上代码定义了一个“启动定位”按钮，id 为 mystart；一个 id 为 myinfo 的 TextView 组件，用于显示经度、纬度值。

③ 主应用文件 MainActivity.java。

```
public class MainActivity extends AppCompatActivity implements View.
OnClickListener,LocationListener{
    LocationManager locationManager;
    public void onLocationChanged(Location location) {
        String s = "纬度="+location.getLatitude()+"\n 经度=" +location
.getLongitude();
        TextView tv = (TextView)findViewById(R.id.myinfo);
        tv.setText(s);
    }
    public void onStatusChanged(String provider, int status, Bundle extras)
{ }
    public void onProviderEnabled(String provider) {  }
    public void onProviderDisabled(String provider) {  }
    protected void onDestroy() {
        super.onDestroy();
        locationManager.removeUpdates(this);
    }
    public void onCreate(Bundle savedInstanceState) {
        super.onCreate(savedInstanceState);
        setContentView(R.layout.main);
        ActivityCompat.requestPermissions(this, new String[]{Manifest
.permission.ACCESS_COARSE_LOCATION, Manifest.permission.ACCESS_FINE_
LOCATION},1);
        Button b = (Button)findViewById(R.id.mystart);
        b.setOnClickListener(this);
    }
    public void onClick(View v) {
        try {
            LocationManager locationManager = (LocationManager)
getSystemService(LOCATION_SERVICE);
            Criteria criteria = new Criteria();
            criteria.setAccuracy(Criteria.ACCURACY_FINE);   //设置为最大精度
```

```
            criteria.setAltitudeRequired(false);         //不要求海拔信息
            criteria.setCostAllowed(true);               //是否允许付费
            criteria.setPowerRequirement(Criteria.POWER_LOW);
                                                         //对电量的要求
            criteria.setBearingRequired(false);
            String bestProvider = locationManager.getBestProvider
(criteria, false);
            locationManager.requestLocationUpdates(bestProvider, 2000, 1,
this);
        }
        catch(Exception e){    }
    }
}
```

该类实现了 OnClickListener、LocationListener 接口，表明按钮响应函数、定位回调函数均在本类中实现；定义了定位管理器 LocationManager 成员变量 locationManager。在回调函数 onLocationChanged()中获得当前经度、纬度值，将其显示在界面中。onCreate()函数中首先完成了定位精度的动态申请；然后利用 Criteria 创建了定位请求标准，利用 getBestProvider()函数获得了最佳的提供定位服务的特征串；最后利用 requestLocationUpdates()函数完成了定位申请。onDestroy()函数主要实现撤销定位申请服务功能。

注意，代码行 locationManager.requestLocationUpdates(bestProvider, 2000, 1, this)，表明每 2s 或位置变化 1m 后就会回调 LocationListener 定义的接口函数一次。

由于手机定位功能涉及 GPS 或基站，因此定位功能与网络信号息息相关。有时在屋里信号弱，运行本示例时会发现界面上根本不显示经度、纬度值。因此，手机定位功能一般在室外应用，运行本示例，会发现经度、纬度值每 2s 刷新一次，并显示在界面上。

习题 11

一、简答题

1. AudioManager 类的作用是什么？
2. MediaRecorder、MediaPlayer 类的功能分别是什么？
3. SurfaceView 类在摄像中的作用是什么？
4. 传感器编程的一般步骤是什么？

二、程序题

1. 完善 11.1 节的麦克风示例功能，使之既能录音，又能播放声音。

2. 完善 11.2 节的摄像头功能，使之既能照相，又能录影。

3. 将 11.3.3～11.3.5 中的加速度传感器、磁场传感器、计步传感器功能，利用 Fragment 片段实现，集成在同一个 Activity 中。

第12章

第三方开发包

手机功能日益丰富，与众多服务商提供的开发包息息相关，人们可以利用高德地图开发包开发自己的地图应用，可以利用天气预报开发包开发自己的天气应用等。总之，应用第三方开发包可以缩短开发周期，这是今后开发的一大趋势。

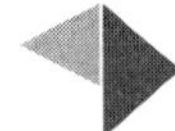

12.1 签名信息

12.1.1 重要性

包名对 App 工程而言相当重要，工程包名不同，则功能不同。但开发 App 的人员众多，无法排除工程包名相同的情况，这将引起歧义，对 App 维护和升级都存在不小的隐患。因此，必须寻求一个标识值，对每个 App 而言是唯一的，就像人的身份证一样。Android 提出一种数字签名技术，由 20 字节组成，每字节包含 2 个十六进制数，而数字签名值是由系统生成的，保证了全球唯一性。这样，当升级某 App 时，首先检查数字签名是否相同，若相同，才能升级，否则即使包名相同，也不会升级。

利用 Android Studio 开发手机 App 时，数字签名有两种：调试版和发布版。调试、发布数字签名区别图如图 12-1 所示。

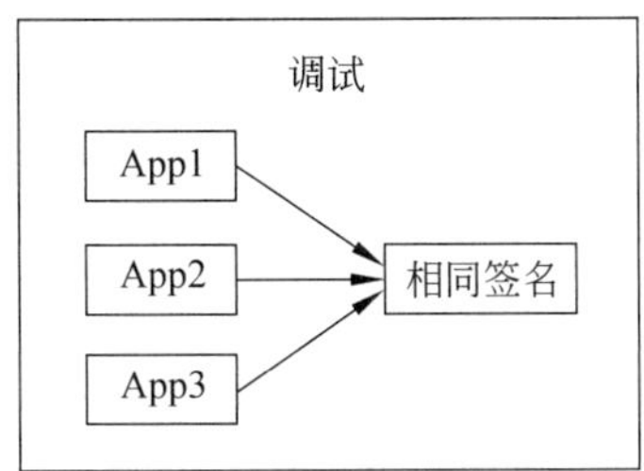

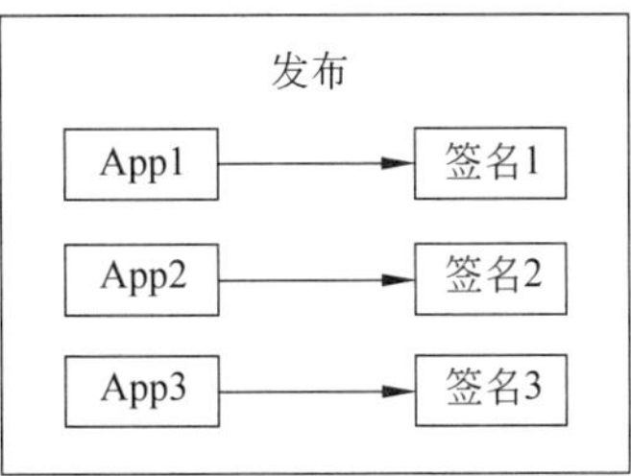

图 12-1 调试、发布数字签名区别图

可以看出，3 个不同功能的 App 调试时可以共享同一个数字签名。当真正作为商品发布时，数字签名必须是唯一的，互不相同。

12.1.2 签名查看

在 Android Studio 中如何查看调试版、发布版的数字签名？下面一一说明。

1. 调试版数字签名的查看过程(假设已经创建了工程，名称为 We)

① 选中主界面最右侧的 Gradle 选项，如图 12-2 所示。

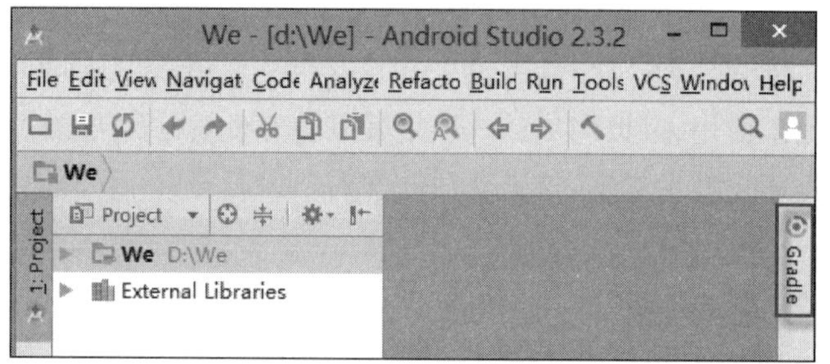

图 12-2 调试版数字签名界面操作 1

② 单击 Gradle，在出现的界面中依次选择"We → app → Tasks → android → signingReport"，界面如图 12-3 所示。

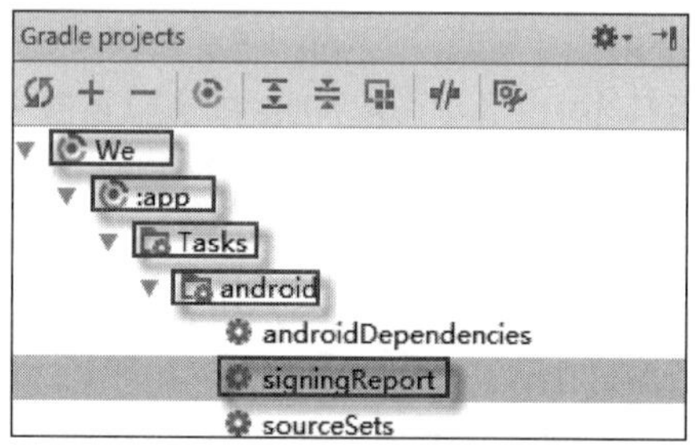

图 12-3 调试版数字签名界面操作 2

③ 单击 signingReport，Gradle Console 控制台上出现签名信息，如图 12-4 所示。可以看出，SHA1 即 App 调试版的数字签名，数字签名保存在文件 C:\Users\dqjbd\.android\debug.keystore 中。

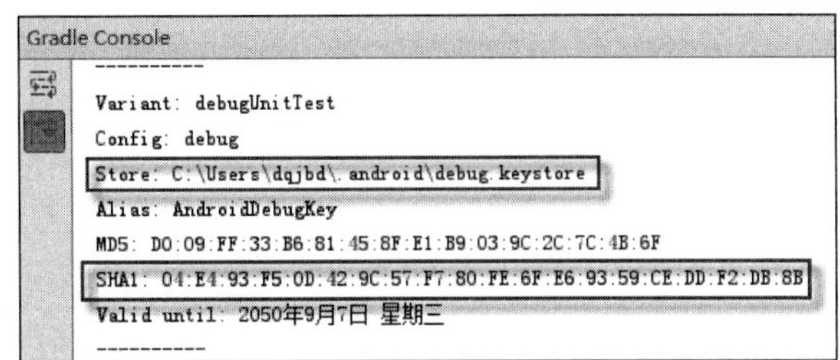

图 12-4　调试版数字签名界面操作 3

2. 发布版数字签名的查看过程

① 从主界面中依次选择"Build→Generate Signed APK",出现如图 12-5 所示的界面。

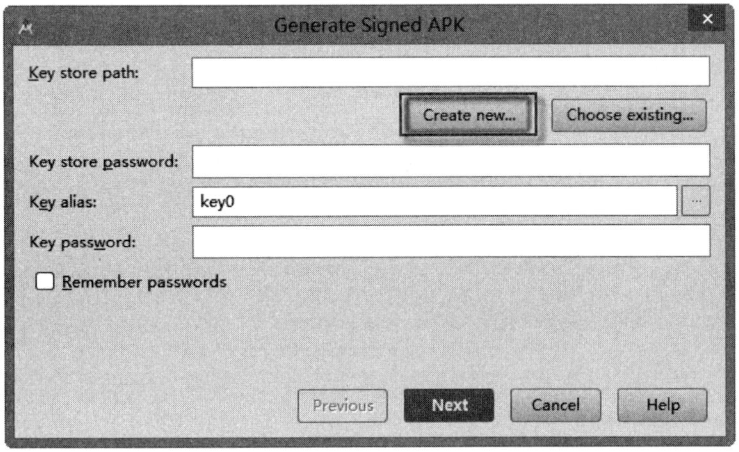

图 12-5　发布版数字签名界面操作 1

② 单击 Create new 按钮,出现如图 12-6 所示的界面。

必须输入的信息在图 12-6 中的方框内,两个密码一定要记住,保存的文件路径没有特殊要求,可以不在 App 工程内。本示例生成的数字签名保存在文件 d:\myfirst.keystore 中。单击 OK 按钮后,就生成了数字签名文件 myfirst.keystore。

③ 查阅数字签名文件 myfirst.keystore。

在主界面中依次选择"View→ToolsWindow→Terminal",打开终端 Terminal 窗口,输入命令行 keytool -v -list -keystore d:/myfirst.keystore,输入图 12-6 中输入的第 1 个密码,按 Enter 键后,则出现数字签名信息,如图 12-7 所示。

图 12-6　发布版数字签名界面操作 2

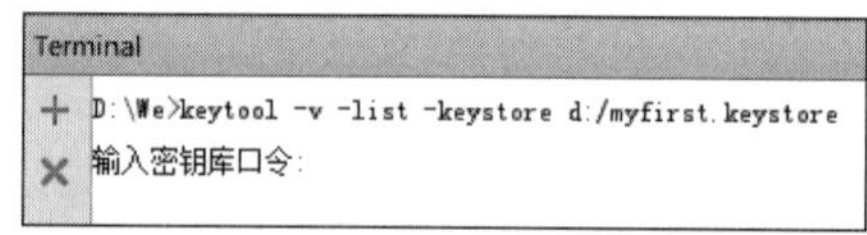

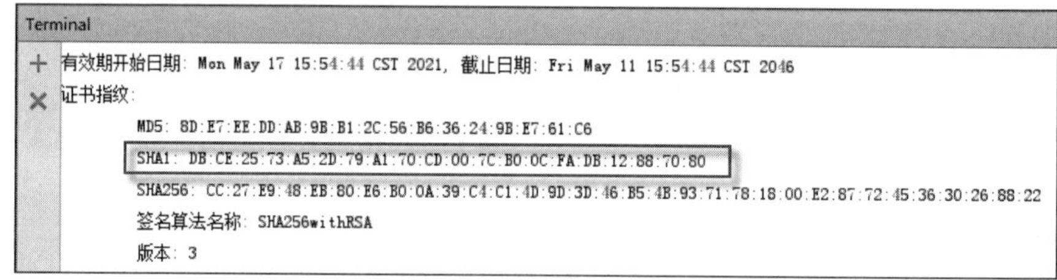

图 12-7　发布版数字签名界面操作 3

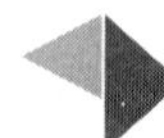

12.2　构建自定义高德地图工程环境

1. 创建名为 We 的 App 工程，包名为 com.example.dqjbd.We，主界面文件为 main.xml，主应用文件为 MainActivity.java。

2. 进入高德地图网站 http://lbs.amap.com，下载所需第三方开发包。此部分内容是开放的，无须注册。本书下载了地图功能文件 AMap_Android_SDK_All.zip 及定位功能文件 AMap_Android_Location_SDK_All.zip。解压缩这两个 ZIP 文件，将其中的 Amap_2DMap_VXXX.jar、AMap_Search_VXXX.jar、AMap_Location_VXXX.jar 包文件复制到 We 工程的 libs 目录下。

3. 将 We 工程与上述 3 个 JAR 文件关联

通过"View→Open module settings"运行主界面文件，选择"app→Dependencies"，出现如图 12-8 所示的界面。

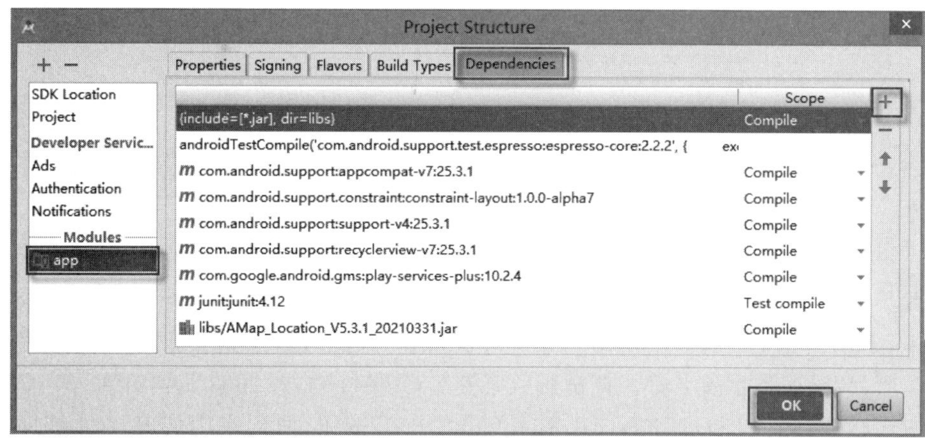

图 12-8　将 3 个 JAR 文件添加到工程中

单击"＋"，选择 libs 目录下已添加的 3 个 JAR 文件，按步骤操作即可，添加成功后，3 个 JAR 文件会显示在图 12-8 的列表中。

此步完成后，就可在工程 Java 文件中应用 3 个 JAR 文件中所包含的各种功能类了。

4. 在高德地图网站进行注册，角色选个人开发，完成注册、登录后，进入"控制台"，选择"我的应用→创建新应用"，界面如图 12-9 所示。

图 12-9　创建新应用界面 1

当单击“创建新应用”按钮时，按步骤操作，就到了图 12-10 所示的界面。

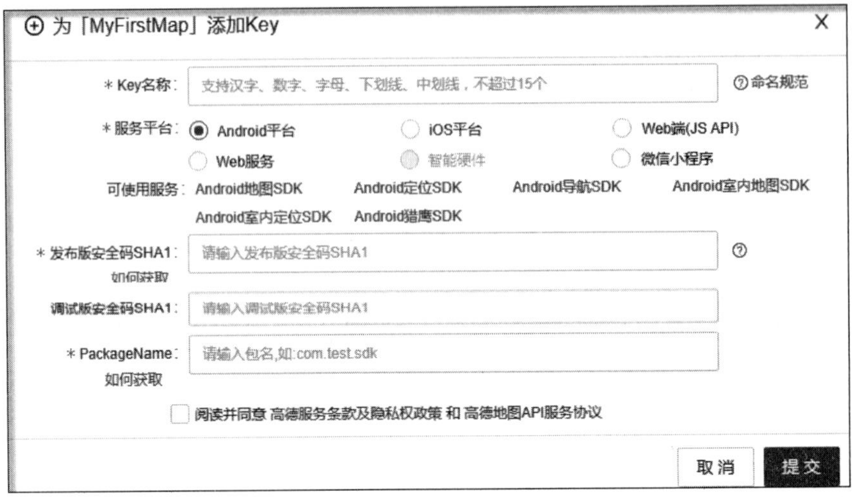

图 12-10　创建新应用界面 2

此页面主要输入 Key 名称(无特殊要求，按命名规范输入即可)；packageName 包名字符串，本示例是 com.example.dqjbd.We；发布版、调试版数字签名。为了方便，本示例的发布版、调试版数字签名都用调试版数字签名，是 04:E4:93:F5:0D:42:9C:57:F7:80:FE:6F:E6:93:59:CE:DD:F2:DB:8B。单击“提交”按钮，回到应用列表页面，此时测试应用下面多了一条刚添加的 Key 记录，如图 12-11 所示。记下这里的 Key 值(bc0f9581b3bb90d5c7f00e415ef0a4a0)，后面会用到。

图 12-11　创建新应用界面 3

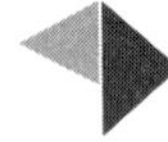

12.3　最简单的高德地图程序

在 12.2 节知识的基础之上，编制最简单的高德地图应用：主界面上有一个“开始”按钮，单击“开始”按钮，显示获得的默认的地图图像，涉及的功能文件如下所示。

① 配置文件 Androidmanifest.xml：主要增加 3 部分内容，如下所示。

• 增加网络、定位及其他相关权限。

```
<uses-permission android:name="android.permission.INTERNET" />
<uses-permission android:name="android.permission.WRITE_EXTERNAL_STORAGE" />
<uses-permission android:name="android.permission.ACCESS_COARSE_LOCATION" />
<uses-permission android:name="android.permission.ACCESS_NETWORK_STATE" />
<uses-permission android:name="android.permission.ACCESS_FINE_LOCATION" />
<uses-permission android:name="android.permission.READ_PHONE_STATE" />
<uses-permission android:name="android.permission.CHANGE_WIFI_STATE" />
<uses-permission android:name="android.permission.ACCESS_WIFI_STATE" />
<uses-permission android:name="android.permission.CHANGE_CONFIGURATION" />
<uses-permission android:name="android.permission.WAKE_LOCK" />
<uses-permission android:name="android.permission.ACCESS_BACKGROUND_
LOCATION" />
<uses-permission android:name="android.permission.WRITE_SETTINGS" />
```

• 在<application>标签中增加如下的<meta-data>子标签。

```
<meta-data android:name="com.amap.api.v2.apikey"
    android:value="bc0f9581b3bb90d5c7f00e415ef0a4a0" />
```

android:name 即键值，是固定的，由高德地图开发包决定；android:value 值是图 12-11 中的 Key 值。

• 增加高德地图用到的 service 配置。

```
<service android:name="com.amap.api.location.APSService" />
```

② 主界面文件 main.xml。

```
<?xml version="1.0" encoding="utf-8"?>
<LinearLayout xmlns:android="http://schemas.android.com/apk/res/android"
android:layout_width="match_parent"
android:layout_height="match_parent"
android:orientation="vertical">
    <Button
        android:id="@+id/mystart"
        android:layout_width="wrap_content"
        android:layout_height="wrap_content"
        android:text="开始"/>
    <com.amap.api.maps2d.MapView
        android:id="@+id/map"
```

```
        android:layout_width="match_parent"
        android:layout_height="match_parent">
    </com.amap.api.maps2d.MapView>
</LinearLayout>
```

以上代码定义了一个 Button 组件,id 为 mystart,用于启动默认地图显示功能;一个 MapView 组件,id 为 map,它是高德地图开发中的功能类,用于显示地图视图。

③ 主应用文件 MainActivity.java。

```
public class MainActivity extends AppCompatActivity{
    private MapView mapView;
    void permissionProcess(){
        String u[] = new String[]{Manifest.permission.ACCESS_COARSE_LOCATION,
                Manifest.permission.ACCESS_FINE_LOCATION};
        ActivityCompat.requestPermissions(this, u, 1);
    }
    /*权限动态申请函数,本函数目前申请了定位权限。若还需其他权限,只在数组 u[]中增加
相应的权限定义即可*/
    void showMap(final Bundle savedInstanceState){
        mapView.onCreate(savedInstanceState);
    }
    /*地图显示函数,savedInstanceState 传过来的是地图图像信息,调用 mapView.
onCreate()函数完成了地图的创建与显示*/
    protected void onCreate(final Bundle savedInstanceState) {
        super.onCreate(savedInstanceState);
        setContentView(R.layout.main);
        mapView = (MapView)findViewById(R.id.map);
        permissionProcess();
        Button b = (Button)findViewById(R.id.mystart);
        b.setOnClickListener(new View.OnClickListener() {
            public void onClick(View v) {
                showMap(savedInstanceState);
            }
        });
    }
}
```

注意:onCreate()函数的形参 savedInstanceState 包含的是地图图像信息。该函数首先调用 permissionProcess()函数完成权限申请,然后获得地图视图对象 mapView,最

后注册开始按钮的 OnClickListener 事件，在响应函数中实现地图显示功能，界面如图 12-12 所示。

图 12-12　最简单的地图程序结果图

12.4　定位功能

12.4.1　相关类及接口

利用高德地图实现定位是一个重要功能，其涉及的类与常用接口如下所示。

1. AMap 类

该类用于定义 AMap 地图对象的操作方法与接口。其常用方法如下所示。

- Polyline addPolyline(PolylineOptions options)：加一个多段线对象到地图上。
- Text addText(TextOptions textOptions)：加一个文字(Text)到地图上。

- Circle addCircle(CircleOptions options)：添加圆形(Circle)覆盖物到地图上。
- Polygon addPolygon(PolygonOptions options)：在地图上添加一个多边形对象。
- Marker addMarker(MarkerOptions options)：加一个 Marker(标记)到地图上。Marker 的图标会根据 Marker.position 位置渲染在地图上。单击 Marker，可视区域将以这个 Marker 的位置为中心点。如果 Marker 设置了 title，则地图上会显示一个包括 title 文字的信息框。如果 Marker 被设置为可拖曳的，那么长按此 Marker 可以拖动它。
- void setMyLocationEnabled(boolean enabled)：设置定位层是否显示。如果显示定位层，则界面上将出现定位按钮，如果未设置 Location Source，则定位按钮不可单击。
- void moveCamera(CameraUpdate update)：按照传入的 CameraUpdate 参数移动可视区域。这个方法为瞬间移动，没有移动过程，如果调用此方法后在回调方法 onCameraChangeFinish()中调用 getCameraPosition()，则将返回移动后的位置。
- CameraPosition getCameraPosition()：返回可视区域的当前位置(包含中心点坐标、缩放级别)。在可视区域变换时，此返回会自己更新。

2. AMapLocationClient 类

该类为定位服务类，此类提供单次定位、持续定位、最后位置相关功能。其相关函数如下所示。

- void setLocationOption(AMapLocationClientOption option)：设置定位参数。
- void setLocationListener(AMapLocationListener listener)：设置定位回调监听。
- void unRegisterLocationListener(AMapLocationListener listener)：移除定位监听。
- void startLocation()：开始定位。
- void stopLocation()：停止定位。

3. AMapLocationClientOption 类

该类用于定位参数设置，通过这个类可以对定位的相关参数进行设置。其相关函数如下所示。

- AMapLocationClientOption setInterval(long interval)：设置发起定位请求的时间间隔。
- AMapLocationClientOption setOnceLocation(boolean isOnceLocation)：设置是否单次定位。

- AMapLocationClientOption setNeedAddress(boolean isNeedAddress)：设置是否返回地址信息，默认返回地址信息。
- AMapLocationClientOption setLocationMode(AMapLocationMode locationMode)：设置定位模式。locationMode 的默认值为 Hight_Accuracy(高精度模式)。

4. AMapLocationListener 接口

该接口仅包含一个函数，如下所示。

- void onLocationChanged(AMapLocation location)：定位回调监听，当定位完成后调用此方法，location 包含定位的结果信息。

5. LocationSource 接口

该接口定义了两个函数，如下所示。

- void activate(OnLocationChangedListener onLocationChangedListener)：定位激活时调用该函数。
- void deactivate()：定位停用时调用该函数。

12.4.2 定位实现

实现最基本的手机定位功能：手机位置变化，显示的地图也随之变化。所涉及的文件如下所示。

① 配置文件 Androidmanifest.xml：同 12.3 节一致。

② 主界面文件 main.xml。

```
<?xml version="1.0" encoding="utf-8"?>
<LinearLayout xmlns:android="http://schemas.android.com/apk/res/android"
android:layout_width="match_parent"
android:layout_height="match_parent"
android:orientation="vertical">
    <TextView
        android:id="@+id/mytxt"
        android:layout_width="match_parent"
        android:layout_height="wrap_content" />
    <Button
        android:id="@+id/mystart"
        android:layout_width="wrap_content"
        android:layout_height="wrap_content"
```

```
        android:text="开始"/>
    <com.amap.api.maps2d.MapView
        android:id="@+id/map"
        android:layout_width="match_parent"
        android:layout_height="match_parent">
    </com.amap.api.maps2d.MapView>
</LinearLayout>
```

以上代码定义了一个 TextView 组件,id 为 mytxt,用于显示定位地址信息;一个 Button 组件,id 为 mystart,用于启动默认地图显示功能;一个 MapView 组件,id 为 map,它是高德地图开发中的功能类,用于显示地图视图。

③ 主应用文件 MainActivity.java。

```
public class MainActivity extends AppCompatActivity implements
AMapLocationListener {
    private AMap aMap;                                          //地图对象
    private MapView mapView;                                    //地图视图对象
    private AMapLocationClient mLocationClient = null;          //地图定位客户端对象
    void permissionProcess(){
        String u[] = new String[]{Manifest.permission.ACCESS_COARSE_LOCATION,
                    Manifest.permission.ACCESS_FINE_LOCATION};
        ActivityCompat.requestPermissions(this, u, 1);
    }
    void showMap(final Bundle savedInstanceState){
        mapView.onCreate(savedInstanceState);
    }
    void initLocation(){
        if (aMap == null) {
            aMap = mapView.getMap();
            mLocationClient = new AMapLocationClient(getApplicationContext());
            mLocationClient.setLocationListener(this);
            mLocationClient.startLocation();
        }
    }
    /* 此函数是本示例手机定位的核心,由代码 mLocationClient.setLocationListener
    (this)知,MainActivity 类必须实现 AMapLocationListener 接口,重写 onLocationChanged()
    函数。由于没有写回调函数时间间隔,因此系统会周期性地默认时间间隔调用
    onLocationChanged()函数。但是必须运行代码 mLocationClient.startLocation(),即设置开
    始定位标识,onLocationChanged()函数才会响应。*/
```

```
    public void onLocationChanged(AMapLocation aMapLocation) {
        if (aMapLocation != null) {
            if (aMapLocation.getErrorCode() == 0) {
                aMap.moveCamera(CameraUpdateFactory.zoomTo(17));
                aMap.moveCamera(CameraUpdateFactory.changeLatLng
                    (new LatLng(aMapLocation.getLatitude(), aMapLocation
.getLongitude())));
            }
        }
    }
```

/* 保证正确定位的前提条件是:aMapLocation 不为空以及 getErrorCode()函数的返回值是 0。若满足上述两个条件,则显示定位经度、纬度下的地图图像。随着手机的不断移动,该图像也不断变化。*/

```
    protected void onCreate(final Bundle savedInstanceState) {
        super.onCreate(savedInstanceState);
        setContentView(R.layout.main);
        mapView = (MapView) findViewById(R.id.map);
        permissionProcess();
        Button b = (Button) findViewById(R.id.mystart);
        b.setOnClickListener(new View.OnClickListener() {
            public void onClick(View v) {
                showMap(savedInstanceState);
                initLocation();
            }
        });
    }
    protected void onDestroy() {
        super.onDestroy();
        mapView.onDestroy();
        mLocationClient.stopLocation();                    //停止定位
        mLocationClient.onDestroy();                       //销毁定位客户端
    }
```

/* 释放手机定位资源,包括地图视图对象 mapView、手机客户端定位对象 mLocationClient 及停止定位设置等,避免该应用已经销毁,但手机定位服务 Service 仍在运行的情况。*/

```
    protected void onResume() {
        super.onResume();
        mapView.onResume();
```

```
        }
        protected void onPause() {
            super.onPause();
            mapView.onPause();
        }
    }
```

onResume()函数调用了 mapView.onResume(),onPause()函数调用了 mapView.onPause()。主要作用是避免不必要的 onLocationChanged()函数响应,造成手机资源浪费。

该示例的运行结果如图 12-13 所示。

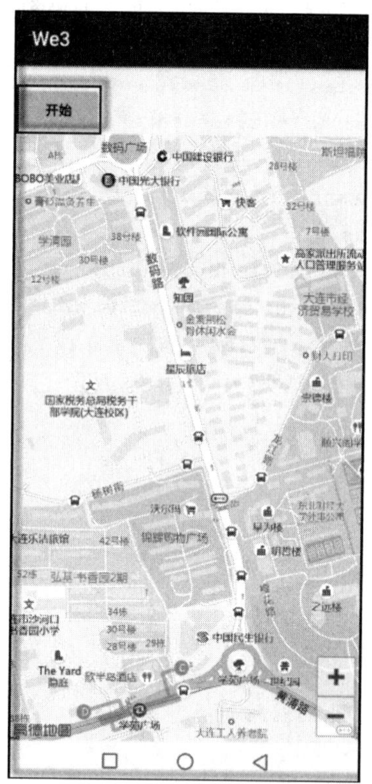

图 12-13　简单手机定位图

由图 12-13 可知,简单的手机定位功能实现了,但手机所在的中心位置在图中并没有标注,也没有显示出所在的省、市、街道等信息。为了实现这些功能,还需要对 MainActivity 类加以修改,如下所示。

```
public class MainActivity extends AppCompatActivity implements
AMapLocationListener,LocationSource {
    private AMap aMap;
    private MapView mapView;
    private AMapLocationClient mLocationClient ;
    public AMapLocationClientOption mLocationOption;  //新增定位选项对象
    private LocationSource.OnLocationChangedListener mListener;
                                                      //新增定位帧听对象
    void permissionProcess(){                         //同前文}
    void showMap(final Bundle savedInstanceState){    //同前文}
    void initLocation(){
        if (aMap == null) {
            aMap = mapView.getMap();
            //设置显示定位按钮 并且可以单击
            UiSettings settings = aMap.getUiSettings();
            settings.setMyLocationButtonEnabled(true);
            aMap.setLocationSource(this);             //设置了定位的监听
            aMap.setMyLocationEnabled(true);
                                    //显示定位层并且可以触发定位,默认是 flase
        }
        location();
    }
    /* 与前文的该函数相比,增加了地图 UI 相关参数设置,使能了显示定位按钮,并设置了对定位
源的帧听。由 setLocationSource(this)可知,MainActivity 类必须实现 LocationSource 接
口,重写 activate()、deactivate()函数。 */
    private void location() {
        //初始化定位
        mLocationClient = new AMapLocationClient(getApplicationContext());
        //设置定位回调监听
        mLocationClient.setLocationListener(this);
        //初始化定位参数
        mLocationOption = new AMapLocationClientOption();
        //Hight_Accuracy 为高精度模式,Battery_Saving 为低功耗模式,Device_
Sensors 是仅设备模式
        mLocationOption.setLocationMode(AMapLocationClientOption
.AMapLocationMode.Hight_Accuracy);
        //设置是否返回地址信息(默认返回地址信息)
        mLocationOption.setNeedAddress(true);
```

```
        //设置是否只定位一次,默认为 false
        mLocationOption.setOnceLocation(false);
        //设置是否强制刷新 WiFi,默认为强制刷新
        mLocationOption.setWifiActiveScan(true);
        //设置是否允许模拟位置,默认为 false,不允许模拟位置
        mLocationOption.setMockEnable(false);
        //设置定位间隔,单位为 ms,默认为 2000ms
        mLocationOption.setInterval(2000);
        //给定位客户端对象设置定位参数
        mLocationClient.setLocationOption(mLocationOption);
        //启动定位
        mLocationClient.startLocation();
    }
    /* 该函数主要是对定位属性进行设置,包括定位精度、地址信息、定位一次还是多次、定位时
间间隔等信息,最后通过 startLocation()函数开启定位功能。*/
    public void activate(OnLocationChangedListener
onLocationChangedListener) {
        mListener = onLocationChangedListener;
    }
    /* OnLocationChangedListener 对象不是我们产生的,是系统产生的,但我们可以截获,
本例就赋值给了 mListener 对象。*/
    public void deactivate() {
        mListener = null;
    }
    /* 释放资源函数,本例将 mListener 对象置位 null。*/
    public void onLocationChanged(AMapLocation aMapLocation) {
        if (aMapLocation != null) {
            if (aMapLocation.getErrorCode() == 0) {
                aMap.moveCamera(CameraUpdateFactory.zoomTo(17));
                aMap.moveCamera(CameraUpdateFactory.changeLatLng(
                  new LatLng(aMapLocation.getLatitude(), aMapLocation
.getLongitude())));
                mListener.onLocationChanged(aMapLocation);
                StringBuffer buffer = new StringBuffer();
                buffer.append(aMapLocation.getCountry() + ""
                        + aMapLocation.getProvince() + ""
                        + aMapLocation.getCity() + ""
                        + aMapLocation.getDistrict() + ""
```

```
                    + aMapLocation.getStreet() + ""
                    + aMapLocation.getStreetNum());
                TextView tv = (TextView)findViewById(R.id.mytxt);
                tv.setText(buffer.toString());
            }
        }
    }
    /* 此函数显示手机当前所在位置的地图,并将具体地址包括省、市、区、街道、门牌号显示在
TextView 组件中。 */
    protected void onCreate(final Bundle savedInstanceState) { //同前文 }
    protected void onDestroy() { //同前文 }
    protected void onResume() { //同前文 }
    protected void onPause() { //同前文 }
}
```

本示例运行后界面如图 12-14 所示,标识了与图 12-13 不同的地方。

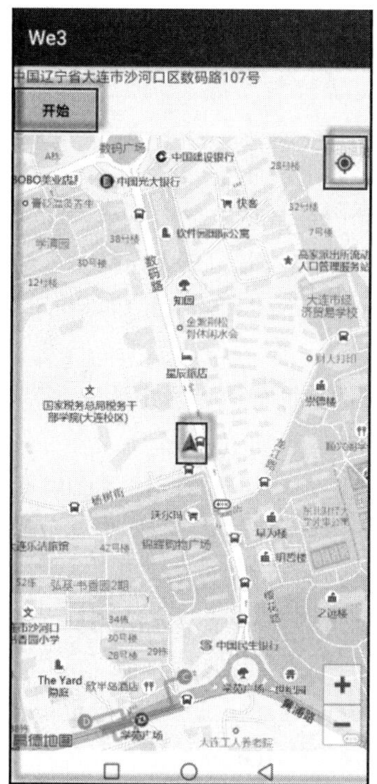

图 12-14　改进的手机定位图

12.4.3 基本搜索

高德地图最基本的功能就是搜索，与搜索相关的类或接口如下所示。

1. PoiSearch.Query 类

该类是查询条件及结果约束类，其相关函数如下所示。

- Query(String query,String ctgr,String city)：构造方法，根据给定的参数构造一个 PoiSearch.Query 的新对象。query 及 ctgr 至少需要定义一个，并且参数 city 必须定义，不能为空。query，查询字符串，多个关键字用"|"分割。ctgr，类型的组合，例如定义如下组合：餐馆|电影院|景点。city，待查询城市(地区)的城市编码 citycode、城市名称(中文或中文全拼)、行政区划代码 adcode，必设参数。
- void setPageNum(int pageNum)：设置查询第几页的结果数目，从 0 开始。
- void setPageSize(int size)：设置查询每页的结果数目。默认值为 20 条，取值范围为 1～30 条。
- int getPageSize()：获取设置的查询页面的结果数目。

2. PoiSearch 类

该类是搜索类，其相关函数如下所示。

- PoiSearch(Context context,PoiSearch.Query query)：根据给定的参数构造一个 PoiSearch 的新对象。context，对应的 Context。query，查询条件。
- void setOnPoiSearchListener(PoiSearch.OnPoiSearchListener listener)：设置查询监听接口。
- void searchPOIAsyn()：查询 POI 异步接口。

3. PoiResult 类

该类是 POI(Point Of Interest，兴趣点)搜索结果分页显示类。PoiResult 封装了此分页结果，并且会缓存已经检索到的页的搜索结果。此类不可直接构造，只能通过调用类 PoiSearch 的 searchPOI()、searchPOIAsyn()方法得到。该类常用的方法如下所示。

- int getPageCount()：返回该结果的总页数。
- ArrayList<PoiItem> getPois()：返回当前页所有的 POI 结果。如果当前页无 POI 结果，则返回长度为零的 list。
- List<String> getSearchSuggestionKeywords()：返回搜索建议。如果搜索的关键字明显为误输入，则通过此方法得到搜索关键词建议。如搜索"早君庙"，搜索引擎给出的建议为“皂君庙”。建议按照与查询关键词发音近似的原则给出。如果有搜索结果,则此方法必返回空。

- List<SuggestionCity> getSearchSuggestionCitys():返回建议的城市。如果搜索关键字无返回结果,则引擎将建议此关键字在其他城市的搜索情况,返回的详细内容见 SuggestionCity 类。如果有搜索结果,则此方法必返回空。

4. PoiItem 类

该类是搜索地址的具体信息类,其相关函数如下所示。

- String getAdCode():返回 POI 的行政区划代码。
- String getAdName():返回 POI 的行政区划名称。
- String getCityCode():返回 POI 的城市编码。
- String getCityName():返回 POI 的城市名称。
- int getDistance():获取 POI 距离中心点的距离。
- LatLonPoint getLatLonPoint():返回 POI 的经、纬度坐标。
- String getPostcode():返回 POI 的邮编。
- String getProvinceName():返回 POI 的省/自治区/直辖市/特别行政区名称。
- String getSnippet():返回 POI 的地址。
- String getTitle():返回 POI 的名称。
- String getTypeCode():返回兴趣点的类型编码。
- String getTypeDes():返回 POI 的类型描述。

考虑如下应用,仅在大连市内进行 POI 搜索。基本搜索功能简图如图 12-15 所示。

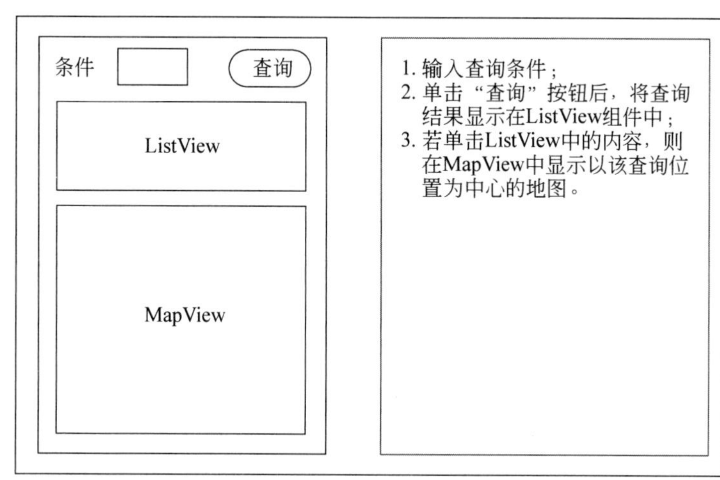

图 12-15　基本搜索功能简图

完成图 12-15 所示功能，所需文件如下所示。

① 配置文件 Androidmanifest.xml：同 12.3 节一致。

② 主界面文件 main.xml。

```
<?xml version="1.0" encoding="utf-8"?>
<LinearLayout xmlns:android="http://schemas.android.com/apk/res/android"
android:layout_width="match_parent"
android:layout_height="match_parent"
android:orientation="vertical">
    <EditText
        android:id="@+id/myfactor"
        android:layout_width="match_parent"
        android:layout_height="wrap_content"
        android:textSize="50sp"/>
    <Button
        android:id="@+id/mypoi"
        android:layout_width="wrap_content"
        android:layout_height="wrap_content"
        android:text="查询"/>
    <ListView
        android:id="@+id/mylist"


        android:layout_width="match_parent"
        android:layout_height="100dp"></ListView>
    <com.amap.api.maps2d.MapView
        android:id="@+id/map"
        android:layout_width="match_parent"
        android:layout_height="match_parent">
    </com.amap.api.maps2d.MapView>
</LinearLayout>
```

以上代码定义了一个 EditText 组件，id 为 myfactor，用于输入查询条件（模糊查询）；一个查询按钮，id 为 mypoi，用于启动查询；一个 ListView 组件，id 为 mylist，用于显示查询结果；一个 MapView 组件，用于显示选中 ListView 组件中具体地址的地图。

③ 主应用文件 MainActivity.java。

```
public class MainActivity extends AppCompatActivity implements PoiSearch.
OnPoiSearchListener,AdapterView.OnItemClickListener{
    MapView mapView;
```

```
ArrayList<PoiItem> aryPois = new ArrayList();  //POI 查询结果集合
void permissionProcess(){
    String u[] = new String[]{Manifest.permission.ACCESS_COARSE_LOCATION,
                Manifest.permission.ACCESS_FINE_LOCATION};
    ActivityCompat.requestPermissions(this, u, 1);
}
public void search(){
    try {
        EditText et = (EditText)findViewById(R.id.myfactor);
        String factor = et.getText().toString().trim();
        PoiSearch.Query query = new PoiSearch.Query(factor, "","大连");
        query.setPageSize(20);
        query.setPageNum(0);
        PoiSearch poisearch = new PoiSearch(this, query);
        poisearch.setOnPoiSearchListener(this);
        poisearch.searchPOIAsyn();
    }catch(Exception e){   }
}
```

/*查询功能函数。首先从 EditText 对象 et 中获得查询条件字符串 factor;然后进行查询参数设置。由代码 new PoiSearch.Query(factor, "","大连"),可得仅查询大连市满足 factor 条件的地点。factor 是模糊查询条件,可输入的值如公园、宾馆、电影院等。setPageSize(20)表明每页最多返回 20 条查询记录,setPageNum(0)表明仅显示首页的记录。当然,可增加代码控制,以显示其他页的记录,由于这不是本示例讨论的重点,因此读者可自行完成;最后设置回调响应函数(由代码 setOnPoiSearchListener(this)知响应函数在 MainActivity 类中,该类必须实现 OnPoiSearchListener 接口中的函数)及启动 POI 异步查询。*/

```
public void onPoiSearched(PoiResult poiResult, int errcode) {
        ArrayList<String>  datas = new ArrayList();
        aryPois = poiResult.getPois();
        for (int i = 0; i < aryPois.size(); i++) {
            datas.add(aryPois.get(i).getTitle()+"-"+aryPois.get(i)
.getSnippet());
        }
        ArrayAdapter<String> ad = new ArrayAdapter<String>(
                    this,android.R.layout.simple_list_item_1,datas);
        ListView lv = (ListView)findViewById(R.id.mylist);
        lv.setAdapter(ad);
    }
```

```
    }
    /* POI 查询结果最主要的回调重写函数。本示例一方面获得了当前页查询结果,并将其赋值给成员变量 aryPois,它是每个 PoiItem 具体地点的对象集合;另一方面形成了"名称+具体地址"的 ArrayList 对象 datas,将其通过 ArrayAdapter 绑定并显示在 ListView 组件中。*/
    public void onPoiItemSearched(PoiItem poiItem, int i) {    }
    /* POI 查询结果的另一个回调函数,一般不用重写。*/
    public void onItemClick(AdapterView<?> parent, View view, int position, long id) {
        PoiItem pi = aryPois.get(position);
        LatLonPoint lp = pi.getLatLonPoint();
        AMap aMap = mapView.getMap();
        aMap.moveCamera(CameraUpdateFactory.zoomTo(17));
        aMap.moveCamera(CameraUpdateFactory.changeLatLng
                (new LatLng(lp.getLatitude(), lp.getLongitude())));
    }
    /* 本函数是 ListView 组件 OnItemClickListener 事件的响应函数,表明已在 ListView 组件中选择了一个具体位置,要显示以其为中心的地图图形。在函数中,根据 position 及成员变量 aryPois,就能获得 position 位置对应的 PoiItem 对象 pi,进而获得具体位置的经度、纬度值,从而完成相应的地图显示功能。*/
    protected void onCreate(final Bundle savedInstanceState) {
        super.onCreate(savedInstanceState);
        setContentView(R.layout.main);
        permissionProcess();
        mapView = (MapView)findViewById(R.id.map);
        Button b = (Button)findViewById(R.id.mypoi);
        b.setOnClickListener(new View.OnClickListener() {
            public void onClick(View v) {
                mapView.onCreate(savedInstanceState);
                search();
            }
        });
        ListView lv = (ListView)findViewById(R.id.mylist);
        lv.setOnItemClickListener(this);
    }
}
```

onCreate()函数的功能包括：权限检测;获得地图对象 mapView;为 POI 查询按钮及 ListView 组件设置事件响应。

12.4.4　公交查询

利用高德地图可以很方便地进行公交查询。为了简化问题规模，仍以查询大连公交为例，其功能简图如图 12-16 所示。

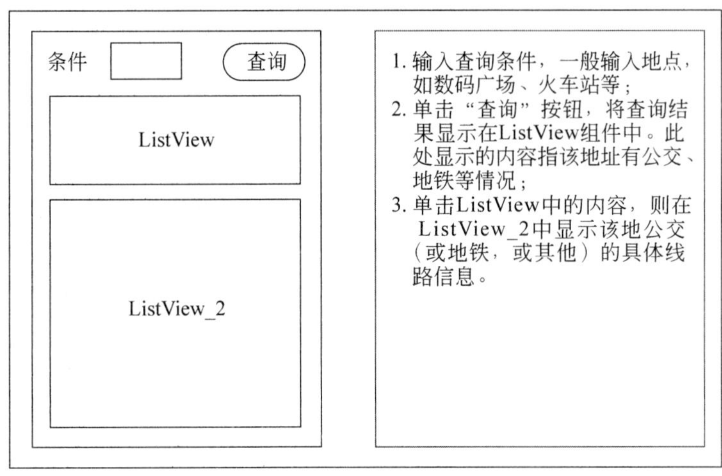

图 12-16　大连公交查询功能简图

完成图 12-16 所示功能，所需的文件如下所示。

① 配置文件 Androidmanifest.xml：同 12.3 节一致。

② 主界面文件 main.xml。

```
<?xml version="1.0" encoding="utf-8"?>
<LinearLayout xmlns:android="http://schemas.android.com/apk/res/android"
android:layout_width="match_parent"
android:layout_height="match_parent"
android:orientation="vertical">
    <EditText
        android:id="@+id/myfactor"
        android:layout_width="match_parent"
        android:layout_height="wrap_content"
        android:textSize="30sp"/>
    <Button
        android:id="@+id/mybus"
        android:layout_width="wrap_content"
        android:layout_height="wrap_content"
```

```
        android:text="公交查询"/>
    <ListView
        android:id="@+id/mylist"
        android:layout_width="match_parent"
        android:layout_height="150dp" />
    <ListView
        android:id="@+id/mysublist"
        android:layout_width="match_parent"
        android:layout_height="match_parent" />
</LinearLayout>
```

以上代码定义了一个 EditText 组件,id 为 myfactor,用于输入查询条件(一般指地址);一个公交查询按钮,id 为 mybus,用于启动查询;一个 ListView 组件,id 为 mylist,用于显示查询结果(该地有公交、地铁或其他信息);一个 ListView 组件,id 为 mysublist,用于显示该地有哪些具体的公交(或地铁,或其他)信息。

③ 主应用文件 MainActivity.java。

```
public class MainActivity extends AppCompatActivity implements
BusStationSearch.OnBusStationSearchListener,
  AdapterView.OnItemClickListener{
    List<BusStationItem> aryStat;
    void permissionProcess(){
        String u[] = new String[]{Manifest.permission.ACCESS_COARSE_LOCATION,
                                  Manifest.permission.ACCESS_FINE_LOCATION};
        ActivityCompat.requestPermissions(this, u, 1);
    }
    public void search(){
        try {
            EditText et = (EditText)findViewById(R.id.myfactor);
            String factor = et.getText().toString();
            BusStationQuery query = new BusStationQuery(factor,"大连");
            BusStationSearch bs = new BusStationSearch(this, query);
            bs.setOnBusStationSearchListener(this);
            bs.searchBusStationAsyn();
        }catch(Exception e){     }
    }
    /*读者会发现本示例的搜索函数与 12.4.3 节的基本搜索函数是相似的,由此可以总结出关
于高德地图 XXX 搜索的一般规律:定义 XXXQuery 对象,设定搜索条件;定义 XXXSearch 对象;注
```

```
册 XXXSearchListener 事件;启动 XXXAsyn()异步搜索。*/
    public void onBusStationSearched(BusStationResult result, int code) {
        aryStat =result.getBusStations();
        ArrayList<String> datas = new ArrayList();
        for(int i=0; i<aryStat.size(); i++){
            BusStationItem items = aryStat.get(i);
            datas.add(items.getBusStationId()+"-"+items.getBusStationName());
        }
        ArrayAdapter<String> ad = new ArrayAdapter<String>(this, android.R.
layout.simple_list_item_1,datas);
        ListView lv = (ListView)findViewById(R.id.mylist);
        lv.setAdapter(ad);
    }
    /*公共交通搜索结果回调函数。相当于一级搜索,指出搜索地址包含的交通类型:如地铁、
公交车等。由 List 成员变量 aryStat 保存,遍历 aryList,将公交类型 ID 及名称合并成一个字
符串,添加到 ArrayList 局部变量 datas 中。将 datas 通过 ArrayAdapter 与 id 为 mylist
的 ListView 组件绑定,完成显示工作。*/
    public void onItemClick(AdapterView<?> parent, View view, int position,
long id) {
        BusStationItem items = aryStat.get(position);
        ArrayList<String> datas = new ArrayList();
        List<BusLineItem> list = items.getBusLineItems();
        for(int i=0; i<list.size(); i++){
            BusLineItem item = list.get(i);
            datas.add(item.getBusLineId()+"-"+item.getBusLineName()+"-"+
item.getOriginatingStation()+
                                "-"+item.getTerminalStation());
        }
        ArrayAdapter<String> ad = new ArrayAdapter<String>(this, android.R.
layout.simple_list_item_1,datas);
        ListView lv = (ListView)findViewById(R.id.mysublist);
        lv.setAdapter(ad);
    }
    /*id 为 mylist 的 ListView 组件 OnItemClickListener 事件回调函数,表明我们要查
询某具体地址某公交类型(如公共汽车、地铁等)有哪些具体线路,始末站点是什么等信息。其原
理与 onBusStationSearched()函数中的列表显示原理是一致的。*/
    protected void onCreate(final Bundle savedInstanceState) {
        super.onCreate(savedInstanceState);
```

```
        setContentView(R.layout.main);
        permissionProcess();
        Button b = (Button)findViewById(R.id.mybus);
        b.setOnClickListener(new View.OnClickListener() {
            public void onClick(View v) {   search();   }
        });
        ListView lv = (ListView)findViewById(R.id.mylist);
        lv.setOnItemClickListener(this);
    }
}
```

onCreate()函数的功能包括：权限检测；公交查询按钮及 ListView 组件设置事件响应。

12.4.5 天气查询

利用高德地图不但可以进行地图相关操作，而且可以获得天气预报信息。获取城市的实时天气、今天和未来 3 天内的预报天气，可结合定位和逆地理编码功能使用，查询定位点所在城市的天气情况。

天气查询的请求参数类为 WeatherSearch，city（城市）为必设参数，type（气象类型）为可选参数，其包含两种类型：WEATHER_TYPE_LIVE 为实况天气；WEATHER_TYPE_FORECAST 为预报天气。

本示例所实现的天气预报查询功能简图如图 12-17 所示。

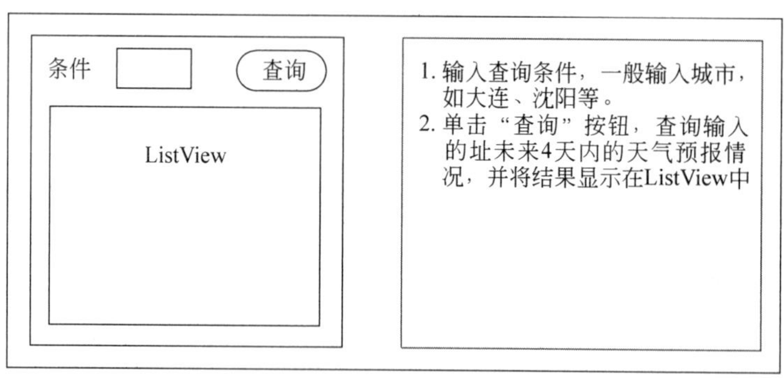

图 12-17　天气预报查询功能简图

完成图 12-17 所示功能需要的文件如下所示。

① 配置文件 Androidmanifest.xml：同 12.3 节一致。

② 主界面文件 main.xml。

```
<?xml version="1.0" encoding="utf-8"?>
<LinearLayout xmlns:android="http://schemas.android.com/apk/res/android"
android:layout_width="match_parent"
android:layout_height="match_parent"
android:orientation="vertical">
    <EditText
        android:id="@+id/myfactor"
        android:layout_width="match_parent"
        android:layout_height="wrap_content"
        android:textSize="30sp"/>
    <Button
        android:id="@+id/myweather"
        android:layout_width="wrap_content"
        android:layout_height="wrap_content"
        android:text="天气查询"/>
    <ListView
        android:id="@+id/mylist"
        android:layout_width="match_parent"
        android:layout_height="match_parent" />
</LinearLayout>
```

以上代码定义了一个 EditText 组件，id 为 myfactor，用于输入查询条件（一般指地址）；一个 Button 组件，id 为 myweather，用于启动查询；一个 ListView 组件，id 为 mylist，用于显示该地址 4 天内的天气预报。

③ 主应用文件 MainActivity.java。

```
public class MainActivity extends AppCompatActivity implements WeatherSearch.
OnWeatherSearchListener{
    List<BusStationItem> aryStat;
    void permissionProcess(){
        String u[] = new String[]{Manifest.permission.ACCESS_COARSE_LOCATION,
                                Manifest.permission.ACCESS_FINE_LOCATION};
        ActivityCompat.requestPermissions(this, u, 1);
    }
    public void search(){
        try {
            EditText et = (EditText)findViewById(R.id.myfactor);
```

```
            String factor = et.getText().toString();
            WeatherSearchQuery mquery = new WeatherSearchQuery(factor,
                        WeatherSearchQuery.WEATHER_TYPE_FORECAST);
            WeatherSearch mweathersearch=new WeatherSearch(this);
            mweathersearch.setOnWeatherSearchListener(this);
            mweathersearch.setQuery(mquery);
            mweathersearch.searchWeatherAsyn();     //异步搜索
        }catch(Exception e){    }
    }
    public void onWeatherLiveSearched(LocalWeatherLiveResult result, int i) {   }
    /* 该函数是 OnWeatherSearchListener 事件回调函数之一,用于获得某地当天的天气预
报情况。*/
    public void onWeatherForecastSearched(LocalWeatherForecastResult result,
int code) {
        LocalWeatherForecast forecast = result.getForecastResult();
        List<LocalDayWeatherForecast> list = forecast.getWeatherForecast();
        LocalDayWeatherForecast wea = null;
        ArrayList<String> datas = new ArrayList();
        for(int i=0; i<list.size(); i++){
            wea = list.get(i);
            String s = "";
            s += wea.getDate()+ "\t";
            s += wea.getDayWeather()+"\t";
            s += wea.getDayTemp()+"度\t";
            s += wea.getDayWindPower()+"级";
            datas.add(s);
        }
        ArrayAdapter<String> ad = new ArrayAdapter<String>(this,android.R.
layout.simple_list_item_1,datas);
        ListView lv = (ListView)findViewById(R.id.mylist);
        lv.setAdapter(ad);
    }
        /* 该函数是 OnWeatherSearchListener 事件的又一回调函数,用于获得某地当天天
气及未来 3 天内的天气预报情况。*/
    protected void onCreate(final Bundle savedInstanceState) {
        super.onCreate(savedInstanceState);
        setContentView(R.layout.main);
        permissionProcess();
```

```
        Button b = (Button)findViewById(R.id.myweather);
        b.setOnClickListener(new View.OnClickListener() {
            public void onClick(View v) {
                search();
            }
        });
    }
}
```

习题 12

一、简答题

1. 简述 App 签名的重要性。

2. 请从开发高德地图公交查询、天气查询中，总结开发高德地图其他应用的一般步骤。

二、程序题

将高德地图天气预报、公交查询利用 Fragment 片段实现，并集成在一个 Activity 内。

参 考 文 献

[1] 欧阳燊. Android Studio 开发实战从零基础到 App 上线[M]. 2 版. 北京：清华大学出版社，2020.

[2] 李冉，李敏. 基于案例的 Android 程序设计教程[M]. 北京：清华大学出版社，2020.

[3] 刘韬，郑海昊. Android Studio App 边做边学-微课视频版[M]. 北京：清华大学出版社，2020.

[4] 刘玉红，蒲娟. Android 移动开发案例课堂[M]. 北京：清华大学出版社，2019.

[5] 仲宝才，颜德彪，刘静. Android 移动应用开发实践教程[M]. 北京：清华大学出版社，2018.

[6] 宋三华. 基于 Android Studio 的案例教程[M]. 2 版. 北京：电子工业出版社，2021.

[7] 白喆. Android 移动应用程序开发[M]. 北京：电子工业出版社，2020.

[8] 赖红. Android 应用开发基础[M]. 北京：电子工业出版社，2020.

[9] 吴绍根，罗佳. Android Studio 移动应用开发基础[M]. 北京：电子工业出版社，2019.

[10] 丁山. Android Studio 程序设计教程[M]. 北京：机械工业出版社，2020.

图书资源支持

感谢您一直以来对清华版图书的支持和爱护。为了配合本书的使用，本书提供配套的资源，有需求的读者请扫描下方的“书圈”微信公众号二维码，在图书专区下载，也可以拨打电话或发送电子邮件咨询。

如果您在使用本书的过程中遇到了什么问题，或者有相关图书出版计划，也请您发邮件告诉我们，以便我们更好地为您服务。

我们的联系方式：

地　　址：北京市海淀区双清路学研大厦A座714

邮　　编：100084

电　　话：010-83470236　010-83470237

客服邮箱：2301891038@qq.com

QQ：2301891038（请写明您的单位和姓名）

资源下载：关注公众号“书圈”下载配套资源。

资源下载、样书申请

书圈

获取最新书目

观看课程直播